3D CADASTRE IN AN INTERNATIONAL CONTEXT

LEGAL, ORGANIZATIONAL, AND TECHNOLOGICAL ASPECTS

3D CADASTRE IN AN INTERNATIONAL CONTEXT

LEGAL, ORGANIZATIONAL, AND TECHNOLOGICAL ASPECTS

JANTIEN STOTER

PETER VAN OOSTEROM

CRC Press
Taylor & Francis Group
Boca Raton London New York

CRC Press is an imprint of the
Taylor & Francis Group, an **informa** business

A TAYLOR & FRANCIS BOOK

CRC Press
Taylor & Francis Group
6000 Broken Sound Parkway NW, Suite 300
Boca Raton, FL 33487-2742

First issued in paperback 2020

ISBN 13: 978-0-367-57789-6 (pbk)
ISBN 13: 978-0-8493-3932-5 (hbk)

Library of Congress Card Number 2005052715

**Visit the Taylor & Francis Web site at
http://www.taylorandfrancis.com**

**and the CRC Press Web site at
http://www.crcpress.com**

Library of Congress Cataloging-in-Publication Data

Stoter, Jantien E., 1971-
  3D cadastre in an international context : legal, organizational, and technological aspects / Jantien E. Stoter and Peter van Oosterom.
      p. cm.
  Includes bibliographical references and index.
  ISBN 0-8493-3932-4 (alk. paper)
  1. Land use--Data processing. 2. Land tenure--Data processing. 3. Land titles--Registration and transfer--Data processing. 4. Cadastres--Data processing. 5. Information storage and retrieval systems--Land use. 6. Three-dimensional display systems. I. Title: Three Dimension cadastre. II. Oosterom, Petrus Johannes Maria van. III. Title.

HD108.15.S762 2006
333.73'13'0285--dc22   2005052715

The Authors

Jantien Stoter, Ph.D., a 1995 graduate in Physical Geography at Utrecht University, began her career as a GIS specialist with the District Water Board of Amsterdam (1995–1997). From 1997 to 1999 she worked as a GIS consultant at the Engineering Office of Holland Rail Consult, where she applied GIS analyses to support the planning of large infrastructure projects. Stoter's university career started in 1999 as an assistant professor in GIS applications, section GIS technology, in the Department of Geodesy, Delft University of Technology, the Netherlands. In 2000, she started her Ph.D. research on 3D cadastre, which she finished in September 2004. She was a member of the program committee of Registration of Properties in Strata, International FIG Workshop on 3D Cadastres, 28–30 November 2001, Delft University of Technology. In February 2004, she received the Prof. J.M. Tienstra research award for her work. This award, which is given every 5 years by the Netherlands Geodetic Commission (NCG) of the Royal Netherlands Academy of Arts and Sciences (KNAW), was established to promote geodetic research in the Netherlands. Since April 2004 she has held the position of assistant professor at the International Institute for Geo-Information Science and Earth Observation, ITC, Enschede, the Netherlands. Her main research and education responsibilities are generalization of geo-information and multiscale databases.

Peter van Oosterom, Ph.D., was the first student to receive the M.Sc. degree in Computer Science from the Delft University of Technology (1985). In 1990, he obtained a Ph.D. degree from Leiden University based on the thesis "Reactive Data Structures for GIS." Oxford University Press published a revised and updated version of the thesis. He is the co-author of more than 100 publications. From 1985 to 1995 he was a computer scientist at the TNO Physics and Electronics Laboratory, The Hague, doing research in the area of spatial databases in the context of several GIS applications. An important project has been the design and development of a GIS based on the open DBMS Postgres (University of California, Berkeley). The system is called GEO++ and was developed in cooperation with TNO Human Factors Laboratory. From 1995 to 1999, he was senior information manager within the Dutch Cadastre, headquarters, Apeldoorn. He was responsible for the design of the renewed cadastral information system. In 2000, van Oosterom was appointed full professor of GIS technology at the Delft University of Technology. The Geo-Database Management Center (GDMC) opened on November 15, 2000 within the Delft University of Technology; its research program concentrates on the theme of geo-DBMS in cooperation with the Geo-ICT industry. In the first half of 2003 he was visiting professor at the University of Queensland, Australia. Van Oosterom has been a member of the subcommittee Geo-Information Models of the Netherlands Geodetic Committee (NCG of the KNAW) since 1989 and a member of the Scientific Council of the ITC, Enschede, since 2000. He is a member of the editorial boards of several journals and research monographs. He is a member of the program committees of nearly all major GIS-related conferences. He has been a regular reviewer for many international GIS-related journals: *IJGIS*

(*International Journal on GIS*), *CaGIS* (*Cartography and GIS*), *GeoInformatica*, *CEUS* (*Computers, Environment and Urban Systems*), etc. He has been guest editor (together with Chrit Lemmen) of several special issues on cadastral systems in *CEUS*. He was the chairman of the program committee of Registration of Properties in Strata, International FIG Workshop on 3D Cadastres, November 28–30, 2001, Delft University of Technology, the co-chairman (together with Henk Scholten) of the program committee of the 10th International IGU Symposium on Spatial Data Handling (SDH2002), May 2002, Ottawa, Canada, and the general chair of the First International Symposium on Geo-information for Disaster Management (Gi4DM), March 21–23, 2005, Delft University of Technology. His main research interests cover spatial databases, GIS architectures, spatial analysis, generalization, querying and presentation, Internet/interoperable GIS and cadastral applications.

Acknowledgments

First, we thank the Netherlands Kadaster because it cooperated in the research presented in this book by providing us with data and by discussions on data models and on research developments. The Kadaster also supported the Ph.D. research that was the basis for this book (see Reference 190).

We are grateful to the companies Laser-Scan, Oracle, ESRI, and Bentley for their collaboration in this research and because we were able to use their software. We appreciate the contribution of the project team of the HSL-Zuid and the Bouwdienst van Rijkswaterstaat because they provided us with 3D data on physical constructions for the case studies. AGI (Rijkswaterstaat, Adviesdienst Geo-informatie en ICT) provided us with point heights of case study areas.

Rod Thompson, Alex Smith, and Diego Embra contributed to this book by providing us with the extensive material for, respectively, Queensland (Australia), Richland County (South Carolina, USA), and Argentina case studies. This included taking pictures and patiently answering our neverending stream of questions. Thank you all!

We express our sincere gratitude to our colleagues and M.Sc. students for their cooperation in various parts of the research described in this book: Sisi Zlatanova, Hendrik Ploeger, Henri Aalders, Calin Arens, Elfriede Fendel, Jitske de Jong, Friso Penninga, Wilko Quak, Axel Smits, Theo Tijssen, Marian de Vries, and Jaap Zevenbergen.

The research presented in this book is part of the research program Sustainable Urban Areas (SUA) carried out by Delft University of Technology. This book was finalized in collaboration with the International Institute for Geo-Information Science and Earth Observation (ITC), Enschede.

Abbreviations

ADT	Abstract Data Type
BLOB	Binary Large Object
CAD	Computer-Aided Design
CSG	Constructive Solid Geometry
DBMS	Database Management System
DDL	Data Definition Language
DML	Data Manipulation Language
FIG	Fédération Internationale des Géomètres/International Federation of Surveyors
GII	Geographic Information Infrastructure
GIS	Geographic Information System
OCL	Object Constraint Language
SDI	Spatial Data Infrastructure
SQL	Structured Query Language
TIN	Triangular Irregular Network
UML	Unified Modeling Language

Contents

1 Introduction

During the last two centuries population density has increased considerably, making land use more intense. This trend has lent a growing importance to ownership of land, which has changed the way humans relate to land. This changing relationship has necessitated a system in which right to land is clearly and indisputably recorded. In this book such a system is referred to as a "cadastre," although many systems with different names that fulfill a similar function, such as cadastral registration, cadastral system, land registry, land registration, land administration, property register, and land book, exist worldwide.

No unique form of cadastre exists. Reference 34 notes:

> It is impossible to give a definition of a Cadastre which is both terse and comprehensive, but its distinctive character is readily recognized and may be expressed as the marriage of (a) technical record of the parcellation of the land through any given territory, usually represented on plans of suitable scale, with (b) authoritative documentary record, whether of a fiscal or proprietary nature or of the two combined, usually embodied in appropriate associated registers.

In principle, this book follows the description of a cadastre as it is given in the FIG (International Federation of Surveyors) Statement on the Cadastre:[48]

> Cadastre is normally a parcel based, and up-to-date land information system containing a record of interests in land (e.g. rights, restrictions and responsibilities). It usually includes a geometric description of land parcels linked to other records describing the nature of the interests, the ownership or control of those interests, and often the value of the parcel and its improvements. It may be established for fiscal purposes (e.g. valuation and equitable taxation), legal purposes (conveyancing), to assist in the management of land and land use (e.g. for planning and other administrative purposes), and enables sustainable development and environmental protection.

Although the aim of this book is not to focus on one cadastre in particular, the Dutch Cadastre will be used as the main example. In the Netherlands the Cadastre, which is the responsibility of the Netherlands Cadastre and Land Registry Agency (Kadaster), comprises both cadastral registration and land registration. Land registration (in the Netherlands) is a public register in which documents describing interests in land are kept. In some countries land registration refers to the ordered and recorded legal documents as in the Netherlands, also called a deed registration, whereas in other countries land registration refers to a property register, also called a title registration.

The cadastral registration in the Netherlands is a record of the rights that are registered on land. In the cadastral registration essential information from documents recorded in the land registration is linked to a location (parcel). Cadastral registration (or cadastre) as used in this book refers to both the active process of registration and the result of registration (also called register).

Basic entities of cadastral registration are "real estate," "(real) property," or "property" and "subject." In general, land and buildings on the land are referred to as real estate, while various rights associated with land are called real property (or property).[48] The subjects are persons or organizations that are entitled to real estate through property rights.

Originally, cadastral registration was often introduced to assist in land taxation. Today cadastral registration also provides relevant information for land transactions and helps to improve the efficiency of those transactions and security of tenure in land in general. It provides governments at all levels with relevant information for taxation and regulation. Cadastral registration is increasingly used by both private and public sectors in land development, urban and rural planning, land management, and environmental monitoring and is no longer related to cadastral surveying and mapping alone.[48,203]

To be able to meet all these requirements, the main tasks of current cadastres can be defined as:

- To register the legal status of and governmental restrictions on real estate; the persons who have interests in land; what the interests are (nature and duration of rights, restrictions, and responsibilities); on what land the interests are established (information on parcels such as location, size, value).
- To provide information on the legal status of and governmental restrictions on real estate.

To perform these tasks adequately, cadastral registration must maintain correct and consistent information, consisting of a complete set of cadastral parcels as well as a record containing interests on the parcels. Moreover, cadastral registration must be organized in such a way that the legal status of real estate becomes clear when querying the cadastral registration.

Individualization of property started originally with subdividing the surface into property units using 2D boundaries. For this reason the basic entity of current cadastral maps is the "parcel," which makes the cadastral map a 2D map. To ensure completeness and consistency, 2D parcels may not overlap and gaps may not occur (forming a planar partition). Although parcels are represented in 2D, someone with a right to a parcel always has been entitled to a space in 3D; i.e., a right of ownership on a parcel relates to a space in 3D that can be used by the owner and is not limited to just the flat parcel defined in 2D without any height or depth. If the right of ownership only applied to the surface, use of the property would be impossible. Consequently, from a legal point of view cadastral registration always has been 3D. In the context of this book the question is posed: Is traditional cadastral registration, which is based on the concept of a 2D parcel, adequate for registering all kinds of situations that occur in the modern world or does cadastral registration need to progress to a 3D approach?

The FIG Bathurst Declaration[50] concluded that "most land administration systems today are not adequate to cope with the increasingly complex range of rights, restrictions and responsibilities in relation to land." Because many existing cadastres are still based on a paradigm that has its origin centuries ago, this paradigm needs to be reconsidered and adjusted to today's world. This book reconsiders the central paradigm of cadastral registration with respect to the issue of dimensions (2D and 3D).

This chapter presents the topic of this book and sets the outlines of the research described in this book. The chapter starts with a description of the need for a 3D cadastre

(Section 1.1). In Section 1.2 the scope of this research is presented; in Section 1.3 the research objectives are described. Related research to this book is presented in Section 1.4. The overall contribution of this work is described in Section 1.5. This chapter ends with an overview of this book.

1.1 NEED FOR A 3D CADASTRE

Pressure on land in urban areas and especially their business centers has led to overlapping and interlocking constructions (see Figure 1.1). Even when the creation of property rights to match these developments is available within existing legislation, describing and depicting them in the cadastral registration pose a challenge. This is not surprising in light of the FIG description of a cadastre in which the parcel is the basic entity. The challenge is how to register overlapping and interlocking constructions in a cadastral registration that registers information on 2D parcels. Although properties have been located on top of each other for many years, it is only recently that the question has been raised whether cadastral registration should be extended into the third dimension. The growing interest in 3D cadastral registration is related to a number of factors:

- A considerable increase in (private) property values
- The number of tunnels, cables, and pipelines (water, electricity, sewage, telephone, TV cables), underground parking places, shopping malls, buildings above roads/railways, and other cases of multilevel buildings, which has grown considerably in the last 40 years
- An upcoming 3D approach in other domains—3D GIS (Geographical Information Systems), 3D planning—which makes a 3D approach of cadastral registration technologically realizable

FIGURE 1.1 (Color figure follows page 176.) Example of complex property situations. Business district La Defense in Paris; a road and a metro in the subsurface intersect buildings and plazas.

The core terms used in this book are 3D cadastre, 3D property unit, 3D (property) situation, and parcel. A **3D cadastre** is a cadastre that registers and gives insight into rights and restrictions not only on parcels, but also on 3D property units. A **3D property unit**, also abbreviated as 3D property in this book, is that (bounded) amount of space to which a person is entitled by means of real rights. In fact, the traditional parcel, with only one person using the parcel, is also a 3D property unit (often not explicitly bounded). However, this has never caused any problems with respect to the third dimension, as current cadastral registration is adequate to give insight into these traditional property situations. The problems arise in 3D property situations.

3D property situations (in this book also abbreviated as 3D situations) refer to situations in which different property units (with possibly different types of land use) are located on top of each other or constructed in even more complex structures, i.e., interlocking one another (Figure 1.2). In this book 3D property situations are also referred to as **stratified properties**. In 3D property situations several users are using an amount of space (volume), which is bounded in three dimensions. These volumes are positioned on top of each other, either all within one base parcel (the volumes are located in the same parcel column defined by the boundaries on the surface) or crossing base parcel boundaries. Real rights are established to entitle the different persons to the different volumes. A **parcel** is a separated piece of land, to which a person (or persons) is (are) entitled with a real right, such as right of ownership. Although the ownership of land is not explicitly bounded in the third dimension, in most countries the ownership reaches as far as the owner has possible interest, while other persons are allowed to use space above and below a parcel as long as the user cannot reasonably object to this use (Figure 1.3).

Consequently, the geological subsurface may be very important for the factual demarcation of the third dimension of ownership. In areas with a solid geological subsurface, e.g., in most Scandinavian countries, a tunnel 25 meters below the surface will not cause any inconvenience to the owner of the surface parcel. Therefore, such a construction may be allowed according to the concept of the right of ownership, while in countries with a "soft" subsurface, the space below the surface may be of much more interest for the owner of the surface parcel because subsurface activity may damage surface property.

To register 3D property situations in current cadastres, the legal status of 3D situations has to be translated in such a way that it can be registered in the current cadastral registration (see Figure 1.4).

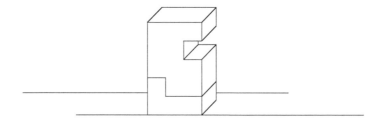

FIGURE 1.2 Example of 3D property situation.

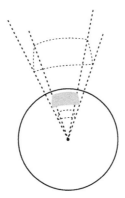

FIGURE 1.3 Illustration of the spatial extent of the right of ownership to a parcel.

FIG Commission 7 (Cadastre and Land Management) produced a vision of where cadastral registrations might be in 2014 taking current trends into account, such as the changing relationship of humankind to land, the changing role of governments in society, the impact of technology on cadastral reform, the changing role of surveyors in society, and the growing role of the private sector in the operation of the cadastre.[49] The study resulted in the following six statements on Cadastre 2014 based on a 4-year process involving input from many countries worldwide:

1. Cadastre 2014 will show the complete legal situation of land, including public rights and restrictions.
2. The separation between "maps" and "registers" will be abolished.
3. Cadastral modeling will take over cadastral mapping.
4. Paper-and-pencil cadastre will disappear.
5. Cadastre 2014 will be highly privatized and public and private sector will work closely together.
6. Cadastre 2014 will be cost recovering.

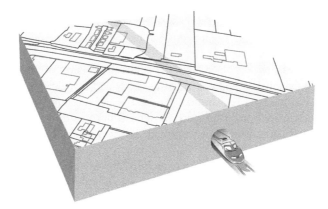

FIGURE 1.4 (Color figure follows page 176.) How to register 3D situations in a 2D cadastral registration.

Although the statement on Cadastre 2014 does not mention 3D cadastre explicitly, the report emphasizes that cadastres in the future will no longer be based on or restricted to 2D cadastral maps. Future cadastres will show the complete (thus, also in all dimensions) legal situation of land, including public rights and restrictions. Also demands from practice will have growing influence on cadastral registration in the future. These aspects motivate the study of the 3D issues of cadastral registration in a broad, integrated view. The result of such a broad integrated approach is that all rights, restrictions, and responsibilities related to land, often overlapping, are considered. This includes many more aspects than would traditionally be of interest and be recorded in a cadastral registration.[203]

Are current registration methods sufficient to fulfill the main tasks of a cadastral registration, i.e., to *register* the legal status of real estate and to *provide* insight into the legal status of real estate?

Since a few situations have occurred (and more are expected in the future) that could not be registered unambiguously and clearly in today's cadastral registrations, discussion started on how to handle 3D situations. To support this discussion the Netherlands Kadaster and the Delft University of Technology (TU Delft), the Netherlands, took the initiative to start research on 3D cadastral registration to study the needs, constraints, and possibilities of a 3D cadastre. This book is the result of this research, which was carried out at TU Delft in collaboration with the Netherlands Kadaster. The main part of the research presented in this book was published earlier.[190]

1.2 RESEARCH SCOPE

The scope of research on 3D cadastre is demarcated by three frameworks that determine the needs, constraints, and possibilities for 3D cadastral registration. These frameworks are linked to each other in a hierarchical order:

1. Juridical framework: How can the legal status of stratified properties be established? How to establish property boundaries other than traditional 2D parcel boundaries? What rights can be used and how can these rights be used?
2. Cadastral framework: Once the legal status of property in 3D situations has been established and described in deeds and in field works that are archived in the land registration, the next issues are how to register the rights and restrictions to property (bounded in three dimensions) in the cadastral registration and how to provide information on the legal status of 3D property situations.
3. Technical framework: What system architecture (computer hardware, software, data structures) is needed to support cadastral registration in 3D situations? What architecture is technologically possible?

This book focuses mainly on the cadastral and technical framework.

1.2.1 TOPICS WITHIN THE SCOPE OF THIS BOOK

Several fundamental considerations outline the scope of this book:

- Current cadastral registration (in combination with current land registration) serves its purposes well in most 2D situations and has a good foundation in today's society based on long history. It is therefore not feasible to visualize a 3D cadastre totally outside the current legal and cadastral framework. This does not mean that feasible adjustments in the framework cannot lead to improvements. Therefore, the precondition of this book is to start with current cadastral registrations and to see where these registrations suffice and where they need improvements or extensions in case of 3D situations. This precondition imposes special demands on a 3D cadastre, as the 3D cadastre should fit within current legal, cadastral, and technical frameworks.
- In this book cadastral registration in the Netherlands is used as the main example. Since cadastral registration abroad has similar fundamental characteristics, the main conclusions drawn in this book may be extended to other cadastral registrations. However, it should be noted that many minor differences are present between cadastral registrations in different countries due to different legislation and different implementation history.
- Cadastral registrations in other countries is considered as well, in order to examine the international need for 3D registration, to see if and how countries throughout the world solve the problem of 3D cadastral registration, and to come to more general (not valid only for the Dutch situation) conclusions.
- Both cadastral and technical issues are addressed. 3D cadastral issues deal with the main tasks of the cadastre in 3D situations and technical issues determine how these cadastral issues can be implemented.
- Since a DBMS (database management system) is an essential part of the architecture that is capable of maintaining large amounts of (spatial) data such as in cadastral registration, a main issue of this book is how to model 3D geo-objects (topologically and geometrically) in a DBMS.
- The cadastral registration must provide access to a wide spectrum of users (citizens, real estate agents, notaries, GIS/CAD specialists). Therefore, another major issue is how the cadastral DBMS can be made accessible for users.
- With respect to 3D GIS, efficient methods for geometric construction, data structuring, organization of 2D and 3D data in one environment, database creation and updating have yet to be developed. This book presents considerations and preliminary solutions for these issues.
- The main focus of this book is to provide technical solutions and technical recommendations to implement a 3D cadastre. For this purpose the needs for a 3D cadastre in general are studied and translated into technical needs. Current (commercially available) techniques are tested to evaluate if they are able to meet these needs. If fundamental solutions are not provided

by commercially available techniques, concepts are designed, which are tested by translating the concepts into prototypes.

1.2.2 TOPICS OUTSIDE THE SCOPE OF THIS BOOK

Topics that are not within the scope of this book can be described as:

- Juridical issues are addressed in this book, but are merely used as preconditions. It is not the aim of this book to give recommendations on major changes of legal systems. However, the experiences and findings in this book may lead to recommendations for developments and further research on legal issues.
- This book does not intend to develop an operational 3D cadastral registration, since this is not considered feasible at this stage, in which many issues still need to be resolved and in which choices need to be made on where to go to. This book first aims at a clear definition of the problem, development of concepts, and validation and evaluation of the concepts by prototyping key aspects.
- Functionality of 3D cadastral registration is the main topic of this research. Performance testing and benchmarking with respect to 3D cadastral registration or other information systems are therefore not part of this research.
- This book addresses cadastral registration in particular and therefore does not address topographical or other registrations.

1.3 RESEARCH OBJECTIVES

The main objective of this book is to answer the question of how to record 3D situations in cadastral registration in order to improve insight into 3D situations. The emphasis of this book is on the technical aspects of cadastral registration. To realize this objective, this book concentrates on four different topics:

1. **Analysis of the background.** This part focuses on identifying problems of current cadastral registration concerning 3D situations both to gain insight into the needs and requirements for 3D cadastral registration and to structure the international discussion on 3D cadastre.
2. **Framework for modeling 2D and 3D situations.** In this part techniques are explored that are needed for a 3D cadastre:
   - How to model 2D and 3D geo-objects in a DBMS, which is the core of the new generation GIS architecture.
   - What is the state-of-the-art of 3D GIS.
   - How to access and analyze 3D geo-objects organized in a geo-DBMS.
   - How to combine 2D parcels and 3D geo-objects in one environment.
3. **Models for a 3D cadastre.** In this part conceptual models are designed based on current registration and on available techniques in order to improve 3D cadastral registration. Also, considerations are given for translating the conceptual models of a 3D cadastre into logical models.

4. **Realization of a 3D cadastre.** The proposed conceptual models are evaluated by translating conceptual alternatives into prototype implementations using techniques explored and developed as part of this book and by performing functional tests. Performance tests are not part of this book.

1.4 PREVIOUS AND RELATED RESEARCH

Research related to this book, which focuses on cadastral and technical aspects of a 3D cadastre, can be classified into research on 3D cadastral registration and research on 3D tools and 3D modeling.

1.4.1 RELATED RESEARCH ON 3D CADASTRES

Israel is one of the countries that faces high pressure on the use of land. This has promoted development of a 3D cadastre. Therefore, in Israel for the past 5 years several studies have started on 3D cadastres.[10,52,53,59,60,184] Mid-European countries such as Ukraine,[112] Hungary,[154] Czech Republic,[77] and Slovenia[164] are in the phase of examining the current cadastre for potential registration of 3D property units, including apartments.

International marine cadastres traditionally have a 3D approach, as the use of the marine environment is volumetric by nature and involves rights to the surface, water column, seabed, and subsoil. The University of New Brunswick (Canada), Department of Geodesy and Geomatics Engineering is developing a 3D marine cadastre to support effective and efficient decision making associated with marine governance.[121,222] In Reference 55 the framework issues that must be considered in the development of marine cadastral data and the use of these data in a marine information system for the United States are discussed. In this discussion 3D aspects are also addressed.

Some other countries and states have already solved part of 3D cadastral registration (Norway, Sweden, Queensland, the United States, Argentina, and British Columbia), as is seen from the international study on 3D cadastral registration (Chapter 3).

1.4.2 RELATED RESEARCH ON 3D TOOLS AND 3D MODELING

3D registration deals with maintaining spatial and non-spatial information on 3D objects, which are core topics of 3D GIS. Therefore, developments in 3D GIS are important when examining a 3D registration.

The main characteristic of research on 3D models intended for 3D GIS and 3D geo-DBMSs is that they are extensive and that the results of this research are fragmented. Examples of 3D models intended for 3D GIS and 3D geo-DBMSs exist.[51,90,91,114,162,163,177] Implementations of 3D models in user-developed systems can be found in several publications.[20,139,174,217]

Research on spatial querying and 3D visualization of geo-objects using Web technologies has resulted in several prototype systems.[14,31,35,93,100,229] Research on spatial querying and 3D visualization of geo-objects organized in a DBMS has not yet resulted in any publications, apart from publications that were written as part of this research.

Since developments in 3D GIS are important when studying the possibilities for a 3D cadastre, a section is included in this book that describes the current state of the art of 3D GIS (Section 7.1).

1.5 CONTRIBUTION OF THE WORK

The main contributions of this work can be summarized as follows:

- Enabling a complex registration addresses many issues in a variety of disciplines (technical, cadastral, legal, organizational). This book is the first extensive research on 3D cadastres in which the problem of registration in complex situations has been studied using an integrated approach. Therefore, this book has strong explorative characteristics resulting in a clear analysis as well as a distinct definition of the essential problems of registering 3D situations in current cadastres taking all involved disciplines into account.
- This book structures the international discussion on the need for 3D cadastre by providing a universal overview of the basic and fundamental needs for a 3D cadastre, considered from different points of view (legal, cadastral, technical) and by providing insight into country-specific aspects that influence the need for and possibilities of 3D cadastral registration.
- This book gives solution directions for a 3D cadastre. Several models for a 3D cadastre are introduced and translated into prototype implementations. Experience with the prototypes will result in concrete recommendations. Based on these recommendations, decision makers will be able to base choices on if, when, and how to implement a 3D registration on fundamental considerations.
- In technical respects this book describes 3D GIS issues in general, i.e., how to model and maintain 3D geo-objects in a DBMS, how to access and query these objects by front ends, and how to combine 3D geo-objects and 2D geo-objects in one 3D environment. With respect to improving 3D GIS functionalities, an extension of a geo-DBMS has been built to support 3D primitives. Further, a study was carried out to generate an appropriate integrated height model in a TIN (triangular irregular network) structure based on both the 2D planar partition of parcels and point heights.
- This work contributes to supporting the demand for 3D geo-information in today's society in general. Other organizations responsible for (spatial) registrations and for spatial data sets can use the outcomes of this work to see the possibilities and constraints to extend their systems into the third dimension (e.g., registrations for cultural heritage, for buildings, for zoning plans, for cables and pipelines, and databases of topographical mapping agencies).

1.6 ORGANIZATION OF THE BOOK

Chapter 1 (this chapter) presents the need for a 3D cadastre, specifies the objectives, the scope, and the contributions of this research, and describes related research. This book is divided into four major parts corresponding with the four main research topics of this book as described in Section 1.3.

1. Part I: Analysis of the Background (this chapter and Chapters 2, 3, and 4)
2. Part II: Framework for Modeling 2D and 3D Situations (Chapters 5, 6, 7, and 8)
3. Part III: Models for a 3D Cadastre (Chapters 9 and 10)
4. Part IV: Realization of a 3D Cadastre (Chapter 11)

The book ends with conclusions (Chapter 12). Readers who are familiar with cadastral registration with respect to the 3D component and are less interested in a detailed study on needs for 3D cadastral registration may skip Part I. Readers who are familiar with spatial modeling in DBMSs both in 2D and 3D and with accessing this information with front ends or readers who are not interested in technical issues of 3D cadastral registration may skip Part II. The introduction and evaluation of new conceptual data models for 3D cadastral registration are described in Parts III and IV.

In Chapter 2 the scope of this research on 3D cadastre is further outlined. The chapter starts with a description of the elementary cadastral model since this model is the basis of all cadastral registrations. The way a cadastre is currently implemented and organized in a country influences the requirements for a 3D cadastre. Thus, Chapter 2 contains a description of common alternatives of cadastral registrations. In the remainder of Chapter 2 an overview is given of the types of cadastral recordings with a potential 3D component. The aim of this overview is to achieve a clear view on which domain the 3D cadastral research should focus. For what types of cadastral recordings should a 3D approach of registration be considered? Cadastral recordings in this chapter are subdivided into cadastral recordings according to private law and cadastral recordings according to public law.

3D property situations, in which different property units are located on top of each other, have been common for decades. How are these situations recorded in cadastral registrations and do these recordings give sufficient insight into the real situation? In other words, what is the actual need for a 3D cadastre? To answer these questions seven countries and states were examined: the Netherlands, Norway, Sweden, British Columbia (Canada), Queensland (Australia), the United States, and Argentina. The results of these international studies are reported in Chapter 3; this chapter also includes case studies that were carried out in the Netherlands, Queensland, the United States, and Argentina.

Chapter 4 elaborates on the needs and opportunities for a 3D cadastre based on the findings described in Chapters 2 and 3.

Chapter 5 aims at clarifying some basic terms and concepts concerning spatial data modeling that are used and applied in this book. Data models and spatial data models in particular are described, followed by a description of the basic phases of data modeling. UML (unified modelling language) is used in this book to describe the data models for a 3D cadastre. The basic characteristics of UML are explained.

How the relationship between spatial data modeling and DBMSs has evolved is also discussed. The chapter ends with a description of standardization initiatives.

As is concluded in Chapter 5, geo-DBMSs are the core of the new-generation GIS architecture. Thus, Chapter 6 discusses the state of the art of geo-DBMSs: how spatial objects can be maintained in 2D and 3D in a geo-DBMS using both geometrical primitives and a topological structure. Spatial analyses on both geometrical primitives and a topological structure are considered as well. The chapter also contains a section describing the implementation of a 3D primitive in a DBMS, a study that was carried out as part of the research presented in this book.

3D GIS is a basic instrument to deal with 3D geo-information in general. Therefore, the state of the art of 3D GIS aspects other than geo-DBMSs is discussed in Chapter 7. Chapter 7 reports also the results of a research that was carried out to access (query and visualize) 3D objects that are organized in a geo-DBMS. For this research three front ends were studied: a CAD oriented system, a GIS system, and a self-implemented system using Web-based techniques.

Chapter 8 deals with the fundamental issue of combining 3D geo-objects (3D cadastral objects) and 2D geo-data (parcels) into one system: how to relate the two data sets in space. A case study was carried out to show possibilities and problems of integrating a 3D geo-object (pipeline) and surface parcels in one environment. TINs, representing integrated height models of point heights and parcels, that were created during this case study are described, together with their data structure and their results. The TINs are inserted in the DBMS, which makes it possible to perform spatial analyses on height surfaces of individual parcels. To obtain a more effective height model, a generalization method was developed and is described in this chapter. This method has been partly implemented in a prototype. The prototype selects only the significant TIN-nodes while removing the nonsignificant TIN nodes. Results of the prototype are also reported.

Chapter 9 introduces three concepts for a 3D cadastre, each with different alternatives, which were designed as part of this research. Based on both cadastral and technical considerations, two of these concepts are indicated in this chapter as the most optimal solution for a 3D cadastre: a hybrid 3D cadastre (short to medium term) and a full 3D cadastre (medium to long term).

Chapter 10 considers issues that accompany translating the conceptual models that were introduced in Chapter 9 into logical models: issues concerning the spatial data model, the administrative data model, as well as the process of data collection to obtain data that can be inserted into the spatial data models. Further, 4D requirements (history) of a 3D cadastre that need to be taken into account in the phase of logical modeling are considered.

Chapter 11 evaluates the proposed conceptual models from Chapter 9 by applying prototypes that contain the key aspects of the conceptual and logical models, to a selection of the case studies introduced in Chapter 3.

Chapter 12 summarizes this book and concludes the major findings of this research. In this chapter it is concluded that a full 3D cadastre is a feasible solution to solve 3D cadastral issues at a fundamental level, taking legal, cadastral, as well as technical aspects into account. The chapter also contains recommendations for future research.

Part I

Analysis of the Background

2 Scope of a 3D Cadastre

Since the current cadastral data model is the starting point of the 3D discussion, this chapter first describes the data model underlying current cadastral registration (Section 2.1). Many types of cadastres exist based on country-specific characteristics, such as local cultural heritage, physical geography, land use, technology, etc. The type of a cadastre (organization, technical implementation, etc.) influences needs as well as possibilities for 3D registration. This chapter contains an overview of different types of cadastral registrations and a discussion on how the different types influence the discussion on 3D cadastre (Section 2.2).

A clear view of the cadastral domain on which the 3D cadastral research should focus is needed to streamline the 3D cadastre discussion. In Sections 2.3 and 2.4 the types of cadastral recordings that have a 3D component are described. In this description a distinction has been drawn between cadastral recordings according to private law (Section 2.3) and cadastral recordings according to public law (Section 2.4). We have tried to give a general, non-country-specific overview. This was not always possible, because existing cadastral recordings are a consequence of legal systems, which are country specific. Where it was not possible to explain the cadastral recordings in general because of differences between legal systems, we used the Netherlands as our main example. This chapter ends with conclusions.

2.1 ELEMENTARY CADASTRAL MODEL

Traditionally, cadastral registrations consisted of a set of cadastral maps containing cadastral parcels with unique parcel numbers and a paper archive in which property information on parcels was maintained. Since the last decennia, cadastral registrations in developed countries were converted from analog cadastral registrations into digital registrations. Spatial information on parcels is no longer maintained on paper maps but in GIS and CAD or even more sophisticatedly in spatial DBMSs. Information on property and other information that is nowadays registered in cadastral registrations (mortgage, soil pollutions, monuments) is no longer maintained exclusively in paper archives but in cadastral databases. A link is maintained between the digital cadastral map and the cadastral administrative database. The link provides the possibility to query the spatial part and administrative part of cadastral registration and combine the results. In more advanced systems it is possible to query the spatial and administrative part of cadastral registration in one integrated environment.

The current administrative cadastral data model is based on three key types: real estate object, person (subject), and right or restriction. The UML class diagram of the data model is shown in Figure 2.1 (see also Reference 98). Real estate objects are (part of) parcels and apartment rights (linked to a "mother" parcel, not shown in Figure 2.1). Persons are natural persons or organizations (non-natural persons) with rights on parcels. Beside rights, there can also be a "restriction" relationship

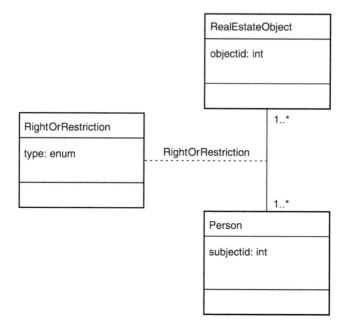

FIGURE 2.1 The current administrative cadastral data model in a UML class diagram.

between a real estate object and a person. Real estate objects and persons have n : m relationships via rights (and restrictions); a person can have rights related to more than one real estate object (e.g., a person owns three parcels) and one real estate object can be related to more than one person (e.g., one person is bare owner of a parcel and another person has the right of superficies on the parcel).[130] Every person in the registration should be associated with at least one real estate object, and vice versa; every real estate object should be associated with at least one person (indicated with the multiplicity of "1..*"; for a description of UML, see Section 5.5). A more extensive description of this data model is described elsewhere.[98]

2.2   TYPES OF CADASTRAL REGISTRATIONS

Cadastres can be classified in many ways, based on different criteria, e.g., as proposed in Reference 48:

- Primary function (e.g., supporting taxation, conveyancing, land distribution, or multipurpose land management activities)
- The types of rights recorded (e.g., private ownership, use rights, mineral leases, public law restrictions)
- The degree of responsibility in ensuring the accuracy and reliability of the data (e.g., complete state mandate, shared public and private responsibility)
- Location and jurisdiction (e.g., urban and rural cadastres; centralized and decentralized cadastres)

- The many ways in which information about the parcels is collected (e.g., ground surveys tied to geodetic control, uncoordinated ground surveys and measurements, aerial photography, digitizing existing historical records, etc.)

All these factors determine the required resolution and scale of spatial data, the type and characteristics of data recorded in both thematic and geometrical attributes, and the organizational and professional responsibility for managing the data. Consequently, these factors also influence the need for 3D cadastral registration in a specific country and how the 3D issue is or will be approached.

Different classifications have been proposed to describe most common alternatives for cadastral (and land) registration.[226,227] These classifications are based on the most essential criteria. Because these classifications form a good overview of the differences that may exist between different cadastral registrations, the classifications are summarized below.

2.2.1 DEED VS. TITLE REGISTRATION

The classification of deed registration vs. title registration is the classification used most often. The most basic difference is that "deed registration is concerned with the registration of the legal fact itself and title registration with the legal consequence of the fact."[69] However, mostly also other factors are taken into account when distinguishing between titles and deeds. A complete definition is given in Reference 69:

Deed registration: A deed registration means that the deed itself, being a document, which describes an isolated transaction, is registered. This deed is evidence that a particular transaction took place, but is in principle not itself proof of the legal rights of the involved parties and, consequently, it is not evidence of its quality. Thus before any dealing can be safely effected, the ostensible owner must trace his ownership back to a good root of title.

Title registration: A title registration means that it is not the deed describing the transfer of rights that is registered but the legal consequence of that transaction, i.e., the right itself (title). So the right itself together with the name of the rightful claimant and the object of that right with its restrictions and charges are registered. With this registration the title or right is created.

In deed registrations (which is common in most of the countries in Western Europe and many of their former colonies, in the United States, and in countries in Latin America falling under Spanish/Portuguese law), the documents filed in the land registration are the evidence of title. The registration itself does not prove title: it only records a transaction between parties. In title registrations (common in the United Kingdom, most of the countries of the Commonwealth and many countries in Central Europe), the register itself serves as the primary evidence. The title is constituted by registration. The registration of title enables a title to be ascertained as a fact. A title registration is an authoritative record kept in a public office. The register is maintained and warranted by the state.

As concluded in Reference 227, the debate on "title versus deeds" is complicated, since no distinct definition can be given. Also technological developments have provided the instruments to decrease the former differences. Generally speaking, there are examples of good and bad title registrations and good and bad deed registrations. The real protection of land ownership is more dependent on the quality of information in the land and cadastral registration and not on the type of land and cadastral registration. To avoid making the 3D cadastral issue more complicated than necessary, this debate is left out of this book. The classification based on titles vs. deeds was mentioned here for completeness.

2.2.2 A Centralized or Decentralized Cadastral Registration

In every country the protection of rights to land is considered a governmental task. However, not every country has a strong national authority. In some cases, financial and technical responsibility lie at the regional or even local level. Therefore, cadastral registration may be the responsibility of local governments, while in other cases it is a state or national responsibility. Apart from the question of whether local or national government is responsible for cadastral registration, cadastral registration can be carried out at different levels (in a central database, in regional or local databases, or at regional and local levels while a centralized database is maintained). The question of the existence of a centralized cadastral database is dependent on three main aspects:

1. State of the art database technology. A cadastral registration consists of an administrative and a spatial part mostly maintained in databases. For decentralized systems many databases have to be maintained, which should be avoided since databases (especially the spatial part) require expensive equipment and expertise. Technical development in the area of databases also motivates concentration at a national level, as DBMS technology favors an approach of one centralized DBMS in which all objects of interest for a specific application are maintained. A centralized DBMS is easier and cheaper to manage.
2. State of the art telecommunication by mobile telephones and Internet facilities. Decentralized systems were set up to bring cadastral information closer to end users. With the modern technologies of telecommunication and the Internet it is no longer as relevant where cadastral information is maintained.
3. The question whether to have a centralized or a decentralized cadastral system is dependent in the way a specific country organizes its whole administration, as a cadastral system is part of the administration of a country.

The ministry responsible for the cadastral registration also differs per country:

- Ministry of Finance. This is mostly the case when a cadastral registration was originally started as a fiscal cadastre.
- Ministry of Agriculture. In some countries this ministry has the responsibility only for rural activities (land consolidation), while in other countries this ministry has the responsibility for the whole national cadastre (e.g., Hungary).

- Ministry of Housing or the Ministry of Public Works. This ministry has the responsibility for the urban cadastre.
- Ministry of Justice. The Ministry of Justice has the responsibility for cadastral registration since land registration originally had a legal nature. Registration takes place in local courts (Austria and Romania).
- Ministry of Interior (Poland).
- A separate authority is responsible to prevent the discussion of the ministerial responsibility.

At what authority level and by which ministry the 3D issue of cadastral registration is approached depends on the organization of cadastral registration.

2.2.3 LAND REGISTRATION WITH SEPARATE OR INTEGRATED CADASTRE

In several countries land registration and cadastral registration are the responsibilities of one organization. This makes it easier to make the contents of both registrations synchronized. In other countries the separation of land registration and cadastral registration has a historical background (e.g., Denmark, Austria, Bulgaria, and Poland). In these countries the land registration and cadastral registration are mostly the responsibility of different ministries. Land registration has generally been the mandate of the courts and the legal profession. Mapping, parcel boundary delimitation, and maintenance of parcel data for fiscal, land use control, and land redistribution purposes are traditionally the responsibility of the surveying profession.[48] In the case of manual registrations, it is difficult to keep two separated registrations up-to-date and identical. In an integrated cadastre, land registration and cadastral registration are better geared to one another synchronized. Improvement of information supply in case of 3D situations can be achieved only by the combined efforts of both land registration and cadastral registration. In a separated system it will be harder to join the two registrations in order to achieve one common goal (i.e., improve insight in 3D situations). A 3D cadastre cannot be achieved without tight collaboration between land registration and cadastral registration.

2.2.4 FISCAL OR LEGAL CADASTRE

Very often cadastral registrations started as a fiscal cadastre for taxation, e.g., the "Napoleontic Cadastre." Such cadastres were based on a full survey of the ownership parcels. After a few decades such fiscal cadastres were changed into legal cadastres. In some countries there are still problems with the old cadastral maps. For example, in the Ardennes, in Belgium, the cadastral maps give the real area of the surface of land parcels that are located on the slope of hills. The transfer of this information to a cadastral map, which is a projection of the terrain onto a horizontal plane, is so expensive that a digital map in this country was never produced (note that this is a nice example of a 3D cadastral aspect). A fiscal cadastre is less complex than a legal cadastre. In the case of a fiscal cadastre a cadastre can be less accurate in maintaining geometry and other attributes if the property tax is based on valuation. In addition, a fiscal cadastre needs an update every year (when following a yearly tax cycle), whereas a legal cadastre needs an update every day. A legal cadastre will therefore impose more conditions on the availability of information in 3D situations.

2.2.5 GENERAL OR FIXED BOUNDARIES

A parcel is defined by indicating its boundaries. General boundaries are boundaries that have to be visible features on the landscape. These features are supposed to coincide with the position of boundaries and can be mapped relatively easily because the features are easy to measure with surveying, with aerial photogrammetry, or from topographic maps. Although these boundaries do not indicate the exact location of parcel boundaries, the parcel is reasonably defined and can be identified beyond doubt. In case of fixed boundaries, all parties involved have to agree fully on the exact position of each boundary point (after which the position of parcel boundaries can be marked on the terrain). The demarcation, measuring, and registration of fixed boundaries require more time. Once 3D property units are defined within cadastral registration, the type of boundaries within the specific cadastral registration (general or fixed) will impose requirements for the boundaries demarcating 3D property units.

2.2.6 FINANCED BY GOVERNMENT OR COST RECOVERY

In general, the maintenance of the cadastral registration is a regular task of the government, which means that normal cadastral activities are generally financed by the government. Cadastral registrations generate income from fees for registration of transactions, mortgages, etc. and supply of information. The income generated by the cadastre goes directly to the state treasury when the cadastre is a regular task of the national government. Consequently, there is no link between the income and the expenses of the cadastral registration. The motivation to take care of user requirements, e.g., to establish a 3D cadastre, is therefore limited[226] unless it is imposed by the government. The alternative is that a cadastre is an independent organization responsible for its own income and expenses, forcing is to listen to changing user requirements.

2.3 3D CADASTRAL RECORDINGS AND PRIVATE LAW

Regarding private law, the main types of cadastral recordings with a 3D component are as follows:

- Right of ownership (Section 2.3.1)
- Limited ownership rights:
  - Right of superficies (Section 2.3.2)
  - Right of long lease (Section 2.3.3)
  - Right of easement (Section 2.3.4)
- Apartment rights (condominium rights) and strata titles (Section 2.3.5)
- Joint ownership via other real estate objects (Section 2.3.6)

In the remainder of this section, these rights are described, together with the cadastral registration of these rights. In this description the cadastral and legal frameworks in the Netherlands are used. Although these frameworks have characteristics that are specific for the Netherlands, the fundamentals can be extended to other countries.

2.3.1 RIGHT OF OWNERSHIP

The most extensive right that a person can have is the full right of ownership. To the exclusion of everybody else, the owner is free to use the thing, provided that its use does not breach the rights of others and that limitations based on statutory rules and the rules of unwritten law are observed (see, for example, Reference 38).

The right of ownership of a parcel has a 3D component. This becomes obvious when the upper and lower boundaries of the right lead to dispute, i.e., when more than one person uses the parcel. Actually, the right of ownership to a parcel (as all other real rights) always relates to a space; otherwise the use of the parcel would be impossible. According to Articles 20 and 21 of Book 5 of the Dutch Civil Code,[38] the right of ownership comprises:

- Exclusive use of the right of space above the parcel
- Ownership of the earthlayers beneath it
- Ownership of buildings and constructions forming a permanent part of the land (directly or by means of other constructions)

This quotation from the Civil Code indicates the ambiguity of the way ownership is defined in the third dimension: the third dimension of ownership is not explicitly bounded. The ownership to a parcel includes the competence to use the land owned. This includes the space above and under the parcel to a height and depth to which the user has (possible) interest. The use of space above and under the surface is permitted to third persons, as long as this is sufficiently high or low that the owner cannot reasonably object to this use, or when this use is regulated by other laws, e.g., by the "Law on Air Traffic," which prescribes regulations for air traffic or by the "Law on Mining," which provides the possibility to extract minerals in the ground of private owners by concession or permit.

Since ownership is not explicitly limited in the third dimension, in principle the right of ownership of land reaches from the middle of the earth up to the sky. Horizontal division of this volume is only possible by establishing rights and limited rights on the surface parcel, such as a right of superficies (Section 2.3.2), right of long lease (Section 2.3.3), right of easement (Section 2.3.4), apartment right (Section 2.3.5), and joint ownership (Section 2.3.6). Horizontal division of the volume enclosing the whole parcel column leads to 3D property units, which are bounded spaces to which persons are entitled by means of real rights.

Restrictions according to public law (Section 2.4) and restrictions imposed by regional and local land use plans, e.g., no more than five floors per building, can also restrict owners in using their parcels (column). Restrictions according to regional and zoning plans are not registered in all cadastral registration and are therefore not considered a focus of this book.

Vertical Accession to Real Estate

The basic rule of accession, derived from Roman law, is that buildings and other constructions that are permanently fixed to the land are considered part of that land. Consequently, constructions below or above the surface that are permanently fixed to the surface are owned by the owner of the land unless other rights or restrictions have been established on the surface parcel (*"superficies solo cedit"*). However, this is not always a strict rule. The owner of a construction below or above the surface is not necessarily always the same person as the owner of the land parcel.

Horizontal Accession to Real Estate

Constructions fixed to the land are part of the property by vertical accession, *unless* the construction is part of another property. In that case the parts encroaching another parcel are part of the main part by the rule of horizontal accession (Figure 2.2). Consequently, these parts do not belong to the encroaching parcel, as would be the case using the rule of vertical accession. Therefore, the owner of the main construction is the owner of parts of the construction that encroach another parcel. For example, where the ownership of a tunnel is not explicitly established and registered on an intersecting parcel, the owner of this component of the tunnel could be found by finding the point, and thus the parcel, where the main part of the tunnel is fixed to the surface, which is presumably where the entrances are (see also discussion in References 66 and 67). The owner of the parcels containing the entrances can in this case be seen as the owner of the entire construction, including the components that run below the surface parcels against which no rights have been established.

With a horizontal accession to real estate, a factual horizontal division in ownership takes place. The legal status follows from the factual situation. Consequently, the legal status may change if the factual situation changes. The disadvantage of horizontal accession to real estate is that the legal status of the situation is not registered and therefore not clear in the cadastral registration.

The horizontal accession to real estate might conflict with the definition of the right of ownership (vertical accession to real estate). According to the Civil Code, the right of ownership contains all constructions that are permanently fixed to the parcel while in the case of horizontal accession to real estate the owner of a parcel

FIGURE 2.2   An illustration of "horizontal accession to real estate." The part of the gray house that is situated under the white house (cellar) belongs to the owner of the parcel under the gray house because this part is a component of the gray house.

that intersects with a construction is not the owner of the construction. In principle, vertical accession always gets priority *unless* horizontal accession can be applied.

It should be noted that the horizontal accession to real estate does not justify the factual situation. It is, for example, not allowed to build a construction encroaching another parcel without permission of the owner of the encroached parcel.

2.3.2 RIGHT OF SUPERFICIES

The right of superficies is a real right to own or to acquire buildings, works, or vegetation in, on, or above an immovable thing owned by another. A construction may also intersect the surface level (located partly below and partly above the surface). The holder of this limited right is the owner of the construction. As a limited real right it restricts the original owner of the land: the owner has to tolerate the existence of the construction in, on, or above his or her land. A right of superficies can be used when the owner of the construction is not the same as the owner of the parcel. By means of this right, a horizontal division in ownership takes place.[85] In most countries no geometry is maintained in the cadastral registration to reflect the spatial extent of the ownership of the buildings nor of the right itself, at least not in 3D. Mostly, the complete parcel is affected with the right of superficies. Although not visible on cadastral maps, it is often possible to add a drawing to the deed or title to clarify the situation. The establishment of a right of superficies provides the possibility to dictate restrictions to the owner of the land in order to avoid damage to the construction.

2.3.3 RIGHT OF LONG LEASE

The legal status of constructions below or above the surface can also be established with a right of long lease (*emphyteusis*). Right of long lease is a legal instrument that is sometimes used in 3D situations; however, this right is not specifically meant for 3D situations. A right of long lease gives the long leaseholder the permission to hold and use the parcel of the bare owner, as if the leaseholder were the owner. The deed of establishment may impose an obligation upon the leaseholder to pay a sum of money to the owner every year. This deed also contains an end date of the lease. Generally, it is not possible to impose a right of long lease to just a part of a parcel or a part of a "parcel column": i.e., no (legal) horizontal division in ownership takes place by a right of long lease. The right of long lease includes the surface parcel as well as space below and above the parcel including the buildings that are fixed to the parcel.

When the legal status of constructions above or below the surface is established with a right of long lease, usually the bare owner of the parcel is the "user" of the 3D construction. The long leaseholder has the right to use the parcel above (or below) the construction. By means of conditions imposed on the leaseholder (described in the deed), the use and protection of the construction can be arranged and also the dimensions to which the right of long lease applies (which causes a factual horizontal division in ownership). The geometry of the space to which a right of long lease applies is not maintained in the cadastral registration and can only be specified in a drawing attached to the deed.

2.3.4 RIGHT OF EASEMENT

An easement (servitude) is a charge (encumbrance) imposed on a parcel (the serving parcel) in favor of another parcel (the dominant parcel). An example of this is when owner A of a parcel can reach the public road easier by crossing the parcel of neighbor B rather than crossing his or her own parcel. An easement can be imposed on the parcel of B in favor of the parcel of A, which makes it possible for owner A to cross the parcel of B.

In general, the deed establishing an easement may impose an obligation on the owner of the dominant property to pay to the owner of the serving property a sum of money. The easement must be exercised in a way that causes the least inconvenience to the serving property. In the example this means that A has to take the shortest path across the parcel of B. When the dominant property is divided, the easement continues to exist for the benefit of each part to which it may be beneficial. The easement is linked to a parcel (establishing a parcel-to-parcel relationship): when the parcel is sold, rights and restrictions of an easement are taken over by the next parcel owner.

In some countries the geometry of a right of easement is explicitly indicated in the cadastral geographical data set (cadastral map) while in other countries, as in the Netherlands, only a drawing can be added to the deed or title specifying the spatial extent of the easement. There is no cadastral registration that maintains the 3D geometry of easements. To obtain insight into the 3D geometry of easements one must look in the deeds that established the easement. The vertical dimension of a right of easement can be relevant in 3D situations, for example, when a right of easement is established for a bridge above the serving parcel or for a pipeline that crosses a parcel. It is also possible to establish a right of easement for a building on a serving parcel. In all these cases 3D insight into the real situation can be improved by a 3D cadastral map indicating the space to which the right applies.

In some countries, it is also possible to establish a right similar to a right of easement without linking it to a dominant parcel. This can be used when a right of easement is established for a pipeline that has no clear dominant parcel. The restriction in this case is established on the serving parcel in a deed and registered in the administrative part of the cadastral registration. Also in this case the geometry of the easement is not added to the cadastral geographical data set. The restriction is linked to the subject who causes the restriction. It can be seen as a contract that imposes an obligation on an owner of land to tolerate, for example, a pipeline. This obligation is also binding for future owners.

2.3.5 APARTMENT RIGHT AND STRATA TITLE

The most frequently occurring 3D situations are apartment complexes. Most countries have introduced legal instruments to establish the ownership of apartment units, such as apartment rights (condominium rights) and strata titles. In Germany, France, and most other European countries legislation on apartment ownership is based on the so-called dual system.[2] Every apartment owner has the full ownership of a part of the building (apartment). The communal areas of the building, such as staircases and elevators, are held in co-ownership. This can be described as compulsory co-ownership, or an accessory restricted co-ownership — "accessory" because it cannot be separated from the ownership of the apartment, "restricted" because during the time the building is divided into apartments, the separation and division of the common areas is not possible.

Some European countries have adopted the "unitary system," e.g., Norway, Austria, Switzerland, and the Netherlands.[2] It is important to note that in this system the apartment ownership is based on co-ownership of the whole complex — consisting of ground parcel(s) and buildings on the parcels(s).

Article 106 of Book 5 of the Dutch Civil Code[38] describes apartment ownership or apartment right (*appartemensrecht*) as follows:

1. An apartment right means a share in the ownership of the property involved in the division which also comprises the right to exclusive use to certain parts of the building which, as indicated by their lay-out, are intended to be used as separate units. The share can also include the right to exclusive use of certain parts of the land pertaining to the building.
2. An apartment owner means a person entitled to an apartment right.

The owners of the apartment units are joint owners of the entire building and the ground below. The underlying ground may consist of several parcels, which can be disjoint. The co-ownership includes the right to have the exclusive use of a certain part of the building: the apartment unit. This means that the persons do not legally own a separate apartment unit, although the apartment ownership can be mortgaged.

The division in apartment rights is based on a notarial deed, the so-called deed of division. A plan (mostly) obliged in this deed is maintained in paper or scanned format in the land registration. This plan gives an overview of the building and a detailed plan of each floor. Thick, dark lines indicate the borders of every apartment, i.e., the area of exclusive use. How apartment units are registered in the current cadastral registration is described in more detail by a case study in Section 3.1.3.

Kenya, South Africa, Australia, and England (basically all Common Law countries) use strata titles[185] in the case of 3D situations. Different terms (e.g., sections) are used in these countries for more or less the same concept. In all those countries analog or scanned drawings of the situation are archived in the land registration. These drawings include an overview of the complete parcel divided into individually owned units and common property, augmented by the cross sections of the different buildings (Figure 2.3). These drawings are maintained in separate archives and not incorporated in cadastral registrations (cadastral map).

Although an apartment right or a strata title is the best way to establish multilevel ownership on one parcel, the registration of the property units involved can still be improved. In current registrations only the ground parcel of the building complex is maintained as part of the cadastral geographical data set, and therefore the individual units cannot be recognized on the cadastral map. Consequently, individual units cannot spatially be queried, although ownership information on the individual units is mostly available in the administrative part of the registration. Another complication that can be mentioned is that an analog (or scanned) drawing is used to clarify the cadastral situation in the deeds and titles. Spatial information available in vector format and in

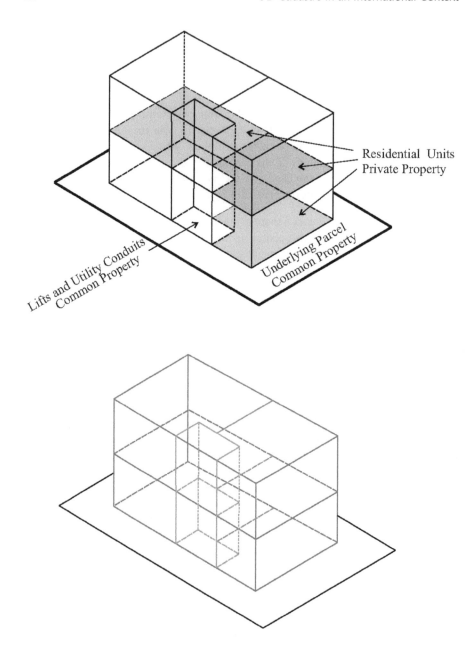

Residential Units
Private Property

Lifts and Utility Conduits
Common Property

Underlying Parcel
Common Property

FIGURE 2.3 Example of drawing in strata title. (Courtesy of Michael Barry.)

real-world coordinates would make it possible to integrate the information from the drawings with the cadastral geographical data set.

Basic characteristics of property units in building complexes are that the units within one complex have a legal relationship with each other (e.g., they share common area in a building) and the building complex is concentrated on one or several parcels.

However, apartment rights and strata titles are also used in the case of independent stratified properties crossing parcel boundaries, e.g., for shops and dwelling units in one building or for public underground parking, which motivates the search for a more general solution for 3D situations. This solution should better reflect the nature of independent multilevel ownership on one parcel or ownership crossing several parcels.

2.3.6   JOINT OWNERSHIP VIA OTHER REAL ESTATE OBJECTS

A special type of joint ownership is called in the Dutch land law *mandeligheid* (compare the French *mitoyennet*). This is a right to land and/or a construction that can be registered similarly to common areas in condominiums. This immovable thing arises when an immovable thing is jointly owned by the owners of two or more properties and where it is designated by them for the common benefit of those properties by a notarial deed between them, which is then recorded in the public registers.[38] Joint ownership comprises the obligation of each joint owner to give the other joint owners access to the thing held in joint ownership. Things held in joint ownership must be maintained, cleaned, and, if necessary, renewed at the expense of all joint owners. A specific cadastral characteristic of joint ownership is that it is registered only on a parcel and is not linked to a subject. The 3D characteristic of an immovable thing held in joint ownership can be important in cadastral registration when the whole parcel is not held in co-ownership, e.g., underground parking places, swimming pools, tennis courts, aerials, etc.

2.4   3D CADASTRAL RECORDINGS AND PUBLIC LAW

Cadastres also register restrictions in the ownership of parcels as dictated by public laws. For a better understanding of the public law restrictions that are registered in current cadastral registrations, a selection was made of Dutch public laws that enforce cadastral recordings containing a 3D component. These laws are described below:

- *Belemmeringenwet Privaatrecht*: Obligation on the owner of land to tolerate a construction for public good
- Law on Monuments (*Monumentenwet*): Registration in order to protect historical monuments
- Law on Soil Protection (*Wet Bodembescherming*): Registration of severe soil pollution

All the restrictions mentioned here are in the Dutch cadastral registration registered on parcels as object restrictions (*Object belemmering*), i.e., as legal notifications. The parcels are affected with a restriction in the right of ownership, which is stored in the administrative database. The restrictions are registered, not the factual objects, which cause the restriction (monument, cable, pollution, etc.). A legal notification (object restriction) is an indication in the cadastral registration that a restriction is imposed on the ownership of the parcel. Legal notifications are an administrative category that describes rights and restrictions but is not rights themselves.

2.4.1 BELEMMERINGENWET PRIVAATRECHT

According to a special law in the interest of public good (*Belemmeringenwet Privaatrecht*),[36] the owner of land can be obliged to tolerate constructions held by others such as lampposts, electrical cables, water pipes, telecom pipes, tunnels, etc.[188] This restriction is used only when no other agreement can be arranged with the owner (e.g., right of superficies, personal rights described in contract, etc.). In addition, the restriction does not allow the imposition of precisely described limitations on the user of the parcel in order to protect the construction against damage. Therefore, this restriction is rarely used.

Since the objects themselves (cables, pipelines, tunnels) are not registered, only the parcels are known below (or above) which a construction is situated. The exact (horizontal and vertical) location of the construction is not known in the cadastral registration, although it is possible to make the outlines of an underground construction visible on the cadastral map.

The obligation of toleration by law only holds for cables and pipelines for public good. Consequently, for those cables and pipelines for which no toleration can be enforced (when it does not serve the public in total) and for which no right of superficies has been established, nothing is registered.

According to private law, the owner of the intersecting parcel becomes the owner of the cable or pipeline, since the construction is permanently fixed to the surface. If horizontal accession receives priority to vertical accession (as in most cases), the owner of the parcel where the cable or pipeline is permanently fixed to the surface (comes to the surface) becomes the owner of the cable or pipeline.

The establishment of a right of superficies (right according to private law) for cables or pipelines provides the possibility to keep the right of ownership explicitly with the cable or pipe holder.

2.4.2 LAW ON MONUMENTS

The Law on Monuments (*Monumentenwet*),[37] established in the Netherlands in 1961, protects buildings and parts of buildings with monumental value but also earth layers below the surface with archaeological value. According to this law, it is possible to impose restrictions on the owner of a monument, e.g., not rebuild certain parts of a house. The restriction is registered on the whole parcel, even when only the façade of a building or just a part of a building is a monument. It is possible to indicate in the administrative part of the registration that not the whole parcel contains a monument. The geometry (outline of the monument or archaeological site) is not maintained in the cadastral registration (Figure 2.4).

More details on the exact location of the monument on the parcel can be found on drawings added to deeds. To protect monuments, a complete, correct, and clear registration of monuments is necessary. An owner of a monument receives funding from the government. This also requires a correct registration of monuments in order to assign the funds to the right person. Another reason why cadastral registration according to the Law on Monuments is becoming more important is that recently archaeological

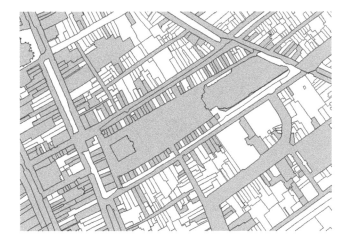

FIGURE 2.4 A selection of parcels (highlighted in the map) encumbered with a notification because of a monument in the city center of Delft, the Netherlands.

sites have received more protection under European agreement.[207] This agreement states that planners of new projects (infrastructure or new city sites) are responsible for the conservation of archaeological treasures in the unexplored subsurface. Cadastral registration can provide the planners of new projects with information on archaeological sites.

Although the actual part that covers the monument is indicated in the deed archived in the land registration, a 2D (or 3D) spatial description of the monument in the cadastral registration would show immediately that it is not the whole parcel (or building) that is a monument. Also the spatial description of underground space with archaeological value in the cadastral registration would provide insight into the exact location of the protected site, without having to look in the land registration.

2.4.3 LAW ON SOIL PROTECTION

According to the Law on Soil Protection (*Wet Bodembescherming*),[39] cases of severe soil pollution have to be registered in the administrative part of the cadastral registration. When (a part of) the subsurface of a parcel is polluted, the parcel is indicated as a polluted parcel. The provinces are obliged to report a severe pollution to the Kadaster. With this report a (2D, analog) drawing of the location of the pollution is archived in the land registration (Figure 2.5). However, since the accuracy of the drawings is not prescribed, the exact locations of pollution are still very unclear in most cases. 3D information on pollution locations is totally lacking. The disadvantage of this registration is that the whole parcel becomes affected by the decision. The exact location (in the horizontal as well as in the vertical dimension) of the pollution is not registered and therefore not known in the cadastral registration.

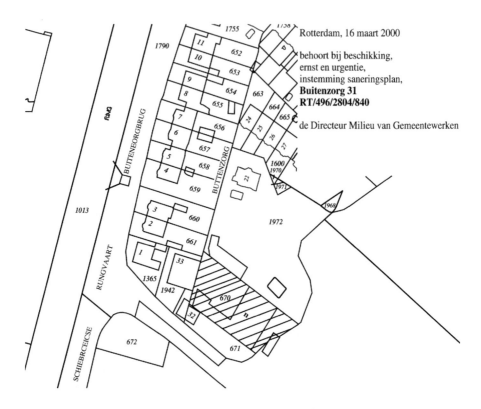

FIGURE 2.5 Drawing added in deed to indicate severe soil pollution. (Note that the polluted area is indicated by hatching.)

2.5 CONCLUSIONS

In this chapter the scope of the 3D cadastre research was outlined. The chapter started with a description of the elementary cadastral model underlying current cadastral registrations. Many countries have met the problems of registering 3D situations within current cadastral registrations, which were originally developed to register the legal status of 2D parcels. The developments on 3D cadastral registration depend on the national legal system, on the state of the art of the cadastral registration, as well as on the type of cadastral registration. In this chapter different types of cadastral registrations were described, including the way the type influences the need for and possibilities of a 3D cadastre. For an overview for which 3D cadastral recordings a 3D cadastre should be considered, this chapter has listed the cadastral recordings with a possible 3D component, distinguishing in cadastral recordings according to private law and cadastral recordings according to public law. Right of superficies, right of long lease, easements (servitudes), apartment rights (strata titles), and public law restrictions are used worldwide to establish 3D property units. These are the rights that should be considered for 3D registration. In most countries the spatial extent of these rights can be described in detail (also in 3D) in the deeds or titles registered in the land registration. However, no cadastral registration exists that reflects the 3D characteristics of these 3D property units as part of the cadastral geographical data set (cadastral map). Consequently, current cadastral registrations are not able to provide 3D insight into the real situation, even though real rights always have entitled persons to volumes and not to flat parcels. In the next chapter the limitations of current cadastral registrations in 3D property situations are studied, by analyzing cadastral registration in seven countries in more detail.

3 3D Recordings in Current Cadastral Registrations

Multilevel use of land is not new. For example, in the Middle Ages cellars below roads along wharfs (*werfkelders*) already existed in Utrecht, the Netherlands (Figure 3.1), and for more than a century stores, workplaces, pubs, and even houses have been situated below railway viaducts. As was mentioned earlier, individualization of property started originally with a subdivision of land using 2D boundaries, causing a 2D parcel to be the base cadastral registration unit. Therefore, cadastres throughout the world had to find solutions to deal with 3D property situations. This chapter tries to find an answer to the following questions: How are 3D property situations recorded in current cadastral registrations and what are the complications of these recordings?

A factor that seems to influence the discussion on 3D registration is the basis of the legal system. For example, in the Netherlands the concept of property rights to real estate is still land (surface) oriented, while other countries, as will be seen in this chapter, have dissolved the complications of 3D cadastral registration at the legal level. The legal system in these countries provided the possibility to establish multilevel ownership that is no longer related to surface parcels.

In this chapter we take a closer look at 3D cadastral issues in seven selected countries and states: the Netherlands, Norway, Sweden, British Columbia (Canada), Queensland (Australia), the United States, and Argentina (Section 3.1 to Section 3.7). The overview of the different countries provides insight into the actual requirements for a 3D cadastre. The Netherlands is used to illustrate problems of cadastral registration of 3D situations that is still very land (surface) oriented since the rights are always registered on ground parcels. The other countries were selected because these countries have introduced solutions that solve part of the problem of 3D cadastral registration. The description of the Netherlands, Queensland, the United States, and Argentina includes several case studies to illustrate current cadastral registration of 3D situations and its problems.

For every country, we examined if the specific country has faced the 3D cadastral problem, and if so, how it has faced the problem. When establishing a 3D cadastral registration, several phases can be distinguished. 3D cadastral registration starts with the ability to establish 3D property units within the legal framework. The next step is to provide insight into the 3D property units, e.g., by scanned drawings included in the land registration (which is the Public Register describing interests in land) or, even better, by integrating the 3D information in the cadastral registration, which links the essential information from documents recorded in the land registration to 3D geometry of real estate objects. In a final phase, regulations could be laid down that define how to prepare and structure the 3D information used to maintain 3D property units in the land registration and/or the cadastral registration.

FIGURE 3.1 (Color figure follows page 176.) Cellars below roads in Utrecht.

This chapter shows that several countries have been able to solve some aspects of 3D cadastral registration, although the approaches differ. The main drawback of these solutions is that they all lack a fundamental approach by taking the legal, the cadastral, as well as the technical framework into consideration: the solutions that were found mainly focus on the legal aspects. Another important finding from the research presented in this chapter is that it is impossible to talk about *the* complications of 3D cadastral registration and it is also difficult to talk with people from different countries about *the* 3D cadastre, because persons from different countries (and different disciplines) may use similar terms with slightly different meanings.

The different countries have been assessed by examining the following questions:

- How can 3D property units be established within the existing legal framework?
- What was the main trigger to establish 3D property units or to start the discussion on how to establish 3D property units?
- Do 3D property units exist as independent properties in the land registration?
- Do 3D property units exist as independent properties (with 3D geometry) in the cadastral registration, and if so, how (e.g., with link to 3D geometry or integrated in the cadastral geographical data set)?
- What are the main shortcomings of current registration of 3D property situations?

This chapter ends with conclusions.

3.1 THE NETHERLANDS

In the Netherlands parcels defined in 2D are the basis for cadastral registration. Constructions and infrastructure below or above the surface are not registering objects themselves. A building registration also does not exist in the Netherlands, although research has been carried out to set up such a registration in the future.[88] Therefore, the legal status of constructions above, on, and below the surface is not registered on the construction itself. The legal status of the construction can be known from the rights that are registered on the intersecting surface parcel or parcels. The notary deed, which has led to registration, may be accompanied by an analog drawing of the physical object, but this is not obligatory. The inclusion of digital 2D and 3D drawings in the cadastral registration is not possible at this time.

The Dutch cadastral geographical data set contains the boundaries of parcels and parcel numbers, outlines of buildings (for reference purposes), street names, and house numbers. The outlines of real-world objects can be incorporated in the topographic part of the cadastral data set (which is not part of the cadastral map), e.g., railways. Apart from the classification code, these lines are encoded with a visibility code. The visibility code indicates the visibility of the topographic line (invisible means "not visible from above").

3.1.1 UNDERGROUND OBJECTS IN THE CADASTRAL REGISTRATION

A special case in the administrative registration is the registration code OB or OBD (*Ondergronds Bouwwerk*: underground construction), which was introduced in 1998. This is just an indication in the administrative database of the existence of an underground object in the subsurface of a parcel. An OB code is linked to a parcel and to a subject (which is the person responsible for the object). The OB code indicates the factual situation but it is not a right or restriction itself. Although it is registered as an object restriction, it has no legal consequences and it does not indicate how the legal status of the construction has been established. To determine the legal status of the underground object, one has to find out what other rights, restrictions, and legal notifications are established on the surface parcel. Recently, it has become possible to add boundaries of underground objects such as transport systems and telecom networks in the topographical part of the cadastral data set using the visibility code.

3.1.2 TELECOM NETWORKS IN THE CADASTRAL REGISTRATION

Quite specific, explicit 3D registrations in the Netherlands (constructions below the surface) are those that fall under the Law on Telecommunication. According to a decision of the Dutch Supreme Court in June 2003,[40] telecom networks are immovable goods and these cables are always owned by the holder of the permit to exploit the cable. This holder has (usually) a right on the parcel where the cable comes to the surface. Since telecom networks are considered immovable goods, the cadastre is obliged to register the transfer of networks as well as the establishment of limited real rights on them. It is expected that in the future this decision will apply to other cables and pipelines (gas and electricity) as well. It should be noted that a lot of

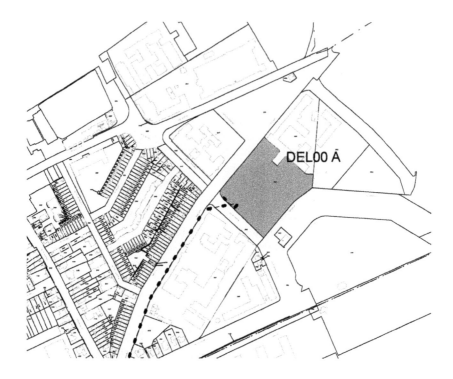

FIGURE 3.2 Example of drawing added to deed in case of a telecom network.

infrastructure objects are or can be used for telecommunication and they may all fall under this law. The registration of telecom networks is done in the following way. If a telecom network is transferred, the holder of a telecom network offers the spatial description (center line) of the network to the cadastre. The network is then registered on at least one "anchor" parcel on which the holder of the network has a real right, e.g., on a parcel where a network substation is located. The other intersecting parcels do not need to be mentioned in the deed but can be found by consulting the drawing archived in the land registration (Figure 3.2).

On all parcels intersecting with the network an object restriction (legal notification) can be registered. The spatial description of the network can only be incorporated in the topographic part of the cadastral data set and not in the cadastral map itself.[104] According to Articles 174 and 175 of Book 6 of the Dutch Civil Code the manager of the cable or pipeline is always responsible for damage caused by a defect in the cable or pipeline or by hazardous material transported through the pipeline whether the manager is the legal owner or not. This also holds if the manager is not explicitly registered as the owner of the cable or pipeline.

3.1.3 CASE STUDIES IN THE NETHERLANDS

To illustrate the way 3D situations are currently registered in the Dutch cadastral registration, five case studies in the Netherlands were selected. The case studies were selected in such a way that they form a representation of the types of 3D situations that currently occur in practice. Another criterion in the selection process was that the cases should be simple in order to illustrate as clearly as possible the constraints of current registration. The case studies are divided into building complexes and subsurface infrastructure objects. Building complexes mostly occur in urban areas and interact with other types of land use. In those cases mostly private parties are involved. Subsurface infrastructure objects are mainly constructions meant to serve the public. Other cases (e.g., soil pollution, archaeological sites, and monuments) are not studied because it is the intention of these case studies to present a picture of the complexity of cadastral registration of 3D situations in general, rather than to analyze all possible cadastral recordings with a 3D component, which are numerous and all have their specific characteristics. Therefore, the most common and basic types of cadastral registration have been selected. It can be expected that types of cadastral registration that are not dealt with in these case studies would show similar basic complications.

Possible future cadastral registration solutions will be shown in Chapter 11, where the prototypes developed as part of this research are applied to a selection of the case studies introduced in this chapter.

Building Complexes

The main characteristics of property units in building complexes are that two or more parties are involved in the ownership of the building and that different property units, often with different functions, are located within one building complex, concentrated on one or several ground parcels. The demand that private persons have concerning the cadastre is that their properties are registered properly. Cadastral query must provide sufficient insight into what persons own, and the location of the property boundaries. Because real estate has significantly gained value during the last decades, it has become more important to register property clearly and unambiguously. Building complexes are therefore relevant objects to study current registration possibilities of 3D situations. How are property units in building complexes registered at the moment? In what way does the cadastral registration provide insight into the property units in building complexes? Does the cadastral registration provide insight into the location of boundaries of the property units, also in the third dimension?

Case Study 1: Building Complex in The Hague

Figure 3.3 shows an example of a 3D situation: a building over a highway in The Hague. The right of property of the building has been established by establishing rights on the three intersecting parcels (Figure 3.4). The cadastral map (Figure 3.4) shows the outlines of the building (on surface level) and the surface parcels. The arrow indicates the view position of the camera in Figure 3.3. The firm ING Vastgoed Belegging BV is holder of the whole building. The rights and restrictions established on the intersecting parcels are as follows. The municipality holds a restricted right of ownership on parcels 1719 and 1720. ING Vastgoed Belegging BV possesses an unrestricted right of ownership on parcel 1718, a right of superficies on parcel 1719, and a right of long lease on parcel 1720. In this example there is one building with one owner ("holder"). However, three parcels are used to establish the legal status of the whole building.

FIGURE 3.3 (Color figure follows page 176.) Building over a road.

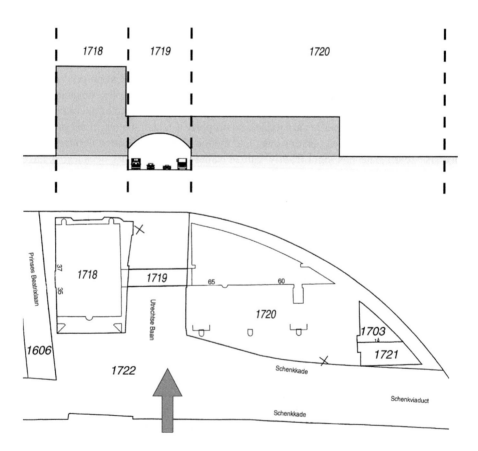

FIGURE 3.4 (Color figure follows page 176.) Cadastral map of the building in Figure 3.3. The arrow indicates the position of the camera.

Case Study 2: The Hague Central Station

The Hague Central Station is a building complex in the city center of The Hague. It is a combination of a multilevel public transport interchange (bus/tram station and railway station), an office center, and shops (Figure 3.5a). All parts of this complex are owned by different governmental and commercial organizations. This is achieved by dividing the high building (office and railway station) into apartment rights, and the establishment of a right of superficies for the bus/tram station.

The use of apartment rights is discussed in more detail in the next case. Here we take a closer look at the right of superficies. A right of superficies is a limited real right that entitles its holder to build and have a building (or an other type of construction) in, on, or above the land owned by another. As a limited real right it restricts the use of the landowner: the landowner has to tolerate the existence of a part of the building on his or her parcel. On the other hand, the holder of the right of superficies is the full owner of the erected building. In the case of The Hague Central Station, the holder of the right of superficies is entitled to build and own the tram/bus station on top of the railway platforms. The cadastral map of this complex is shown in Figure 3.5b. The arrow indicates the position of the camera in Figure 3.5a. The bus/tram station on top of the railway platform is erected on parcel 13295, the business center is on top of the railway station on parcel 12131.

According to the cadastral DBMS, the right of the concerning parcels are as follows:

Parcel	Kind_of_right	Right_owner
12131	VE	VER. VAN EIG. STICHTHAGE
		divided into two apartment untis:
12205A0002	VE	STICHTHAGE TRUST B.V. GEV. TE'S-GRAVENHAGE
12205A0001	VE	NS VASTGOED BV
13288	VE	NS VASTGOED BV
13289	VE	NS VASTGOED BV
13290	VE	NS VASTGOED BV
13291	EVOS	NS VASTGOED BV
13291	OS	Gemeente Den Haag
13292	EVOS	NS VASTGOED BV
13292	OS	Gemeente Den Haag
13293	EVOS	NS VASTGOED BV
13293	OS	Gemeente Den Haag
13294	EVOS	NS VASTGOED BV
13294	OS	Gemeente Den Haag
13295	EVOS	NS Railinfratrust BV
13295	OS	Gemeente Den Haag

VE = full right of ownership
OS = right of superficies
EVOS = right of ownership, restricted by a right of superficies

Analyzing these results, it is clear which persons have a right on the relevant parcels. For example, for parcel 13295, the database shows that NS Railinfratrust BV is owner of the land (with the railway platforms), and that the municipality of The Hague (in Dutch: *gemeente Den Haag*) is holder of the right of superficies (tram/bus station). However, neither these data nor the cadastral map gives insight into how the rights are divided in the vertical dimension on every single parcel. There is also no indication

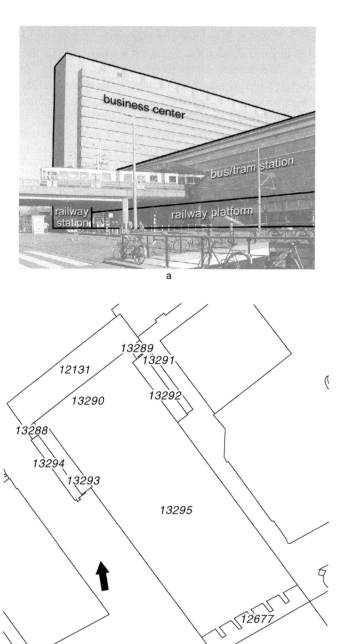

FIGURE 3.5 (Color figure follows page 176.) The Hague Central Station, combination of a business center, a railway station, and a bus/tram station. (a) Overview of situation; (b) Cadastral map.

FIGURE 3.6   Apartment complex used in case study.

in the cadastral registration that the municipality is the factual owner of the bus/tram station. A study in the Public Registers did not reveal much more information. Except for parcel 12131 (divided into apartment rights), the concerning deeds do not contain a spatial description or a clear drawing to clarify the division into 3D property units.

Case Study 3: Apartment Complex

A typical form of multiple use of space, known in Dutch law since 1953, is apartment ownership (condominium ownership). For this case, we used a "simple" apartment complex, consisting of one ground parcel and three apartments. One apartment is located on the ground floor, and the two other apartments are located on the second and third floor, next to each other, with an entrance on ground level (Figure 3.6).

The deed of division of this apartment complex (archived in the land registration) contains a drawing with a cross section and the overview of every floor (Figure 3.7). The individual apartments are numbered. The rights at the location of the apartment complex according to the cadastral registration are as follows:

```
Parcel      Kind_of_right   Right_owner

5238 G0     VE              VER. VAN EIG. I.HOORNBEEKSTRAAT 51-55, DELFT
                            divided into three apartment units:
  6408 A3   VE              PERSON1
  6408 A2   VE              PERSON2
  6408 A1   VE              STOTER

VE = full right of ownership
```

At first glance it seems that there are four owners, the *vereniging van eigenaren* (association of owners) and the holders of each of the three apartments. But this conclusion is incorrect. The parcel 5238 G0 refers to the ground parcel with the apartment complex erected on it. In practice the Kadaster names the *vereniging van eigenaren* (the association of owners) as owner. From a legal point of view, this is not correct. The complex is co-owned by all the apartment owners, not by the association. In Dutch law the association of co-owners is merely a legal body entrusted with the

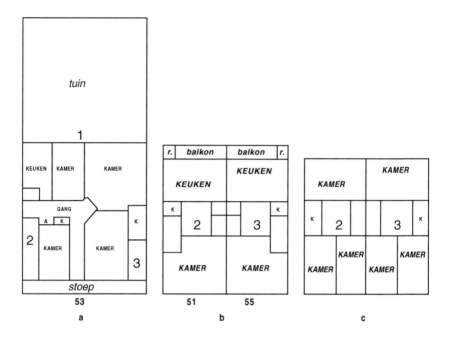

FIGURE 3.7 Drawing added to deed of division. (a) 1st floor; (b) 2nd floor; (c) 3rd floor.

day-to-day administration and management of the complex. All the co-owners of the complex are by definition members of this association, which is not explicitly registered in the cadastral registration.

Apart from the (co-owned) ground parcel, the individual apartments are each indicated by a unique number (6408 A1, 6408 A2, 6408 A3). The suffix A shows that this number refers to an apartment right. The last digit is the same as the apartment number in the deed of division.

Importantly, the individual apartments, the areas of exclusive use, cannot be found on the cadastral map (Figure 3.8). The land registration has to be queried to find the plan of division. Addition of 3D spatial information on the individual apartments in the cadastral registration would enhance insight. Another disadvantage of current apartment registration is that the plans in the notarial deeds are only available on analog (and in the future on scanned drawings) in a local coordinate system (in 2D layers). When spatial information on apartment units is available in vector format in the national reference system, this information could be incorporated as part of the cadastral geographical data set or in other geo-data sets (e.g., topographic data) when requested.

Subsurface Infrastructure Objects

Infrastructure objects are objects that are necessary to transport all kinds of things (cars, trains, electricity, water, data communication). The main characteristics of

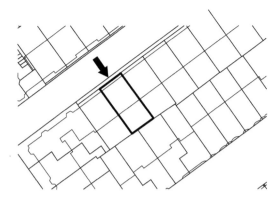

FIGURE 3.8 The cadastral map of the apartment complex in Figure 3.6. The parcel in question has been drawn with a thicker line-style. The front of the building is indicated with an arrow. Note that the parcel is larger than the footprint of the building, since the parcel also includes a garden ("tuin" in drawing of deed of division).

infrastructure objects are their benefit to the public, their linear shape, and the fact that they cross parcel boundaries. From a cadastral point of view, it is important to register the property rights of infrastructure objects and to register public restrictions because of the infrastructure objects, not merely to secure the value of the real estate for the persons involved, but also to indicate who is responsible for the object (for example, in case of damage). In addition, establishment of rights on infrastructure constructions provides a means to protect the construction against damage by specifying conditions in the accompanying deeds. A precise registration is also required, since the holder of the construction is usually obliged to pay the parcel owner a sum of money. Finally, information on the exact location of tunnels and pipelines is indispensable in risk management with regard to the increased attention to calamities in the past 10 years (although it can be questioned if this is a specific cadastral task).

In this section two case studies of subsurface infrastructure objects are described (a railway tunnel in an urban area and a railway tunnel in a rural area) to show the possibility to locate infrastructure objects in current cadastral registration.

Case Study 4: Railway Tunnel and Station in Urban Area

An interesting case of multiple use of space in the Netherlands can be found in the center of Rijswijk, a suburb of The Hague. Some years ago the railway line running through this town was tunneled. On top of this tunnel, buildings were constructed. A small part of the tunnel area is shown in Figure 3.9 and Figure 3.10. In Figure 3.11 the cadastral map of the situation is shown. According to the administrative part of the cadastral registration the following rights have been established on the parcels:

FIGURE 3.9 **(Color figure follows page 176.)** Rijswijk railway station (left) and kiosk (right).

Parcel	Kind_of_right	Right_owner
7854	OS	NS VASTGOED BV
7854	EVOS	NS RAILINFRATRUST BV
7855	OS	NS VASTGOED BV
7855	EVOS	NS RAILINFRATRUST BV
7856	VE	NS VASTGOED BV
7857	OS	NS VASTGOED BV
7857	EVOS	NS RAILINFRATRUST BV
7944	OS	DE GEMEENTE RIJSWIJK
7944	EVOS	NS RAILINFRATRUST BV
7945	VE	NS RAILINFRATRUST BV
7946	VE	NS RAILINFRATRUST BV
7949	EVOS	NS RAILINFRATRUST BV
7949	OS	DE GEMEENTE RIJSWIJK

VE = full right of ownership; OS = right of superficies;
EVOS = right of ownership, restricted by a right of superficies

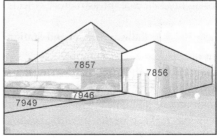

FIGURE 3.10 The location of parcels around the building of the railway station.

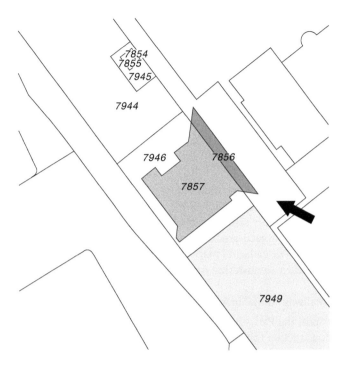

FIGURE 3.11 Fragmented pattern of parcels caused by the projection of 3D objects on the surface. The arrow indicates the position of the camera in Figure 3.10.

In this area there is the following:

- A railway station building, owned by NS Vastgoed BV (parcel 7856 whole parcel column; 7857 ground level)
- A railway tunnel and platforms owned by NS Railinfratrust BV (parcel 7854, 7855, 7857, 7944, 7949 underground; 7945, 7946 whole parcel column)
- Public space owned by *Gemeente Rijswijk* (7944, 7949 ground level)
- A kiosk, owned by NS Vastgoed BV (7855 and 7854 ground level)

In Figure 3.9 the pyramid-shaped object is the building of the railway station (parcels 7856 and 7857), the building on the right is a kiosk (parcels 7854 and 7855), and the railway tunnel is located beneath the buildings.

The cadastral map and the photograph of Figure 3.10 show that the station building, owned by NS Vastgoed, has been built for the major part above the tunnel (assuming that the tunnel is located below the surface parcel 7857) and for a relatively small part next to the tunnel (parcel 7856). For the first part NS Vastgoed holds a right of superficies on the parcel owned by NS Railinfratrust BV; for the second part NS Vastgoed has the full ownership of the parcel. This case also shows that the 3D spatial extent of rights is not available in the cadastral registration, although it is possible to see that more than one person is entitled to a parcel.

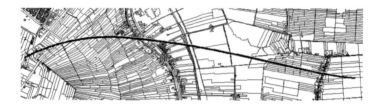

FIGURE 3.12 The railway tunnel in the "Green Heart" of the Netherlands.

This example is a good illustration of how 3D physical objects below and above the ground control the parcel pattern in the cadastral map (e.g., 7856 and 7857 for the railway station building; also the tunnel is identifiable in the patterns of parcels). Moreover, 3D physical objects are "divided" into parts according to the parcel boundaries on the surface. The cadastral map on this location reflects the basic principle of the current cadastre, i.e., registering rights on 2D parcels.

Case Study 5: Railway Tunnel in Rural Area

In the Netherlands, the Paris–Amsterdam High Speed Railway (Figure 3.12) is currently under construction (planned to be finished in 2007). Since this railway passes through unaffected rural land, it was decided to drill a tunnel for this part of the railway. The project team of the tunnel provided us with 3D data for the tunnel, which we then imported as one spatial object (a linear object) into the cadastral DBMS. Therefore, it was possible to query the legal status of the intersecting parcels. Normally this is not possible because physical objects are not maintained within the cadastral registration. The tunnel itself is about 15 meters in width and 8.5 kilometers long: 7160 meters for the actual drilled tunnel and two entrance sections of 660 and 770 meters in length.

In November 2001, the activities for this tunnel started. The drilling of the tunnel was completed in January 2004. We had access to three snapshots of the cadastral database: June 2000, June 2001, and September 2003. Between June 2000 and September 2003, most of the property rights needed by the Ministry of Transport and Public Works were obtained and registered. For this reason we were able to study the differences in the legal status of the parcels that contain the tunnel between the different snapshots. The results of this investigation are shown in Table 3.1.

As can be concluded from this table, at the location of the planned tunnel many changes have taken place between June 2000 and September 2003. Of the original 104 (complete) parcels that intersected with the tunnel in June 2000, 36 are not subdivided in September 2003 (and 50 were not subdivided in June 2001). The other 68 parcels (and 54 in June 2001) are subdivided (without being yet surveyed) because the tunnel has been built below just a part of these parcels. The subdivision of parcels avoids that part of parcels that does not intersect with the tunnel and is encumbered with a right for the tunnel. Most of the subdivided parcels are divided into two parts. A minority of them are divided in three, or even four new parcels.

Of the 104 intersecting parcels, in June 2000 the Ministry of Transport and Public Works had a right on 12 intersecting parcels which are all ownership rights. In June

TABLE 3.1

Results of the Queries on the Legal Status of the Parcels Intersecting with the Railway Tunnel Passing Through the Green Heart of the Netherlands

		June 2000	June 2001	September 2003
1.	Number of parcels intersecting with the projection of the tunnel	104	104	104
2.	Number of intersecting parcels that contains part parcels	0	54	68
3.	Number of parcels of (1) that is encumbered with a right that belongs to the Ministry of Transport and Public Works	12	80	99
4.	Number of parcels (including part parcels) that is encumbered with a right that belongs to the Ministry of Transport and Public Works	12	91	121
5.	Number of rights mentioned in (3) that is a right of ownership	12	44	47
6.	Number of rights mentioned in (3) that is a right of superficies	0	36	52
6a.	Number of parcels affected with an OB notification	0	36	52
7.	Number of rights mentioned in (4) that is a right of ownership (registered both on part parcels and complete parcels)	12	53	60
8.	Number of rights mentioned in (4) that is a right of superficies (all registered on part parcels)	0	38	61

2001, the Ministry had a right on 80 intersecting parcels: 44 ownership rights and 36 rights of superficies. Finally in September 2003, the Ministry had a right on 99 intersecting parcels: 47 ownership rights and 52 rights of superficies. All intersecting parcels affected with a right of superficies are also affected with the legal notification OB (underground construction), with the Ministry as subject. In the snapshot of June 2000, none of the intersecting parcels had an OB notification.

The results based on the cadastral database of September 2003 show that at that moment the Ministry still had to obtain a right on five intersecting parcels.

This case study reveals some complexities of current registration of linear objects that cross several parcel boundaries:

- The information that can be obtained from the cadastre is fragmented because only the rights on the intersecting parcels are registered. It is not possible to query the object itself.
- The location of the object itself is not registered. Even if a right for the object has been established, the exact location of the object (in 2D and 3D) is not known.

- A drawback of the cadastral registration of cross-boundary objects is that there is redundancy: for every parcel crossed by the object, a reference is made to the same subject (holder of the object), which may result in inconsistencies.
- Cadastral registration of infrastructure objects is not uniform. Different rights and restrictions are used to register the legal status of these objects.

3.1.4 EVALUATING 3D CADASTRE IN THE NETHERLANDS

How can 3D property units be established within the existing legal framework?

Property rights in the Netherlands always have to relate to surface parcels. Consequently, the ownership of real estate above and below the surface is always established on surface parcels. Owners can be restricted in using the whole parcel column by limited rights or a parcel column can be divided into different property units by apartment rights (which are also related to the surface parcel).

What was the main trigger to establish 3D property units or to start the discussion on how to establish 3D property units?

The discussion in the Netherlands was started since a few 3D property situations were met that could not be registered unambiguously in current cadastral registration.

Do 3D property units exist as independent properties in the land registration?

3D property units do not exist independently in the land registration, but are always related to 2D parcels. The only exception is an apartment unit, which is known as an individual property unit in the land registration. However, an apartment unit must also always relate to one or more 2D parcels. Recently cables and pipelines meant for telecommunication can be registered in the cadastral database. However, these objects still need to be registered on a surface parcel (the anchor parcel). Information on 3D property units can be obtained only by querying deeds that establish real rights on surface parcels.

Do 3D property units exist as independent properties in the cadastral registration?

Only in the case of apartment rights, the 3D property units exist as separate real estate objects in the administrative part of the cadastral registration. Apart from apartment units, the only real estate objects known in the Dutch cadastral registration are parcels, and recently cables and pipelines meant for telecommunication. The outlines of subsurface objects can only be indicated in the topographic part of the cadastral database by using a specific classification and visibility code.

What are the main shortcomings of current registration of 3D situations?

The main shortcomings of Dutch cadastral registration in case of 3D property situations is that the 3D situation is projected on the surface and that the 3D spatial extent of rights is not available in the cadastral registration. In addition, the real situation is not properly reflected in the cadastral registration, e.g., by showing 3D outlines of physical constructions and legal spaces above and below the surface.

3.2 NORWAY

Norway has a solid subsurface in a geological sense, in contrast to the subsurface of the Netherlands, which consists only of sediments. In Norway tunnels for roads, trains, and water drilled in the subsurface do not influence the economic value of the surface property. Therefore, these subsurface objects are already common practice in Norway without subdivision and formal registration in the cadastre and in the Land Book. The owners of surface properties are compensated financially only if the surface property has been damaged in any way.

At the beginning of the 1990s, providing the possibility for 3D property was listed as an important motivation for the improvement of cadastral legislation in Norway, because the current legal framework does not provide the establishment of 3D property units with separated ownership on one surface parcel. It was expected that investors would be more willing to invest money in registered ownership, rather than in the variety of limited rights that are currently used to establish stratified property.

A committee was established in 1995, which concluded that three types of 3D property should be facilitated:

1. Volumes below the surface of the earth, such as underground parking, shopping areas, tunnels
2. Buildings and other constructions erected on pillars or by other means realized above the original surface of the earth, mostly above roads or railways
3. Constructions on pillars at sea or in fresh water

The findings of the committee led to a proposal for a law on *construction properties*. It is assumed that this law will be enacted in 2006.[126] In this law the surface property is still the basic property object including all land and permanently fixed constructions except what is subdivided from the surface property. It is expected that the chosen legal instruments will have an effect on prices. A 3D construction property has the following characteristics:

- A 3D property construction can be established only by subdivision of the surface property and may cross several parcels.
- It is up to the parties involved to decide whether to use the 3D property construction solution or to use other possible solutions such as servitudes or to remain unregistered in the cadastre.
- The parties involved enjoy much freedom and carry the risk of making bad arrangements. It is expected that the new law will accept construction or building drawings as satisfactory for registration, without additional surveying. Any detailed surveying of the 3D property beyond that level of accuracy would be the choice and responsibility of the parties involved.
- Since a new parcel can only be established when it follows the planning and building acts, a subdivision of a parcel in general is not permitted unless it is likely that the subsequent construction on the parcel is approved. This means that there is a direct link between the new parcel and the building

to be created. 3D construction properties that will remain unused are prevented by this regulation. In addition, the potential for speculation in land and in space is reduced. A 3D construction property will be approved when it is needed to support a particular and approved construction. Therefore, the law on 3D construction property inhibits the free construction of 3D property units.

- A 3D construction property will cease to exist should the actual construction to which it alludes collapse and not be rebuilt within 3 years.
- A 3D construction property can only be established when the surface can still be used for a relevant purpose as part of the property from which the construction property will be subdivided. Therefore, a building standing directly on the surface cannot be established as a 3D construction property.
- A 3D construction property cannot be established for parts of buildings. It is only possible in the case of separate buildings in which the 3D properties have no relationship to the neighboring properties beyond the usual relationship between neighboring surface properties. In the other cases, apartment rights (*eierseksjon*) must be used, for example, in the case where new units are part of a common owned building.

At this moment no specifications for surveying or solutions for the cadastre are part of the proposal. Conditions in this area would only delay the introduction of the law that meets the demand of the market. For the short-term future it is expected that the cadastre will accept rather simple solutions such as visualizing the projection of the 3D property on the surface only, while referring to more detailed information contained in the deeds.

Awaiting the new law, the municipalities (which are the cadastral authorities at local level) have for many years established properties as volumes above and below the surface, subdividing the volume from the surface property. They have extended the existing cadastral law with municipal regulations to be able to divide properties in 3D. The proposed regulations are based on existing practices. An example of this practice is the municipality of Oslo. This city introduced a practical approach to register 3D properties as real property both in the cadastral registration and the title register.[208] These properties have the same rights and restrictions just as surface parcels. The existing law does not provide for these 3D real properties, and hence the Oslo method has mostly been limited to underground facilities. In the case of 2D subdivision, the new parcel boundaries are surveyed and marked. In the 3D case, it is impossible to survey before the actual construction has been built. Therefore, the plans and drawings from the applicant are sufficient. Usually, this drawing is also accepted as the final document against which a survey certificate is issued without any surveying. On the survey certificate each corner is given by coordinates and heights both at floor and ceiling level. The registration number and the survey identify the parcel as a volume, but in the various registers the parcel size is given in square meters and not in cubic meters. This is due to the Land Subdivision Act, which has no provision for 3D parcels. A 3D parcel is identified by a unique parcel number. 3D parcels can be recognized because the parcel number ends with 300. The numbering of the 3D parcels is done in such a way that the relationships with the surface parcel are preserved.

3.2.1 EVALUATING 3D CADASTRE IN NORWAY

How can 3D property units be established within the existing legal framework?

The new law will enable establishment of 3D construction properties that may cross several surface parcel boundaries. Although such construction properties are not yet formally allowed, municipalities and the land registration already accept it, as shown by the Oslo method.

What was the main trigger to establish 3D property units or to start the discussion on how to establish 3D property units?

Currently multilevel ownership can be established by apartment rights or just by virtue of the owner's legal right to use his or her property (unobstructed by legislation). In the latter case, the legal status is not established and not registered, which is always a risk, especially in case of constructions owned by private persons. Therefore, it is required to ensure the legal status of real property in the cadastre. Apartment rights must always relate to a surface parcel on which the related building is erected, while the 3D construction properties are not necessarily related to the surface parcels. The 3D construction property enables 3D ownership for which apartment rights are not an appropriate solution. Examples include independent volumes below the surface crossing several parcels (underground garages, shopping areas, tunnels, etc.), buildings and other constructions erected on pillars or by other means realized above the original surface, frequently built above roads and railways.

Do 3D property units exist as independent properties in the land registration?

The 3D property units that will be established will be known in the land registration. However, there are no requirements for surveying and mapping the 3D property unit. The 3D geometry of the property unit may therefore not be known (in detail) in the land registration.

Do 3D property units exist as independent properties in the cadastral registration?

The 3D property units exist in the administrative part of the cadastral registration. The footprint of 3D construction properties can be drawn in the cadastral map. However, the 3D geometry of the 3D property unit will not be maintained.

What are the main shortcomings of current registration of 3D situations?

The first shortcoming in Norway is that construction property has to relate to real constructions. Furthermore, the cadastral registration can be improved by first setting up regulations to survey 3D property units and second by solving the technical aspects of 3D cadastral registration, which is "how to incorporate the 3D information in the cadastral map."

3.3 SWEDEN

Before January 2004 in Sweden the division of ownership was not possible in the third dimension. This led to remarkable legal structures. For example, the space for the Stockholm underground is granted through an easement. The dominant parcel to which the easements are linked is a small property formed for a lift shaft going down to the underground railway.[111]

The need for 3D property has been influenced by the fact that apartment units in an apartment complex can only be owned totally, e.g., by one housing association. Each member in such association has a flat connected with his or her apartment and when the member's share is sold, the right to the flat follows with the purchase. So, each apartment owner may sell his or her net share of the cooperative association (*bostadsrätt*). Both the association and the member may take a loan secured as mortgage. However, the association loan can be secured in the land registration and then be related to the whole property, while the member loan is secured in a register kept by the association and is then related to the membership. Difficulties can arise when two types of security are in the same property and when different types of use are combined in one building (e.g., apartment units and offices), since this requires different right holders as well as the possibility of mortgaging the parts separately. One of the problems is the nontransparency of the related information, as the property and mortgage information is maintained in different registers. The separation of the right holders would make the apartment units as well as the offices more attractive on the real estate market (the office property will no longer constrain the housing property, and vice versa). Therefore, for financial and administrative reasons, there is a need to divide properties in such a way that the facilities or parts of them can be mortgaged separately and owned as separate properties.

A new law is elsewhere described that facilitates 3D property units.[87,111] The law came into force in January 2004.[199] The law was prepared by a committee appointed by the Swedish government in 1994 to investigate the potentials for solving the problems of different types of use in one building. The main conclusion of the committee was that the most appropriate solution would be the facility to establish 3D property similar to 2D property. 3D properties can then be mortgaged and information on the 3D properties will be accessible through the real property register. The main objection to the proposal was that the fundamental property concept should not be radically altered from 2D since the number of 3D properties will probably be small. Therefore, the new 3D properties had to fit within the structure of 2D properties. The following criteria have been set up for 3D properties (*3D-fastighet, 3D-utrymme*):

- Title must be in perpetuity.
- Title shall, as far as possible, be independent of the (land) property within whose parcel column it is located and shall be separately transferable, without any simultaneous transfer of the surface land.
- A 3D property must be an object for credit; public authorities, credit providers, and other outsiders shall be permitted to obtain information on the rights established on the property.
- The new rules should as far as possible be in accordance with the existing principles of real property law.
- The ultimate aim of 3D property formation is to create better opportunities for 3D property use and also to permit such properties to serve as security for the grant of credits.

Formation of 3D property is permitted only if it accommodates, or intends to accommodate, a building or other construction and if it is assured of the rights necessary to its appropriate use (e.g., rights to joint facilities, easements). To avoid empty, air-space property units, the 3D property has to relate to a real construction. When it relates to a construction to be built, the cadastral authority can set a deadline for the completion of the construction. Unlike in Norway, a construction itself may be divided into different property units with this new law. This is also the main type of ownership situation that the new law aims to facilitate. However, a 3D property for housing purposes must contain at least five apartment units, which means that the new legislation does not afford scope for the creation of individual apartment ownership. The 3D property units may intersect boundaries of surface parcels.

The 3D property is registered in the real property register and therefore accessible by the public. The new law takes care only of the legal issues and then in the same way as 2D properties. The projection of the 3D property units is indicated on the cadastral map. Details describing the boundaries, like marks, are described in scanned files in the cadastral database. Therefore, these files can be checked (separately from each other) in computers.

3.3.1 EVALUATING 3D CADASTRE IN SWEDEN

How can 3D property units be established within the existing legal framework?

The new law enables the establishment of 3D property units that may cross several surface parcel boundaries.

What was the main trigger to establish 3D property units or to start the discussion on how to establish 3D property units?

The main problem of the existing legal system is that parts of multifunctional building complexes cannot be mortgaged independently, which may discourage investors from investing in multipurpose building complexes. In normal cases this is no problem as the situation is instead handled through tenant ownership. However, in cases with mixed land, use problems can arise. Because the new law prescribes that a 3D property unit should contain at least five individual apartment units, the mortgage of individual apartment owners can still not be registered in the cadastral registration.

Do 3D property units exist as independent properties in the land registration?

The 3D property units are registered in the land registration. However, there are no requirements for surveying and mapping the 3D property units. The 3D geometry of the property unit may therefore not be known in detail in the land registration.

Do 3D property units exist as independent properties in the cadastral registration?

Although 3D property units are registered as independent property units in the administrative part of the cadastral registration, it is not yet clear how 3D property units will be documented as part of the cadastral geographical data set. Until now, it was not the goal of the Swedish legislator to regulate the way 3D property units are incorporated in the cadastral database. At this moment the footprint of 3D property units can be drawn on the cadastral map. This means that 3D property units are registered in the same way as 2D property units.

What are the main shortcomings of current registration of 3D situations?

As in Norway, the 3D property units have to relate to built constructions. Consequently, the 3D property units do not cover all 3D situations. Furthermore, cadastral registration can be improved by setting up regulations to survey the 3D property units and by solving the technical aspects of 3D cadastral registration: for example, how to incorporate the 3D information as part of the cadastral geographical data set.

3.4 QUEENSLAND, AUSTRALIA

In Queensland, Australia, 3D registration has also partly been solved. Since 1997, it has been possible to create parcels defined with 3D geometries. The legal framework of Queensland, which originated from Common Law, provided the possibility of establishing 3D property units (which can be both freehold and leasehold estates). However, the cadastre only includes the footprint of these 3D parcels on the cadastral map, and therefore the cadastral issue of 3D property units is not completely solved in Queensland.

The cadastral registration in Queensland is used to illustrate in more detail the possibilities and constraints for 3D registration of a land registration that is already able to define parcels with a bounded volume. In Section 3.4.1 the different types of parcels that can be established in Queensland are described. In Section 3.4.2 a case study from practice will be introduced to show possibilities and constraints of current cadastral registration of a 3D situation in Queensland. Improvements of the cadastral recording of this case are shown in Chapter 11, where the prototypes developed as part of this research are applied to this case study. Section 3.4.3 evaluates the 3D cadastral issues in Queensland.

3D cadastral issues in Queensland have been studied in collaboration with Queensland Government, Department of Natural Resources, Mines and Energy (NRME).

3.4.1 RESTRICTED, BUILDING, AND VOLUMETRIC PARCELS

According to the Land Title Act of Queensland,[168] a standard parcel (defined in 2D, but implying the 3D column) is a lot or a collection of lots that is unlimited in height and depth. Apart from these "unrestricted" parcels, four types of parcels with a 3D component are distinguished:

1. Building parcels, which are parcels that are generally defined by floors, walls, and ceilings
2. Restricted parcels, which are parcels restricted in height or depth by a defined distance above or below the surface or by a defined plane (restricted easements can also be restricted in height **and** depth); the boundaries of the restricted parcels must coincide with the boundaries of the surface parcel

3. Volumetric parcels, which are parcels that are fully bounded by surfaces and are therefore independent of the 2D boundaries of the surface parcels
4. Remainder parcels, which are parcels that remain after a volumetric parcel or building parcel has been subdivided out of it

The "in strata" parcels that were used before 1997 (and are not applied anymore) included both the volumetric parcels and the restricted parcels.

A standard parcel may be subdivided using three different formats of survey plans: standard, building, or volumetric format. In the document "Registrar of Titles, Directions for the Preparation of Plans,"[169] the conditions for the different plans are exactly described.

A standard format plan defines land using a horizontal plane and references to marks on the ground. A standard format plan is used for standard parcels, restricted parcels, and restricted easements. In case of standard parcels, the drawn parcel refers to the whole parcel column. Restricted parcels (which are restricted in height or depth) are also indicated on standard format plans by *values relative to the surface* (defining horizontal planes), or by a defined plane. For restricted easements, the vertical restriction shall be detailed on the plan with reference to the Australian Height Datum together with details of the Permanent Mark on which this is based (Reference 169, p. 20).

A building format plan defines land using the structural elements of a building, including floors, walls, and ceilings (building parcels). A building format plan is used in situations similar to apartment units in the Netherlands. A parcel is subdivided into a minimum of two building units (lots) and a common property that is shared. The common property is linked to the units and not to the persons owning the units. Lot numbers in buildings shall be numeric and may be made up in the form FL, TFL, or TL, where T is a tower number, F is a floor number, and L is the lot number. The building format plan should include a main plan with the location of each building or structure with respect to the outer boundaries of the base parcel (i.e., the projection of the outermost walls of the building). This plan should include any subsurface basements and a diagram of every level of the building showing the parcels and common property on that level (Reference 169, p. 32). The maximum amount of encroachment (the intersection of this building with any other parcel) permitted is limited to half the width of the wall (Reference 169, p. 36). Consequently, "the boundary of a building format lot may not be projected beyond the boundaries of the base parcel."

A volumetric format plan defines land using 3D points to identify the position, shape, and dimensions of each bounding surface and is used to reflect volumetric parcels. A volumetric parcel is a parcel that is fully limited by bounding surfaces (which may be other than vertical or horizontal) that are above, below, or partly above and partly below the surface of the ground (compare with restricted parcels and notice the difference). Volumetric parcels have been possible in Queensland under the Land Title Act since 1997. The use and purpose of volumetric parcels (not per se related to constructions, e.g., for a panorama) are determined by the local government and other legislation. One volumetric parcel can intersect several surface parcels. All lines on a volumetric format plan are straight and all surfaces are flat unless explicitly stated otherwise; hence any surface that is mathematically definable (so that an intersection can be calculated) can be registered. The height used to define volumetric parcels cannot refer to above or below a depth from the surface (the height cannot be defined as relative height or depth) since "this is subject to change and not capable of mathematical definition."[169] The corners of volumetric parcels should refer to existing structures or marks as much as possible. The vertices of the corners should be given in bearings and distances of existing cadastral corners and the height levels in the Australian Height Datum. Each volume shall be given an area, which is the area of its footprint, and a volume in cubic meters. The plan should show a 3D representation of the parcel. The 3D descriptions are maintained in titles in the land registration while a footprint of the volumetric parcel is shown on the cadastral map.

The cadastral geographic data set of Queensland has a "base layer," which is a complete non-overlapping coverage, and consists of parcels, road, rail, watercourse, and intersection parcels. An intersection parcel is part of a roadway (the intersection of two roads). Volumetric parcels are not part of the non-overlapping coverage, but the footprints of these 3D parcels are drawn on the cadastral base layer, and therefore they are overlapping with the base parcels. Also easements, having their own geometry and survey plans, are drawn on the base layer and may therefore intersect several parcels. Initially, easements are defined on a single base parcel, but the base parcel may become subdivided, leaving the easement whole. Building parcels are not drawn on the cadastral map.

3.4.2 A CASE STUDY IN QUEENSLAND

Since volumetric and restricted parcels are advanced examples of 3D property units, a case study from practice will be used to illustrate the establishment of these parcels: the establishment of 3D property units for the Gabba Cricket stadium in Brisbane. This stadium overlaps two streets: Vulture Street in the north and Stanley Street in the south (Figure 3.13).

Three 3D properties have been established: for the intersection with Vulture Street a stratum with parcel identifier 100 (established before 1997) and a volumetric parcel with identifier 101 and for the intersection with Stanley Street a volumetric parcel with identifier 103. The volumetric parcels were established after 1997. All three parcels are leasehold estates. This means that the holder of the real estate has the right of use and exclusive possession of the property for a specified time, which is comparable to the right of long lease. However, it should be noted that most volumetric parcels are related to freehold estates.

The titles establishing the 3D parcels contain very detailed 3D information imposed by the regulations: cross sections are added in case of the strata title and 3D diagrams are added in the titles for the volumetric parcels (see Figure 3.14 for parcels 101 and 103). All coordinates needed to demarcate the 3D property are present in the titles in bearings and distances. The coordinates are determined only when the information is entered into the cadastral database. The height of all coordinates is defined in the Australian Height Datum.

The footprints of the 3D properties are part of the cadastral geographical data set. Figure 3.15a shows the cadastral map with the footprints of the 3D parcels and Figure 3.15b shows the cadastral base map without the footprints of parcels 100, 101,

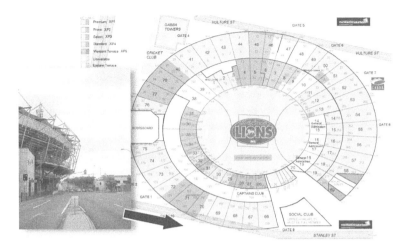

FIGURE 3.13 (Color figure follows page 176.) Overview of Gabba Stadium overhanging Stanley Street in the south and Vulture Street in the north, Brisbane, Australia.

and 103 and without the geometry of easements. Figure 3.15 shows that 3D parcels are not part of the base parcel map and that volumetric parcels and traditional strata parcels exist separately from the base map and may therefore intersect parcels of the base parcel map. For example, the 3D stratum parcel 100 crosses two parcels of the base map.

This example shows the very good potential for establishing 3D properties in the current registration in Queensland. How 3D information, which is part of survey plans and volumetric titles, could be used further to improve cadastral registration, is explained in Chapter 11, where the concepts developed as part of this research are applied to this case study in a prototype.

3.4.3 Evaluating 3D Cadastre in Queensland

How can 3D property units be established within the existing legal framework?

3D parcels, either bounded or unbounded, can be established. The way Queensland has solved the 3D property problem shows that the law introduced in 1997 made it possible to establish 3D property units unrelated to the surface.

What was the main trigger to establish 3D property units or to start the discussion on how to establish 3D property units?

The existence of overlapping and interlocking constructions called for the ability to establish multilevel ownership. The legal system was extended to allow the establishment of 3D property units and the cadastral registration followed the legal practice.

Do 3D property units exist as independent properties in the land registration?

The 3D property units (bounded and unbounded parcels) are known in the land registration. The "Registrar of Titles directions for the preparation of plans" dictates how

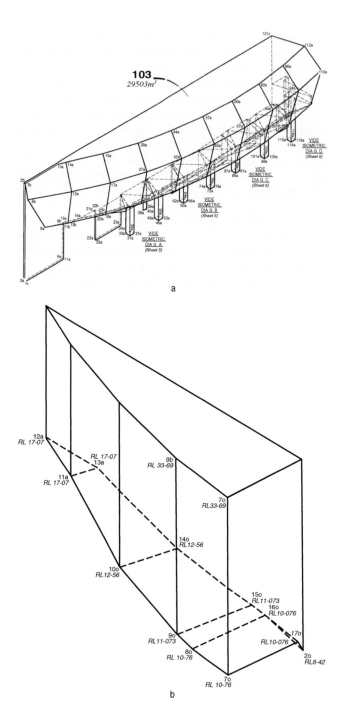

FIGURE 3.14 Examples of 3D diagrams added to volumetric titles. (a) 3D diagram of parcel 101; (b) 3D diagram of parcel 103.

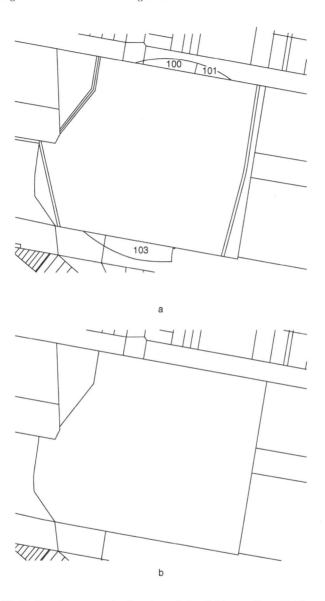

FIGURE 3.15 Cadastral map on the location of the Gabba stadium, Brisbane, Australia. (a) Cadastral map with footprint of 3D parcels (100, 101, and 103) (and easements); (b) Cadastral map without footprints of parcels 100, 101, and 103 (and without easements).

to incorporate 3D information in survey plans. In the case of restricted parcels, the projected parcels with values relative to the surface are sufficient, whereas volumetric survey plans require 3D diagrams, including values in the Australian Height Datum. It should be noted that the survey plans are scanned drawings. It is, therefore, not possible to view the volumetric parcels in an interactive 3D environment.

Do 3D property units exist as independent properties in the cadastral registration?

3D property units exist in the administrative part of cadastral registration. The footprint of the volumetric property is drawn on the cadastral map, and is therefore known in the cadastral registration. However, the 3D geometry is not available in the cadastral geographical data set, and therefore it is not possible to query the 3D situation from the cadastre, nor is it possible to see if two volumetric parcels overlap.

What are the main shortcomings of current registration of 3D situations?

Although, the titles contain detailed 3D information, complications in the registration of the 3D properties arise, for a number of reasons:

- Because the 3D information is laid down on paper or scanned drawings (which are 2D visualizations), the 3D information cannot be interactively viewed. This is a weak point, because the ability to do so would be very helpful in cases of complex volumetric parcels; it would help in interpreting the situation correctly (e.g., parcel 103).
- The 3D properties are described only by coordinates and edges on drawings; i.e., no 3D primitive is used. Therefore, it is not possible to check if a valid 3D property has been established. Is the 3D property closed? Are the faces planar? Are two neighboring 3D properties correctly represented (based on the source documents), that is, no overlap or gaps between these objects in 3D space (similar to the 2D parcel map)?
- The 3D spatial information is not integrated with the cadastral map (linked to legal information) or with other 3D information; e.g., two or more neighboring parcels cannot be visualized in one view in 3D and it is also not possible to check how volumetric parcels spatially interact in 3D (overlap, touch, etc.). Exploration of the 3D situation is limited to a case-by-case approach via documents while the 3D information of several neighboring surveys is not related.

When the 3D information itself would be available in the cadastral information system, the user interface can become 3D (and therefore closer to the real-world situation than with the flat 2D map). In Queensland, the basic improvement for 3D registration would therefore be to incorporate the information on 3D property units, which is already very well described in survey plans in the land registration, into the cadastral registration.

3.5   BRITISH COLUMBIA, CANADA

In British Columbia, Canada, an owner of a parcel has the right to subdivide the land into air-space parcels according to Section 139 of the Land Title Act 1996.[17] The air-space parcel may continue or exist completely below the surface. Only the "fee simple estate," which consists of all ownership rights that can be attached to a certain parcel (complete ownership), can be subdivided and not a leasehold estate (which is an estate created between a landlord and a tenant under a contract, comparable with the right of long lease). For every subdivision, even in 2D, a subdivision plan has to be made. For air-space parcels a special part of the Land Title Act applies.

Every new 3D parcel (air-space parcel) has to be created within an existing conventional parcel. The grant of an air-space parcel does not transfer any easements or restrictive covenant that limits the use of the grantor's land. The title to the ground below and to the air-space above and below the granted air-space parcel, as well as the easements and covenants, remains the possession of the grantor. This means that an easement has to be created separately if access to the newly created air-space parcel is desired or if the existing easements have to apply to the new air-space parcel as well.

The main requirement for creation of an air-space parcel is the provision of an air-space plan on the title.[18] This plan must consist of a 3D drawing to show that the boundaries lie within the boundaries of a single parcel (Figure 3.16). This raises the question what will happen when the surface parcel is subdivided in the future. The plan must further indicate if it is a subdivision of the whole parcel shown on the plan or just a part thereof. A geodetic elevation (in the National Height Datum) is needed, which must be noted on at least one of the corners of the parcel on the ground and for every corner or angle of the subdivided air-space parcel. Air-space parcels can be used for stratified property, but also for the purpose of later granting a right of view to benefit a parcel next to a planned construction.[56]

For a further division of the air-space parcel, the rules of the Condominium Act apply. This divides the air-space into strata lots. The Condominium Act states that a building or land may be subdivided into strata lots by the provision of a building strata plan. The strata lots are coupled with an interest as a tenant in the remaining common areas. It is possible to establish either freehold or leasehold condominiums. The new strata lots have the same status as any land that is registered at the Land Title Office. The strata plan must contain a diagram of the proposed project, showing the boundaries of the land included in the strata plan and the location of the buildings.

In British Columbia the survey plans are registered in the Crown Land Registry and in the Land Titles Office. The Crown Land Registry lists all Crown land converted to private ownership, all private land turned over to the government, all existing Crown land tenures, leases, licences, or other time-limited holdings, and includes maps that record the location of Crown land parcels. In British Columbia the Crown owns 90% of the land. The remaining 10% is privately owned.[56]

In the Land Title System, all titles are given a parcel identifier number, which is part of the legal description and should be included in all land titles documents. A registered title for a "fee simple estate" can either be a conventional parcel or an air-space parcel, which are both considered land under the Land Title Act. It can also be a part of the building, i.e., a strata lot according to the Condominium Act.

There is no general map that covers all existing parcels. There is only a plan that defines the specific area. Therefore, information on the 3D, as well as 2D, properties can be found only in the land registration in the title documents. One has to look in the survey plans to gain insight into the legal situations.

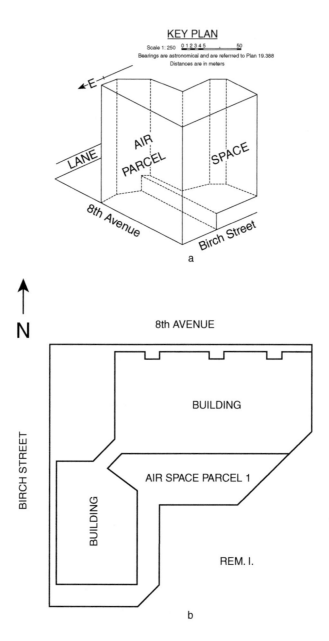

FIGURE 3.16 Drawing in title of air-space parcel. (a) Plano fair-space parcel; (b) cross section. (From J. Gerremo and J. Hannson.[56]).

3.5.1 EVALUATING 3D CADASTRE IN BRITISH COLUMBIA

How can 3D property units be established within the existing legal framework?

3D property units with separate ownership within one parcel are allowed since it is possible to establish air-space parcels, apart from conventional parcels and apart from lots that are the results of subdivision under the Condominium Act. Air-space parcels may not intersect surface parcel boundaries.

What was the main trigger to establish 3D property units or to start the discussion on how to establish 3D property units?

The existence of overlapping and interlocking constructions called for the ability to establish multilevel ownership. The legal system was extended to establish 3D property units. The cadastral and land registration followed the legal practice.

Do 3D property units exist as independent properties in the land registration?

Air-space parcels are known as individual property units in the land registration. The 3D property situations are indicated with 3D diagrams in survey plans and can be known from the documents and records in the land registration.

Do 3D property units exist as independent properties in the cadastral registration?

In British Columbia, cadastral registration is actually the land registration, which includes a title registration. The survey plans are maintained as part of the titles. However, there is no cadastral map in British Columbia. In 2D, neighboring parcels cannot be integrated in one view, by which it is difficult to see an overview of a certain situation and to see if two parcels overlap. Consequently, air-space parcels can also not be shown in one integrated view with other air-space parcels.

What are the main shortcomings of current registration of 3D situations?

Because 3D survey plans are prepared and available in a similar way as in Queensland, basically the same shortcomings apply. In addition, 3D cadastral registration in British Columbia would be improved by two major steps. The first step is to make 2D survey plans digital and to create one parcel map of the plans, with no overlaps and gaps in 2D. The second step is to make 3D survey plans digital (to be able to view the 3D property units interactively and to check the 3D property units) and to include the 3D information that is available in detail in survey plans in the digital cadastral data set. This would make it possible to query the air-space parcels in a combined view with the cadastral geographical data set.

3.6 THE UNITED STATES

The legal system in the United States differs significantly from the countries discussed earlier. This is reflected in the cadastral registration (systems and organization). However, the real world is not that different (very dense population and constructions/ownership above and below each other). A classic 3D property case in the United States is the financing of the construction of Grand Central Station in New York City by the sale of air rights over the railway yards in the early 20th century.

FIGURE 3.17 (Color figure follows page 176.) Pan Am building above Grand Central terminal.

Even after the project was completed the owners continued to exploit the space above the railway and the terminal itself by the sale of construction rights. An example of the latter is the Pan Am Building (Figure 3.17).

Except in the State of Louisiana, real property law in the United States is based on the Common Law. Although the Common Law, like the Roman–law based civil law in continental Europe, knows the Latin maxims *superficies solo cedit* (the owner of the land is owner of the attached buildings) and *cuius est solus, eius est usque ad coelum et ad inferos* (who owns the land, has a right from the sky, down into the depths of the earth), it is undisputed that land (immovable property) may be held apart from the surface, and subject to horizontal division. So, titles to land can be stratified and vested in various owners simultaneously. The result is that each owner holds a different portion (cubic space) either below or above the surface. This stratification of land has more often been achieved through the creation of leasehold rather than freehold estates, but it is not considered impossible for the landowner to carve out of his or her estate a flying freehold. Another way of stratification in the United States is the concept of condominium or apartment ownership.[58]

Land records offices are highly decentralized in the United States. They are typically maintained at the county level. As a result, numerous jurisdictions exist with a wide variety of record-keeping systems. The current system leaves the control and operation of the conveyancing system at the local or county government level.[127]

a   b

FIGURE 3.18 (**Color figure follows page 176.**) 6th Street, double skyways connect 2nd and 4th floor. (From http://www.cgstock.com. Photographer: Chris Gregerson.)

3.6.1 Case Studies in the United States

To show the currently used methods and to reveal complications that are currently met in the United States, two cases are described in more detail (see also Reference 131):

1. Covered pedestrian bridge (skywalk) in Ramsey County, Minnesota
2. Condominium example from Richland County, South Carolina

Case Study 1: Skywalk System in Minneapolis, St. Paul

In the Minnesota Twin Cities Minneapolis and St. Paul the pedestrian skywalks dominate the downtown streetscape and pull pedestrians away from street level. The system consists not only of the covered skyways above the street (and sometimes also below the street via tunnels), but also within the buildings. The construction of this network was motivated by the fact that suburban enclosed shopping malls were attracting shoppers away from the downtown area. Via the skywalk system of enclosed passages a story above the street, the downtown buildings are linked, making it possible to reach shops, restaurants, and offices without going outside. Currently, 100 buildings in Minneapolis and 65 in St. Paul are linked.

Figure 3.18 gives an impression of the skywalk system in St. Paul. This is a real 3D property case as the owners of the buildings, streets, and the parts of the skywalk system (with overlap in the 2D projection) can be different persons.

Figure 3.19 shows a map of part of the system. The case study is situated near the crossing of Seventh St. and Cedar St.

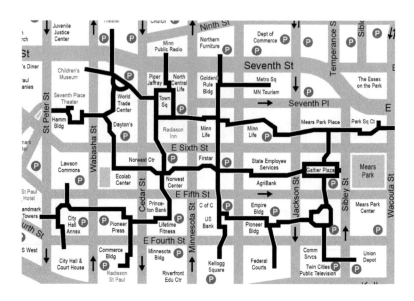

FIGURE 3.19 **(Color figure follows page 176.)** Map of the St. Paul skywalk system.

Legal situation: There are two types of real estate recordings in Minnesota: Abstract Title and Torrens Title. Land, which has been registered pursuant to a District Court Order, is called Torrens land. The owner of registered land has Torrens Title. Land, which has not been registered, is called Abstract land. Abstract refers to the abstract of title, that is not more than a condensed history of all registered deeds, mortgages, and other documents relating to the land from the registry of deed. Before the enactment of the Torrens Act in 1901 (now Minnesota Statues Chapter 508), all land was Abstract land. Creation of condominium property is possible according to the rules provided in Minnesota Statutes Chapter 515 and Chapter 515A (Uniform Condominium Act). A condominium can be registered or unregistered land.

The registration (Torrens Title): The Torrens registration of land titles results in the creation of a Certificate of Title. The registrar must examine the documents presented and determine that they meet the legal requirements to transfer the property. The registrar will issue the Certificates of Title, which are proof of ownership for the purchaser. This certificate is kept in the office of the Registrar of Titles (Minnesota Statutes, Chapter 508.03–508.12)

A Registered Land Survey (RLS) is a survey performed for the identification of registered Torrens lands, according to Minnesota Statutes 508.47. A registered surveyor must certify the RLS to be a correct representation of the parcel. The RLS is filed in the office of the Registrar of Titles. Before this, the county surveyor must approve it. The RLS must correctly show the legal description of the land and the outside measurements of the parcel and all tracts delineated therein. All tracts are lettered consecutively beginning with the letter A. It is interesting to see that the law provides that multilevel tracts can be surveyed. Minnesota Statues 508.47 subd. 4 reads:

A registered land survey which delineates multilevel tracts shall include a map show-ing the elevation view of the tracts with their upper and lower boundaries defined by elevations referenced to National Geodetic Vertical Datum, 1929 adjustment.

We include two fragments of RLS 554 (Figure 3.20). Showing one page with the parcel surveyed on the second (skyway) level of a multistory building at Cedar St. (Figure 3.21), and the page with the elevation data. Observe that the skyway level is between 87.25 and 104.91 feet (where tracts H, I, J, F, and G are situated). However, it is important to note that this building complex itself is not only surveyed in RLS 554, but also in RLS 517 and 518. In this case RLS 517 and 518 give boundaries for parts of the complex that are not only below, but also next to and above the part surveyed in RLS 554.

The Certificate of Title documents are registered at the office of the Registrar of Titles (Minnesota statutes 508.34). The RLS, describing the geometric aspects of the involved real estate objects, is also kept there (Minnesota Statutes 508.47).

Information is available to the public via the Internet through three Web sites:

- http:// gis.ci.stpaul.mn.us/ rcsurvey
- http:// rrinfo.co.ramsey.mn.us/ public/ characteristic/
- http:// maps.metro-inet.us/ RamseyCoGIS/ Viewer.htm

The last Web site offers the following services: access Certificate of Title, Registered Land Survey, and the cadastral map (Figure 3.22).

These Web sites give access to the legal descriptions of the 3D properties. A search in the http://rrinfo.co.ramsey.mn.us/public/characteristic/index.asp shows, for instance, a legal description of a property in the Cedar Street complex (with Parcel Identification Number 06.28.22.12.0130) as:

Tracts B,E,I & J in RLS 554 & in sd RLS 518 tracts F & MM& those parts of tract KKK lying bet plane surface elevations of 173.64 ft and 185.40 ft & lying bet plane surface elevations of 232.38 ft & 267.64 ft & lying above a plane surface elevation of 350.66 ft city of St Paul datum & those part of tract KKK lying bet plane surface elevations of 104.91 ft & 114.97 ft city of St Paul datum & lying swly of a line drawn parallel with & 61.13 ft nely of as measured at right angles to the swly l of sd tract KKK.

This shows that the property itself involves tracts from both 518 and 554 and that the height measures are part of the actual description.

From this example we can conclude that, although the legal system allows the es-tablishment of 3D property units, it is complicated to reconstruct the real 3D situation from the filed documents. 3D information from different surveys must be combined and this information cannot be viewed in one visualization, let alone in a 3D view.

Case Study 2: Condominiums in South Carolina, Richland County

The second case study is more common and is related to a situation that is present all over the world: apartments or condominiums. The case that was studied and is described here is located in South Carolina, Richland County (contains the City

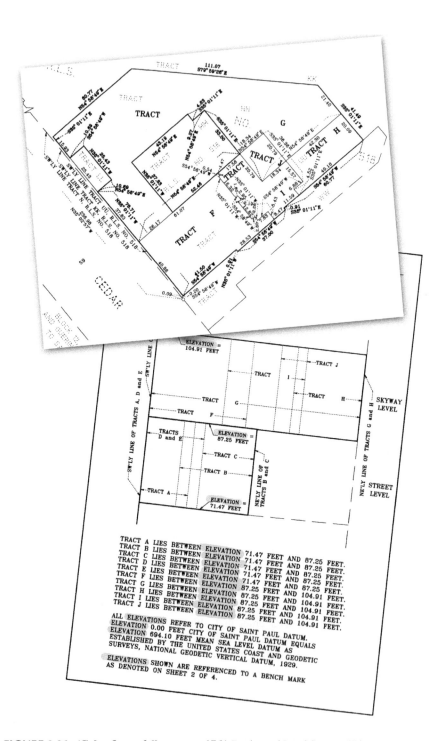

FIGURE 3.20 (Color figure follows page 176.) Registered Land Survey 554.

FIGURE 3.21 (Color figure follows page 176.) Arial photograph of some buildings connected by the skywalk system (Figure 3.20 shows the RLS 554, documenting the second level of the lower left tower of twin towers).

of Columbia, the state capital). Figure 3.23a shows the location of the county and Figure 3.23b shows the specific complex (Middleborough) that is analyzed in more detail here.

Legal situation: Condominium ownership is defined by the South Carolina Horizontal Property Act (South Carolina code of laws Chapter 27), section 27-31-20 as "the individual ownership of a particular apartment in a building and the common

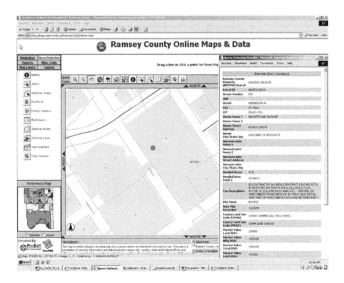

FIGURE 3.22 (Color figure follows page 176.) Base map fragment (showing parcels and structures) of the example used in this section.

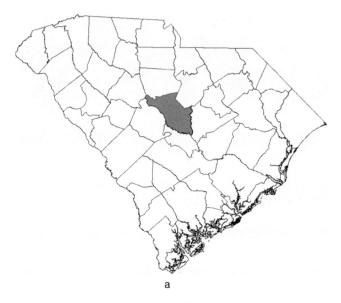

a

b

FIGURE 3.23 Case study from Richland County. (a) Location of Richland county in South Carolina; (b) Middleborough condominiums.

right to a share, with other coowners, in the general and limited common elements of the property."

A building under horizontal property regime contains common elements (like the land on which the building stands, the foundations, main walls, roof, elevators, etc.) and the privately owned apartments. An apartment is a part of the property intended for any type of independent use, including one or more rooms or enclosed spaces in a building, or a delineated place outside a building (e.g., spot for parking a car, or storage of a boat). An apartment owner has the exclusive ownership of his or her apartment, and a common right to a share in the common elements of the property (Horizontal Property Act section 27-31-60). So apartment ownership in South Carolina is based on the so-called dualistic system: exclusive ownership of the apartment, combined with a share in the common elements.

The registration: Horizontal property regime is established by the recordation of a master deed or lease, executed by the owner (or owners) of the real property. This deed must contain *inter alia* (Horizontal Property Act section 27-31-100):

- A description of the land and the buildings
- The general description and number of each apartment, expressing its area, location, and any other data necessary for its identification
- The description of the common elements of the property
- The value of the property and of each apartment and the percentage of the share of each of the co-owners in the expenses and each's rights in the common elements
- A description of the rights and obligations of the co-owners

After establishing the horizontal property regime, each apartment unit can be transferred by recording a deed. This deed will refer to the master deed for a more accurate description of the property and interests.

Attached to the master deed is filed a plot plan and building plan (Horizontal Property Act section 27-31-110). The plot plan is a map showing the horizontal and vertical location of the buildings within the boundary of the property under horizontal property regime. This must be signed and sealed by a registered land surveyor. There must also be attached a plot plan of the construction showing the location of the building and a set of floor plans of the building, which must show graphically the dimensions, area, and location of each apartment, and the dimensions, area, and location of common elements affording access to each apartment (e.g., stairways, corridors). Other common areas must only be shown graphically insofar as possible, but must be described in detail in words and figures. An authorized and licensed engineer or architect must certify the building plans. Each apartment must be designated on the plans by letter or number on the plans (Figure 3.24).

The master deed of the Middleborough condominiums gives the following description of the boundaries of each apartment (source master deed D 593 page 93):

Each Apartment encompasses and includes all that portion of the building designated on the Floor Plans as an Apartment and consisting of all living and storage place bounded by the upper surface of the floor slab, by the unexposed surfaces of the drywall or plastering forming interior walls and ceiling, and by the exterior surfaces of windows and window frames and of exterior doors and the door frames.

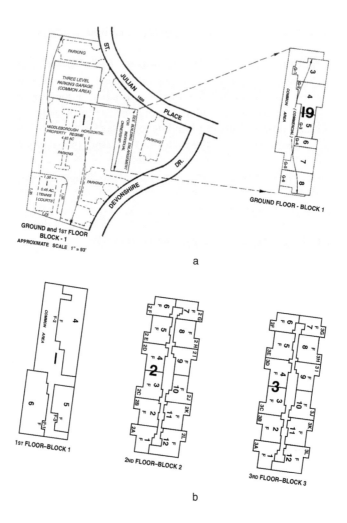

FIGURE 3.24 Fragment of Property Tax Map (R11851) of Middleborough (levels 4 through 18 skipped as they all look the same).

It can be observed that the description refers not only to the floor plans, but also to the constructive elements of the building: walls, ceiling, and floor.

The Web site http://www.richlandmaps.com/maps/maps.html can be used to obtain a map (including parcels, buildings, streets, and aerial photographs) for any area of interest within Richland County. It is possible to locate an address or parcel number on the map (via the dated 1996 parcel layer) and it is also possible to obtain administrative valuation/taxation information via links to the Web site http://www2.richlandonline.com/Assessorearch (information dated December 2003) (for both Web sites, see Figure 3.25). In this process the property tax map (number) plays a linking role as via this number all related administrative information can be obtained, including all condominiums involved, even as they are not directly visible

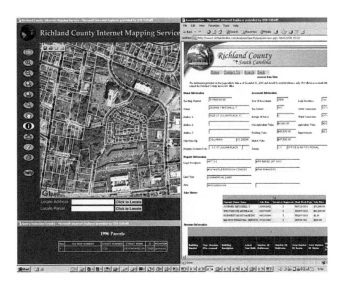

FIGURE 3.25 (Color figure follows page 176.) The Richland County Internet mapping service for locating parcels and obtaining related online taxation information (including sale history of the property); note the tall building, a little above the center of the map on the aerial photograph (Middleborough).

on the parcel map. Via traditional means, which means not via an online information system but via an analog system, it is possible to obtain the involved property tax map itself and the master deed including the attached exhibits (such as building plans) as available for property under the South Carolina horizontal property regime. The building plans maintained in the land registration contain some height information. However, this information can only be presented through 2D means, i.e., paper or scanned drawings or screen, which limits the insight into the real situation. In addition the 3D information is not available on the cadastral map.

Future cadastral registration of this case study is shown in Chapter 11, where a prototype is applied to this case study.

Future system: Investigations are ongoing in Richland County to improve the registration of condominiums by truly including the height information and moving from a 2D map model to a 3D cadastral model. The basic principle is that the whole county is covered with non-overlapping 2D parcels. In certain cases, e.g., when the domain type is 'Condominium,' these objects are further subdivided. Again, these sub objects can be of different types; e.g., Unit, Common Element, Limited Common Element, Lake, Outlot, some of which can be 3D in nature (such as the condominium unit). A prototype system has been developed by Richland County GIS, in which the individual condominium unit has its own 3D geometry (Figure 3.26).

Some aspects currently not covered in the 3D cadastral prototype of Richland are as follows:

- Objects crossing several ground parcels (the solution proposed for condominiums may not be optimal for objects such as tunnels or pipelines).

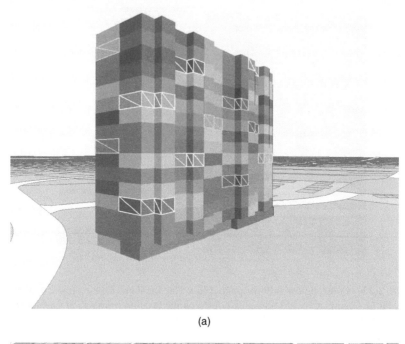

(a)

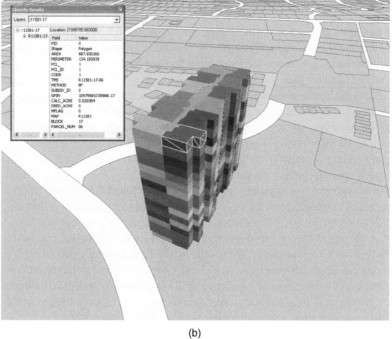

(b)

FIGURE 3.26 (Color figure follows page 176.) The 3D interface (ArcScene) of the 3D representation of the condominiums. (a) Selection of a number of units; (b) selection of a single unit and the display of a number of relevant attributes.

- 3D objects, which cannot be modeled in the level or layer approach (in case of more complex 3D configurations).
- Using absolute heights of 3D objects instead of the relative "unstable" height (see Section 8.1). Absolute coordinates are more stable and they provide unambiguous definitions of 3D objects, especially in mountainous areas. Absolute heights also require that the height of the surface parcels be available (in order to determine the position of a certain 3D object: above or below the surface).

Condominium units are probably the most frequent type of real property unit (and relevant for taxation), so this is a very reasonable start. It may already prove difficult enough to take this step as the users have to make quite a big mental move from a 2D to a 3D model with difficult visualization and interaction aspects associated. For example, one could imagine internal 3D (shared) objects which are not visible unless some other objects are first made transparent or not displayed (e.g., by removing top layers as in Figure 3.26).

3.6.2 Evaluating 3D Cadastre in the United States

How can 3D property units be established within the existing legal framework?

The case in St. Paul showed that real property (parcels) can be defined using elevation information; i.e., parcels can be defined in 3D. These parcels may overlap other parcels when projected on 2D surfaces. The case in Richland County shows that 3D units in a building can be distinguished as separate real estate (apartments).

What was the main trigger to establish 3D property units or to start the discussion on how to establish 3D property units?

The existence of overlapping constructions called for the ability to establish multilevel ownership as in Queensland and British Columbia. In the legal system it is possible to allow the establishment of 3D property units and the cadastral registration follows the legal practise.

Do 3D property units exist as independent properties in the land registration?

The 3D property units in St Paul and in Richland County are known as individual property units in the land registration. For the properties in St. Paul a Registered Land Survey containing a 3D spatial description is maintained in the land registration. In these surveys elevation information is referenced to National Geodetic Vertical Datum. Condominiums are known in the land registration of Richland County and clarified by a lot plan, a plan with the location of the buildings within the property boundaries as well as a set of floor plans, showing the locations of the common property and the individual apartment units. It should be noted that the plans in both case studies are scanned drawings. It is therefore not possible to view the 3D property units in an interactive 3D environment.

Do 3D property units exist as independent properties in the cadastral registration?

The U.S. systems are currently able to handle a 2D cadastral map representation of the situation, which can be used as an entry to related information, such as the legal/administrative information (extract, but also the source document) and the RLS, which can contain 3D information. Although constructions are also drawn on the cadastral map, the cadastral map does not contain any 3D information, and therefore it is not possible to query the 3D situation from the cadastre.

What are the main shortcomings of current registration of 3D situations?

Cadastral registration in the United States is characterized by its highly decentralized and, as a result, nonhomogeneous nature. This has some drawbacks for the nationwide and uniform access to real property information. As a result, this may have some negative effects on the transparency of the market and the legal security of real property. However, the United States as a nation seems to be able to cope with this situation (via a work-around such as insurance to deal with the lower level of legal security). There are states/counties in which this cadastral registration is quite advanced and locally a higher level of legal security can be provided. As in British Columbia and Queensland it turns out that the registration could be improved through the application of the principles of a true 3D cadastral registration. A future system might consider taking the 3D information out of the surveys and including it in a 3D GIS. Technology investments and required knowledge levels for a full 3D cadastral system is much higher than the current 2D-based systems. Instead of the distributed approach in which each county builds and operates its own system, which will be very expensive in the case of a 3D cadastral system, it will be more effective to set up such a system jointly. This could also result in a more uniform cadastral registration with all its positive side effects. The data ownership and maintenance responsibility can remain at the current local level and does not have to be changed. In a way, the nation will then have the best of both worlds: decentralized local expertise, but at the same time harmonized (and centralized/clustered, cost effective) data management and access to the information. This will then include the increasing need of 3D functionality without the need for every county to make large/impossible investments.

3.7 ARGENTINA

Argentina[103] is a federation of 23 provinces, plus the Federal Capital District (Buenos Aires City). The agencies responsible for the land registration and for cadastral survey depend on provincial government. Cadastral registrations are related to different ministries depending on the specific province. The implementation of information systems within the cadastres has increased their potential to provide different services to society. Cadastral registrations in Argentina cover the following aspects: legal (land transfer, land market), fiscal (land valuation, land tax), and multiple purpose use (planning, local government). As each province has its own cadastre the importance of each role mentioned above varies according to the province, especially the multipurpose role. The multipurpose role of cadastre is sometimes complemented by municipal cadastres according to the provincial organization and development.

As in most other countries of the world, the parcel is the common registration unit for cadastres in Argentina. All land parcel division, unification, amalgamation, etc. must be registered within the cadastral registration. In addition to parcels, horizontal property units (apartments, condominiums) are also registered. Cadastral maps cover basically urban and rural areas. Urban maps have the following contents: geo-reference information, administrative and parcel boundaries, buildings, and complementary improvements. Rural maps are composed of geo-reference information, administrative and parcel boundaries, and topographic data.

In general, the cadastral principles are quite similar to the Netherlands: they are based on a deed registration. Furthermore, the registration of real estate ownership is compulsory and during the initial establishment of the cadastre the real estate properties are systematically established in the register (and not sporadically). In 2003, there were about 13.5 million parcels, of which 12 million are classified as urban parcels, on a total population of a little more than 36 million inhabitants.[103]

In 2002, an initiative has started to strengthen the Cadastral Federal Council, which has 24 members (23 provinces and the Federal Capital District), in order to better comply with a wider range of future requirements. An important effort in this context is to achieve the new National Law of Cadastre.

3.7.1 A Case Study in Argentina

In this section the practice of horizontal layers is illustrated with an example (Figure 3.27). The apartment complex is located in the city of San Fernando in province of Buenos Aires and is located on the ground parcel with the identification IV A 5 3A, which has the full explanation: Circ IV (Cincurnscription), Secc A (section), MZ 5 (block), and Parcel 3A.

As indicated in the introduction above, in Argentina horizontal property units are registered. This is based on a specific Horizontal Property Law (no. 13.512) dating to 1948.[6] This law recognizes within a building complex the following 3D parts: (1) private ownership units, (2) common areas with exclusive use by specific owners (e.g., balconies), and (3) true common areas (this includes the structure of the building like walls and columns, space of the elevator, etc.). This is based on the "dual system" (or perhaps it should even be called a "ternary system") and is different from the situation in the Netherlands, based on the "unitary system," where the whole complex is in co-ownership with exclusive use of certain parts. It should be noted that the Horizontal Property Law is also used quite often for subdivisions that are at the same level.

In Argentina each private ownership unit has a unique parcel number within the province, assigned in a manner similar to other parcels according to the urban law; e.g., IV A 5 3A 01-01. For this, a true survey plan (and not a building design document) is required, which contains the following three parts:

1. Cover with identification of municipality, owner, and ground parcel
2. Location of original parcel within the block and a projection of the building on the parcel (Figure 3.28)
3. For every layer the measurement of the new 3D parcels with a local identification number (Figure 3.29)

a

b

FIGURE 3.27 Horizontal property in Argentina. (a) Side view; (b) top view; arrow indicates view position side view.

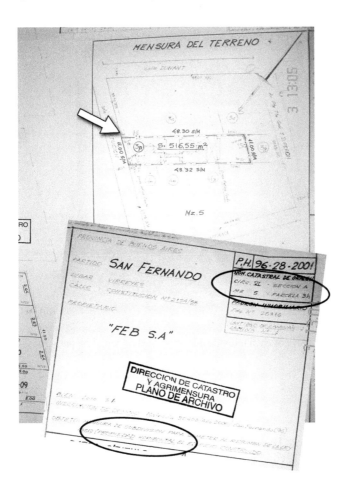

FIGURE 3.28 Identification information and location/projection of horizontal property on survey plan (the arrow indicates the side view direction). Note that the ground level of the construction is indicated with the hatched area and the projection of the construction on the ground surface is indicated with the areas marked with a cross.

Note that this is a survey plan and contains distances to corners (relative position) and does not contain coordinates as in the cadastral map. It should be stressed that the document is not a design or construction plan, but a true survey plan with real measurements. In general, there are no volumes on the survey map; in some provinces the land surveyor puts a mark on the parcel and a number that indicates the height of the floor (relative to the street level).

The example from the city of San Fernando located in the province of Buenos Aires in Figures 3.28 and 3.29 shows the measurements per layer approach, which does not explicitly describe the height dimension as, for example, some of the volumetric descriptions in Queensland do. This is the approach applied in some of the provinces, such as Buenos Aires. In other provinces this can again be different. For example, in

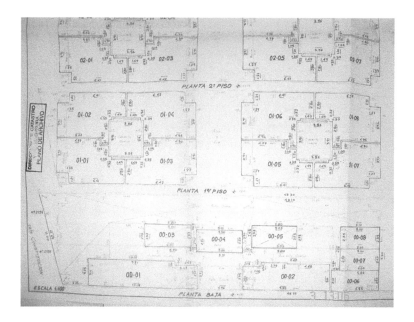

FIGURE 3.29 Surveys of each individual layer in the horizontal property (note the included measurements).

the province of Santa Fee, information is added in the form of cuts or slices. A set of cuts can then be used to obtain a volumetric impression or even an approximated description.

3.7.2 EVALUATING 3D CADASTRE IN ARGENTINA

How can 3D property units be established within the existing legal framework?

Property rights in Argentina must always relate to surface parcels. Consequently, the ownership of real estate above and below the surface is always established on surface parcels. A parcel column can be divided in horizontal layers, which in turn can be subdivided into units according to the Horizontal Property Law. Both the horizontal layers and the units are related to the surface parcels. The resulting units can be considered as true 3D parcels; for example, they get a unique parcel number.

What was the main trigger to establish 3D property units or to start the discussion on how to establish 3D property units?

The main type of 3D real estate property, the apartment, is covered by the Horizontal Property Law. It is expected in the near future that this will not be sufficient for other types of 3D property situations, such as tunnels, underground constructions, pipelines and cables, etc.

Do 3D property units exist as independent properties in the land registration?

The 3D property units, horizontal properties, do exist as real parcels in the land registration. However, they are always related to ground parcels. Other types of 3D property units or rights do not yet occur in the land registration (property registers).

Do 3D property units exist as independent properties in the cadastral registration?

Again, only in the case of horizontal property, the units exist as separate real estate objects in the administrative part of the cadastral registration. The exact geometric representation depends on the actual province, but a formal survey plan is required. On the cadastral map the ground level of the construction is indicated together with a projection of the whole construction within the ground parcel. Other types of 3D property units or rights do not yet occur in the cadastral registration (parcel registration).

What are the main shortcomings of current registration of 3D situations?

The main shortcomings of the Argentina cadastral registration in case of 3D property situations are (1) that it is limited to horizontal property units (apartments), (2) that the 3D situation is projected on the surface, and (3) that the 3D spatial extent of rights is not available in the cadastral map. 3D property units crossing several ground parcels (such as tunnels of infrastructure) are not fully supported as cadastral objects with their own 3D geometry expressed in the same reference system as the cadastral map (parcels and buildings).

3.8 CONCLUSIONS

This chapter presented the 3D cadastral issues in seven countries and states: the Netherlands, Norway, Sweden, British Columbia (Canada), Queensland (Australia), the United States, and Argentina. From this chapter it can be concluded that no complete solution exists for 3D cadastral registration.

In the Netherlands rights and limited rights referring to space or objects above or below the surface are always established on surface parcels. Therefore, to obtain the legal status of a 3D object or of a volume, the intersecting surface parcels need to be queried. The other countries described in this chapter are or will be soon able to establish 3D property units with multilevel ownership within the existing legal framework (with some extensions). These solutions differ per country; e.g., the footprints of 3D property units are limited to the 2D surface parcels (British Columbia, apartments in the United States and Argentina) or not (Norway, Sweden, Queensland, parcels in the United States), the 3D property units have to relate to built constructions (Norway, Sweden, apartments in the United States and Argentina) or not (British Columbia, Queensland), the 3D property units have to be described in survey plans (British Columbia, Queensland, the United States, Argentina) or not (Norway, Sweden).

As can be concluded from this chapter, none of these 3D solutions is a complete solution for 3D cadastral registration. First, a digital 3D description of the 3D property unit in vector format is not maintained (only scanned or paper drawings) in the land registration. The 3D drawing in the land registration is just a 2D visualization of the 3D situation. Therefore, it is not possible to view the 3D situation interactively and the geometry of the 3D property unit cannot be checked. Second, the 3D properties are still not incorporated in 3D (linked to legal information) in the geographical data set of the cadastral registration (cadastral map); hence it is not possible to query the 3D situation. In the best case the 3D property units are incorporated in the cadastral data set in the same way as 2D properties (as footprints). Therefore, these solutions do not address technical issues, such as how to store, query, and visualize 3D property objects in 3D and how to make sure that 3D properties do not overlap (the condition that 2D parcels may not overlap assures complete and consistent registration in current cadastres).

Although the examples of establishing multilevel ownership show good potential for a 3D cadastre, in some countries the step to register 3D properties that are no longer related to surface parcels may be too extensive for the short-term future, e.g., in the Netherlands. An introduction of multilevel ownership requires redefining the 2D cadastral concept. It depends on the legal system if 3D property units that are no longer related to surface parcels are straightforwardly possible within the current legal framework. In countries where the concept of ownership of real estate is still restricted to a surface parcel, the 3D cadastre either has to find solutions to improve cadastral registration using the concept of a surface parcel or has to reconsider the traditional concept of ownership.

4 Needs and Opportunities for a 3D Cadastre

The first part of this book has focused on the basic question: "What are the needs for a 3D cadastre?" An inventory was made of types of cadastral recordings with a possible 3D component. To show the complexities of current cadastral registration of 3D situations, international developments on 3D cadastral registration were studied.

The implementation of a 3D cadastre will only be successful if the considerations for the 3D cadastre reflect on current cadastral frameworks. Current cadastral registrations of 3D situations are therefore summarized in Section 4.1. In this section, a distinction is drawn between countries that are surface oriented and countries that already provide the possibility to establish 3D property units.

The current cadastral registration meets complications in 3D situations, causing a need for a 3D cadastre. This chapter elaborates on the findings of the first part of this book to determine the needs for a 3D cadastre (Section 4.2). In addition, this chapter describes the potentials and opportunities for a 3D cadastre (Section 4.3). When looking at the opportunities for a 3D cadastre, it is relevant to look for applications outside the cadastral domain that may benefit from a 3D approach of cadastral registration, and vice versa, both directly (since 3D information can be interchanged) and indirectly (since they can learn from the experiences of each other). Section 4.4 describes 3D applications outside the cadastral domain. This chapter ends with conclusions.

4.1 CURRENT CADASTRAL REGISTRATIONS AND 3D

Requirements and developments of 3D cadastre are dependent on the type of cadastre as well as on the historical and legal background of a specific country. Cadastral registrations throughout the world are based on the principle that a parcel is the basic registration entity for cadastral registration. This principle of cadastral registration follows the juridical definition of ownership of land, which says that ownership of land is defined by boundaries on the surface and is not explicitly limited in the vertical dimension. In general, the ownership of land includes all space above and below the parcel, as well as all constructions that are permanently fixed to the land. The consequence is that property to land is very well registered in the cadastral registration by means of 2D parcels. The international study (Chapter 3) has shown that 3D property units are currently either established by imposing limited rights on the intersecting surface parcels (Section 4.1.1) or by the explicit definition of the 3D legal space (Section 4.1.2). However, 3D property units are not yet part of the cadastral map.

Establishment of special rights for objects above and below the surface is not necessary from a legal point of view in all cases. For example, many underground situations relate to infrastructure where the owner of the parcel is also the owner of the subsurface object (e.g., a subway tunnel under land owned by the municipality).

In these cases no reference to a subsurface object is made at all in the deed, let alone is a drawing provided. Consequently, this will also not lead to a cadastral recording of the situation. Other cases of underground constructions that do not lead to a cadastral recording are cases of non-registered personal rights (short lease), obligations to tolerate constructions for public good that follow from general laws, and when nothing is registered.

4.1.1 SURFACE-ORIENTED CADASTRAL REGISTRATIONS

In countries that are still surface oriented, 3D property units are established and registered by means of limited rights and other restrictions on intersecting parcels. Although it is possible to register the legal status of these 3D situations administratively, the registration is not satisfactory, for several reasons:

- The right itself is administrated, but not the function of the object to which the rights refer (underground infrastructure, metro station, subterranean parking place).
- 3D spatial information on rights (geometry, location) is not available, e.g., does the right of superficies apply to space above or below the surface?
- The administrative information (by means of restrictions and limited rights) may indicate that something *could* be located above or below the surface. However, this is the only information that the current cadastral registration can provide in 3D property situations.
- No rules or standardization exist for establishing rights and for setting up deeds in 3D situations, leading to diverse solutions. Every notary (or licensed surveyor in some countries) who is confronted with registering rights in 3D situations must decide which rights to use in specific situations and what information to include in deeds (ranging from detailed 3D surveys to a general description).

Cadastral Registration of Property Units in Building Complexes

Property units in buildings are mainly established by means of apartment rights (or strata titles), and sometimes with a right of superficies. In the case of apartment rights spatial information is available in the land registration using the legally prescribed paper or scanned drawings including cross sections. Although not strictly 3D, a drawing of each vertical layer is provided. In case a right of superficies (or other limited right) is established, in general no drawings are available in the land registration. Only in the case of apartment units, the 3D property unit is known in the administrative part of the cadastral registration. In both cases the 3D property unit is not incorporated (either in 2D or in 3D) in the spatial part of the cadastral registration. The deeds and thus the scanned drawings are or will soon be accessible through the cadastral registration, which means that this information becomes available in, but is not integrated with, the cadastral registration. From the cadastral registration, it is possible to see which persons have a right on a parcel or an apartment unit. However, the cadastral registration cannot provide 3D information on how properties are located in the building complex itself.

Cadastral Registration of Infrastructure Objects

In the case of infrastructure objects, the 2D parcel strongly limits the amount of information that can be obtained from the cadastral registration:

- The rights for 3D infrastructure objects are established by means of ownership rights, limited rights, and legal notifications, all established on intersecting parcels and not on the infrastructure objects themselves. These rights are not related to the infrastructure objects.
- There is no uniform way to establish the legal status of infrastructure objects and consequently the registration for infrastructure objects is not uniform.
- The infrastructure object is partitioned over the many parcels it intersects with. No information on the whole infrastructure object is available, not even an identification; i.e., the existence of the object is not known in the cadastral registration. Since the spatial extent of the objects is not known, the following queries cannot be performed. "Which parcels intersect with the 3D object?" "What rights and restrictions are established on the parcels intersecting with the 3D object?" "Are there any 3D objects (tunnels, pipelines) intersecting with a specific parcel?"
- When the parcel is subdivided (e.g., in the case of a transfer of a part of the parcel), it is not always known in which part of the parcel an infrastructure object is actually located. Therefore, the cadastral database can become polluted, as all child parcels will be encumbered with a restriction due to the (potential) presence of a construction. Thus, registration does not necessarily reflect the real situation.

Locating Infrastructure Objects in Current Cadastral Registration

There are basically three possibilities of locating infrastructure objects in current cadastral registration that are based on 2D parcels. The case of the HSL railway tunnel in the Netherlands (see Section 3.1.3) can be used to illustrate these possibilities. We used the parcel boundaries of the intersecting parcels and 3D spatial information on the tunnel to create a fictive cadastral map (see Section 11.1.5) with new parcel boundaries to limit the parts of the parcels that are affected by the tunnel according to Dutch rules. In some countries it is possible to establish a right of superficies for just a part of a parcel. It also should be noted that in the Netherlands it is possible to expropriate only the total parcel column, whereas in other countries it is possible to impose a limited right, e.g., a right of long lease, for a certain part of the parcel column (e.g., to hold a tunnel). Although the actual cadastral geographical data set of 2003 was used, the examples in this section are not intended to show the actual parcel boundaries: they are meant only to clarify the alternatives.

The first map (Figure 4.1a) would be the result if all parcels intersecting with the tunnel were completely affected with a right to build the tunnel. The location of the 3D object is (vaguely) indicated when all parcels that are intersecting the tunnel are selected. This selection is done by finding all the parcels that are encumbered with a right of which the Ministry of Transport and Public Works is the subject. The relationships between the tunnel and limited rights and notifications that are established are not stored (the tunnel itself is not stored). The only information that

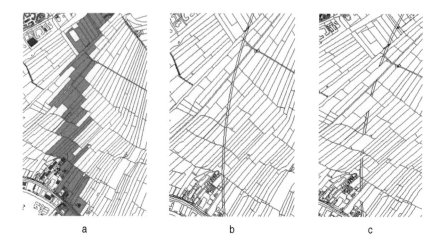

a   b   c

FIGURE 4.1 Three possibilities to register infrastructure objects. (a) Whole parcel is affected; (b) new parcels are generated; (c) as b but now some parcels are not divided.

the cadastral registration can provide is what rights and notifications are established on a parcel and who the subjects are of the rights and notifications. In the case of the HSL tunnel, this subject is the Ministry of Transport and Public Works. Because the ministry owns many other objects as well, this does not give insight into the nature of the 3D object that the ministry keeps on the intersecting parcels: the object could also be a viaduct or a road at surface level. In addition, the result could also be a mix of several different objects (belonging to the same owner).

When the tunnel partially intersects a parcel, normally the ownership or the right of superficies will be obtained for only a part of the land (according to Dutch legislation). This will lead to the creation of new parcels. Figure 4.1b illustrates this situation: the ministry has obtained rights of ownership or superficies for the extent of the tunnel (with a needed safety zone on both sides). New parcels are generated. Still, the relation between the tunnel and all the parcels is not maintained in the database. Because of the pattern of new parcels, the location and direction of the tunnel are clearly visible. But if other constructions are partly built on top of the tunnel and new parcels will be created according to the footprint of these buildings as in the Rijswijk case (Section 3.1.3), this image will be disturbed. Also, the same parcel pattern might be the result in the case of physical objects above the surface, such as roads. The cadastral map is even more disturbed in Figure 4.1c. It is more realistic to suppose that the Ministry is not the owner of only the land right above the tunnel, but also of complete parcels. For example, during the negotiations they may agree to buy all the land from the original owner (and not only the small zone that is actually needed). In this case, there is no need to generate new parcels.

Not all countries require the creation of new parcels to be able to establish the restriction on only a part of the parcel. In those cases the exact location of the restriction can be defined in the deed, but is mostly not maintained in the cadastral registration. Consequently, the location of the restriction is not clear from the cadastral registration.

4.1.2 3D PROPERTY UNITS IN CADASTRAL REGISTRATIONS

Some countries and states have redefined the unlimited ownership of a parcel. In these countries and states it has recently (or will soon) become possible to establish ownership rights related to bounded volumes by defining volumetric parcels (Queensland and the United States), air-space parcels (British Columbia), or 3D construction properties (Argentina, Norway, and Sweden). These 3D property units are the result of subdividing 2D parcels or are actual parcel columns. The solutions fit into the existing legal framework of the specific country or required only small adjustments. The possibilities to establish multilevel ownership have not yet been translated into an appropriate cadastral registration of 3D property units. The 3D property units exist as independent properties in the land registration and are described on 3D survey plans. The 3D property units also exist as independent properties in the administrative part of the cadastral registration. However, it is impossible to view the 3D property units interactively (which is helpful to get insight into complex 3D property units) since the drawings are available only in scanned or paper format. In addition, consistency checks are not possible: Are two 3D properties neighbors, is there a gap, is there overlap? Finally, the 3D properties are not available in 3D as part of the cadastral geographical data set and therefore cannot be examined in 3D.

4.2 BASIC NEEDS FOR A 3D CADASTRE

The complexities described in the previous section are not all new. However, they have become more obvious during the last decades. This is partly because 3D situations occur much more often now than 40 years ago (number of multipurpose buildings has increased, number of cables and pipelines has grown, many tunnels have been built during the later part of the last century). But also as a result of the considerable increase in the value of property during the last few decades, users want to have the legal status of their property clearly ensured in the cadastre. This means that the cadastre should give sufficient insight into property and in the boundaries of property in all dimensions.

From the complexities and limitations summarized in the previous section, conclusions on the basic needs for a 3D cadastre can be drawn. The basic needs for a 3D cadastre can be summarized as follows:

- To have a complete registration of 3D rights (rights that entitle persons to volumes). The current cadastre already registers rights that entitle persons to volumes; however, a 3D cadastre should explicitly register the 3D space to which these rights apply (which is nowadays only available in land registrations).
- To have good accessibility to the legal status of stratified property including (3D) spatial information as well as to public law restrictions.

Whether information that does not directly support the main tasks of a cadastre should be registered and maintained in the cadastral registration, e.g., the exact location of cables and pipelines, is disputable and dependent on the background of a cadastral registration. In addition, it will be more effective (e.g., with respect to data integrity and data consistency) if information on constructions and other objects of interest are maintained at their source and accessible within and from the 3D cadastre.

Based on these considerations, we can conclude that a 3D cadastre should incorporate the following functionalities:

- Register 3D information on rights (what is the space to which the person with a real right is entitled?) and make this information available in a straightforward way.
- Establish and manage a link with external databases containing objects of interest for the cadastre (infrastructure objects, soil pollution areas, forest protection zones, monuments) and incorporate the location (and other information) of these objects in the cadastral registration.
- Use the information on these objects to support registration tasks, i.e., to detect and correct errors or in the process of registering and viewing the legal status of 3D situations. Are all intersecting parcels encumbered with a right for the infrastructure object?

Linking different registrations and linking different databases can be established by setting up a well-functioning national geo-information infrastructure (GII).

The term "spatial data infrastructure" (SDI) or "geo-information infrastructure" (GII) is often used to denote the collection of technologies, policies, and institutional arrangements that facilitates the availability of and access to geo-information to the benefit of many users.[125] The word infrastructure is used to promote the concept of a reliable, supporting environment, analogous to a road or telecom network, that, in this case, facilitates the access to geo-information using a minimum set of standard practices, protocols, and specifications. Like roads and networks, a GII facilitates the conveyance of virtually unlimited packages of geographic information.[125] A GII consists of the following four components.[61]

1. Geographic data
2. Technology for storing, access, distribution, and use of geo-information
3. Standards for describing, exchanging, and linking geo-information
4. Policy and organization

A distributed setup of registrations within a GII provides the possibility to link information maintained in different databases. In this way the geometry of infrastructure objects and other 3D objects of interest can remain and be maintained at their original source (in databases at organizations that are responsible for these objects), while this information can be used to improve cadastral registration in 3D situations.

4.3 OPPORTUNITIES FOR A 3D CADASTRE

A 3D approach to cadastral registration offers improvements for the main tasks of a cadastre for a number of reasons:

- 3D registration provides information on the 3D extent of rights, limited rights, and legal notifications and allows integration of 3D information in the current cadastral geographical data set. In the case of 3D registration, a 3D property unit can be queried in a 3D environment in the same way a parcel can be queried in the current registration (with some other attributes).

- A 3D cadastre will incorporate digital information on 3D situations. In the current registration, scanned or paper drawings clarifying the 3D information can be added to deeds. The availability of deeds in digital scanned form has already improved the accessibility of information. It is now possible to link digital documents to parcels in the cadastral geographical data set (e.g., the document appears after clicking on a parcel). However, a vector representation of the situation in the national reference system (not scanned) instead of a drawing will offer better registration possibilities, as it is easier to integrate vector information with the current cadastral geographical data set to obtain an overview of the whole 3D situation (and not just at the location of the specific parcel). Digital information will also offer better possibilities for quality checks. In addition, digital information facilitates the exchange and integration of information between and within cadastral offices, municipalities, and provinces, and it facilitates viewing of 3D (property) situations interactively.
- When enabling 3D registration, the parties involved have a tool to register 3D situations, which may motivate them to include spatial information in deeds and to establish the legal status of 3D situations in a uniform way. This makes it possible to have uniform, and consequently readily accessible, recordings of 3D property units (it should be noted that coordinates, also in 3D, should always be obtained from cadastral surveying).

A 3D cadastre can interact with other registrations, which offers other opportunities as well:

- If the exact 3D location of infrastructure constructions is available within the cadastral registration (maintained in databases by holders of these objects), the cadastre can use this source for certain cadastral tasks, e.g., during cleanup of registration or to support other cadastral tasks.
- Holders of infrastructure constructions will benefit from a clear registration of the location of infrastructure objects, because they have more legal protection since rights are better maintained and they do not pay compensation for parcels that do not intersect. In current practice, errors occur such as a cable crossing a parcel where no limited right or notification has been established or a limited right or notification has been established when the parcel is not crossed by a cable. By knowing the exact locations, the parcels and thus the persons involved and who need to be compensated can be more accurately determined.
- Linking databases containing infrastructure objects with the cadastral registration can also be used for registering infrastructure objects, such as pipelines. For example, according to a decision of the Dutch Supreme Court, telecom networks are considered immovable goods (see Section 3.1.2). This decision will in the future apply to other cables and pipelines

as well. Consequently, the cadastre must be able to register these networks apart from parcels and apartments. This registration can be improved when a direct link is maintained with the database of the holder of the networks (and of other pipelines in the future).

4.4   3D APPLICATIONS OUTSIDE THE CADASTRAL DOMAIN

To ensure legal security and to support town and regional government in general, 3D geo-information receives more attention in today's society where there is an increasing interest in placing different types of land use on top of one another. Registrations and applications outside the cadastral domain are therefore also confronted with the fact that 3D information becomes more and more important. A 3D cadastre can benefit from other domains that develop toward 3D, and vice versa, because knowledge and experiences can be shared and because 3D data can be interchanged.

Many examples of applications that have a growing interest in 3D information have been cited elsewhere.[137,138] Traditionally, military applications were the first to look for 3D solutions and provided the first elaborated systems for 3D visualization and simulation.[101] Nowadays, more and more civil applications need the third dimension:

- Urban planning is one of the most demanding areas pushing 3D developers to provide rapid modeling approaches, extended visualization and interaction tools, and elaborated spatial functionality.[118,183] The influence of new buildings and infrastructure on the existing environment can be visualized best in 3D environments, which is important in discussions with citizens. In addition, 3D visualizations of planned infrastructure and underground constructions provide better insight into the vertical planning of regions.[86]
- Landscape modeling seeks specific 3D tools for interactive design and simulation.[13,25]
- Road, railway, and canal construction and maintenance benefit greatly from visual 3D environments.[16]
- Maintaining 3D information on real-world objects enables us to deal with 3D characteristics of buildings, e.g., calculating the volume of buildings (for tax purposes) or dictating a maximum construction height and depth.
- 3D geo-information can serve as input for 3D spatial modeling such as modeling noise levels[89] and risk modeling for buildings when a tunnel is being drilled.[119]
- Knowledge about 3D characteristics of natural processes can be used to impose limitations and obligations, e.g., in cases of noise control, odor nuisance, and safety measures.
- In telecommunications the decision on the locations of antennas requires 3D analysis to obtain information on the area that can be covered and on the costs of using a specific location.
- Geological applications (e.g., finding fractures or salt domes) require 3D analysis.[220]
- To predict the consequences of bursting of dikes (flooding), a good terrain model is needed together with 3D software.[221,228]

FIGURE 4.2 (Color figure follows page 176.) To avoid damage to cables, digging first by hand is necessary (*De Volkskrant* Newspaper, July 2000).

- Cables, pipelines, and tunnels can be better protected against damage when their 3D location can be visualized in the real world[175] (Figure 4.2). Based on knowledge of the location of constructions, precisely defined restrictions can be imposed on the owners of the surface land from doing anything that could damage the underground construction.
- Location-based services (LBS) for shopping, tourism, rescue operations, etc. are another area where the use of 3D visualization and 3D GIS is rapidly increasing.[29,76]

A last example with increasing interest in incorporating 3D geo-information is the domain of local land-use plans. At the moment, there are no standards or rules to incorporate 3D information in local land-use plans. Consequently, every local land-use plan that has to regulate different types of land use on top of each other reinvents the way to deal with the 3D component of local land-use planning. Local land-use plans can also differ within one project, as local land-use plans are the responsibility of municipalities and infrastructure objects may cross municipality boundaries, e.g., as in the case of the HSL tunnel.

An example of a local land-use plan that had to deal with 3D information is the *Noord-Zuid lijn* in Amsterdam.

In Amsterdam a metro tunnel is being drilled from north to south (the Noord-Zuidlijn). A local land-use plan was needed in which the use of a tunnel below other types of land use was guaranteed. The tunnel is planned partly below houses.

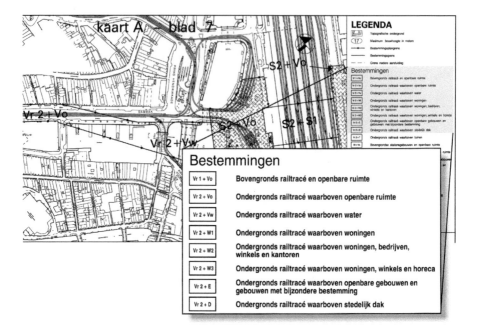

FIGURE 4.3 Local land-use plan of metro tunnel (Noord-Zuid lijn) in Amsterdam. Ondergronds railtracé waarboven means "subsurface metro line on which."

Figure 4.3 shows part of the map that was produced for this local land-use plan. It is a 2D map. The areas on the 2D map are encoded (as "multi-layers") and the 3D information ("tunnel below houses") is added as a description in the legend and not as a 3D spatial description. Consequently, the local land-use plan of the Noord-Zuidlijn does not include 3D spatial information, which also does not occur elsewhere in the local land-use plan.

4.5 CONCLUSIONS

This chapter has summarized the complexities and limitations of current cadastral registration in 3D property situations presented in the previous chapters.

From a legal point of view it does not seem problematic to establish 3D property units. This can be realized either within legal frameworks that still strongly hold to the unlimited concept of ownership that is linked to surface parcels (using right of superficies, apartment rights, and strata titles) or within more flexible legal frameworks that enable establishment of multilevel ownership (e.g., air-space parcels, volumetric parcels, construction properties) as was observed in a few other countries.

In countries where 3D property units are established by means of limited rights on surface parcels, the registration of the legal status of 3D situations has until now been limited to an administrative registration. Apartment units (or strata titles) or in cases when more than one real right is established on one parcel involve a 3D situation. In those cases it is possible to query which rights and persons are involved. However, no 3D overview of the situation can be obtained.

Also in countries in which it is possible to establish 3D property units independent from the surface, no solutions have been found to incorporate the 3D geometry of 3D property units in cadastral registration. Current cadastral registrations all lack a fundamental approach for 3D cadastral registration by combining legal, cadastral, as well as technical aspects with respect to 3D situations.

This chapter has defined the essential elements for a 3D cadastre. A 3D cadastre should be able to:

- Maintain the spatial extent of real rights, and provide information on the spatial extent of real rights
- Establish and manage a link with external databases that contain objects that are of interest for the cadastre (infrastructure objects, monuments, soil pollution zones, etc.)
- Use information on these objects in the work processes of cadastral registration

Registration of 3D situations offers other opportunities as well. Once 3D information on situations is accessible (e.g., from the cadastral registration based on links with other registrations via the GII), this information can be used in other applications, and vice versa. For example, exact information on the location of cables, pipelines, and tunnels offers opportunities to use this information in the management (planning activities) of the subsurface.

The remainder of this book aims at meeting the needs of cadastral registration by studying possibilities and constraints of establishing a 3D registration both from a technical and a cadastral point of view. The proposed solutions for a 3D cadastre should fit to some extent in current legal frameworks in order to be feasible for the near future.

Part II

Framework for Modeling 2D and 3D Situations

5 Theory of Spatial Data Modeling

This chapter presents an overview of the basic concepts and terms in spatial data modeling. The aim is to familiarize the reader with concepts used in this book. First, data models and in particular spatial data models are described (Section 5.1), followed by a description of the different phases in data modeling including their characteristics (Sections 5.2 to 5.4). UML (Unified Modeling Language) has become a standard to represent data models and is used to represent the data models in this book. Therefore, a short introduction UML is also included in this chapter (Section 5.5). DBMSs are essential systems in spatial data modeling and in the new-generation GIS architecture. Section 5.6 describes how the relationship between spatial data modeling in GISs on the one side and in DBMSs on the other side has evolved. Finally, when looking at spatial data models, the standardization initiatives on spatial data modeling are important; these are described in Section 5.7. The chapter ends with concluding remarks.

5.1 DATA MODELS

The term *model* is a frequently used term in many disciplines. Models in general are used to make an abstraction of reality with the aim to make reality understandable. Data models are intended to interpret the world in a way that is understandable to computers.[213] A data model is a generic blueprint (structure); the data model can be populated with instances (data) to come to an abstraction of reality for a specific application. Data models consist of the following:

- Classes
- Attributes
- Relationships
- Constraints
- Operations

Classes and objects: In data models classes are abstractions of phenomena in the real world that can be identified, e.g., parcels, persons, buildings. Objects are instances of classes. An object instance has at least a unique id, which is in principle meaningless but which can be used in references. In data models, the term *object* does not refer to objects as they occur in the real world, but to the representation of the real-world objects, which may be very confusing. The representations can be maintained in a DBMS. For example, a road can be referred to as an object, i.e., the representation of the road. The object, containing both spatial and non-spatial attributes, can be maintained in the DBMS (e.g., line, with attributes such as owner, type of asphalt, etc). Objects are basic elements in object-oriented modeling (see Section 5.3.2).

Attributes: Objects have attributes in which the property of the objects is described, e.g., a land parcel can have "area" or "land use" as an attribute.

Relationships: In the data model, relationships exist between the objects, identifying how the objects are related. For example a land parcel has a relationship with person: a parcel is owned by a person. There are three kinds of relationships with respect to cardinality: one-to-one, many-to-one, many-to-many. The objects can be structured in a class hierarchy. Objects that are derived from other objects have either a "is-part-of" or a "is-a" relationship with the objects they are derived from. The first type of relationship is called "aggregation" and the second type is called "specialization."

Constraints: A constraint is a limitation on objects or on relationships in the data model, e.g., "the age of the object person must be more than zero." Consistency constraints can be used to prevent any logical contradiction within a model of reality.[43] This is not the same as correctness, which excludes any contradiction with reality itself. Consistency constraints are used to enforce the logical consistency of the data model. Consistency constraints can be organized into two groups:[204]

1. Inherent constraints, which are incorporated in the definition of the data model. The model can disallow certain objects or limit certain relationships by its definition. For example, if the data model does not define relationships between a parcel and a subject, this relationship cannot be maintained.
2. Explicit constraints, which are not part of the data structure but which need to be explicitly defined, e.g., the constraint that an employee cannot earn more than his manager.

Operations: The operations describe all the actions that can be performed on objects. Here we focus on operations in DBMSs. Four generic DBMS operations on objects using the Data Manipulation Language (DML) are distinguished in the database literature:[204]

1. Retrieve: Make a whole data set available to the user.
2. Insert: Add new data to the database.
3. Delete: Remove data from the database.
4. Update: Change existing data.

Apart from these generic operations, three other supporting operations in DBMSs are distinguished[204] (the first two are used in the DML operation mentioned above, like the third operation is a modeling action part of the Data Definition Language [DDL]):

1. Selection: Retrieve operation under a particular condition (only specific parts of a database become available).
2. Navigation: Operations that permit a logical path on the basis of a selection to be followed.
3. Specialization: Complex operation that allows a new object to be created on the basis of existing ones.

Note that the term *specialization* is also used to denote a special type of relationship in data modeling (as mentioned above).

5.1.1 DATA MODELS IN GIS

In GIS, a data model is the structure used to identify and represent objects referenced by space relative to the earth surface.[179] Models of spatial information are usually grouped into two broad categories: field-based models (raster) and object-based models (vector).

In the field-based model, the world is modeled as a regular tessellation (raster), which is sampling based. For example, height can be modeled in a field-based approach in which each point in space has exactly one value of height. Field-based models are often used to model continuous spatial trends such as elevation, temperature, and soil. In object-based models, the focus is on abstracting spatial information into distinct, identifiable, and relevant things or entities called objects. Individual objects are modeled together with their attributes. Object-based models are often used for human-made objects and are common in modeling transportation networks (roads), land parcels for property tax, and legal ownership-related applications.

Objects in GIS

Traditionally geo-sciences focus only on real-world phenomena with a spatial extent. It is therefore relevant to distinguish between spatial (or spatial-temporal) objects and non-spatial objects. A spatial object is the representation of a real-world object having spatial (topology, size and shape, position and orientation) and thematic characteristics.[7,109,113,163] A spatio-temporal object has three fundamental components: location (spatial), attributes (aspatial) and time (temporal).[223]

Till recently GIS models maintained only spatial objects, while non-spatial objects, such as subjects or rights in a cadastral context, were maintained in DBMSs or were integrated in GIS as semantic characteristics of spatial objects. However, integrated architectures are evolving in which both spatial and non-spatial objects are maintained in one integrated DBMS (see section 5.6).

Relationships in GIS

In spatial data models, spatial relationships exist. Spatial relationships describe the relationships between the geometric elements of spatial objects. In spatial modeling, spatial relationships serve two main purposes:

1. To find the spatial relationships between two spatial objects (used in querying), e.g., find all parcels that are adjacent to a certain parcel
2. To enforce the consistency of a model by formulating consistency constraints using spatial relationships (used in modeling and editing), e.g., two parcels should not overlap

Spatial relationships can be classified as topological or geometrical. Topological relationships describe the connectivity, containment, and adjacency relationships among spatial objects. These relationships are invariant under transformation such as translation, scaling, and rotation.[224] Geometrical relationships are described in terms of distance and directions and depend on the absolute positions of objects relative to a given reference system.[224]

Constraints in GIS

In spatial data models, consistency constraints can be used to enforce spatial characteristics. For example, topological constraints can enforce that lines intersect only at nodes and parcels shall not overlap. Semantic constraints can enforce spatial characteristics that are dependent on semantics, e.g., a building area should always be adjacent to a street.[26] Semantic constraints are application dependent.

Operations in GIS

Operations on spatial objects can be performed on both the spatial characteristics and the thematic characteristics of the objects or on a combination of these characteristics. Here we focus on operations on spatial objects maintained in spatial models in DBMSs. DBMSs have a strictly defined functionality based on relational algebra and calculus[170] and were originally not designed to manage spatial objects. The traditionally available operations had to be "translated/extended" into the spatial domain to be able to handle spatial objects. As was seen, four basic operations are distinguished in the database literature:[204] retrieve, insert, delete, and update. A similar set of operations (but more elaborated) has to be available for spatial data. The operations related to introducing a new element, deleting and updating an existing one have to be extended with respect to the data structure used. Four groups of operations related to DBMSs are distinguished that use the geometrical characteristics of spatial objects:[179]

1. Update operations: Standard DBMS operations such as insert, delete, modify, etc.
2. Select operations, e.g., "retrieve all parcels that overlap with this pipeline": Three basic groups of selection operations with respect to spatial objects can be defined to be offered at DBMS level:

    (a) Metric operations: Selection operations that require computations of geometrical properties, e.g., compute distance, volume, area, length, and center of gravity. Metric operations need coordinates of the spatial objects and the result is always quantitative. Metric operations are unary operations and should not be confused with metric relationships, which are binary operations.
    (b) Proximity operations: Selection operations related to spatial location, e.g., objects in a certain area, volume, or field of view.
    (c) Relationship operations: Selection operations based on spatial relationships between objects.

3. Spatial join: Like the join operator in relational databases (see Section 5.3.1), the spatial join is one of the more important operators. When two tables are joined on a spatial predicate (intersect, contains, is-enclosed-by, distance, northwest, adjacent, meets), the join is called a spatial join. This is equivalent to the map overlay in GIS. The operations combine two sets of spatial objects to form a new set. An example is "find all natural areas and forest areas that overlap."

4. Spatial aggregate: Retrieve spatial objects based on spatial characteristics of other spatial objects; an example is "find the station closest to this building."

The spatial join and the spatial aggregate are actually complex select operations.

5.1.2 DESIGN PHASES IN MODELING

A data model is a structure to capture an abstraction of reality for a specific application. In designing a data model, three phases are distinguished in the literature that have their own data model associated with them:[204,213]

1. A conceptual model (Section 5.2)
2. A logical model (Section 5.3)
3. A physical model (Section 5.4)

5.2 CONCEPTUAL MODEL

In the conceptual phase all classes that need to be included in the data model are identified, together with the characteristics and relationships of the classes. The aim of the conceptual model is to demarcate the part of the real world that is relevant for the specific application. The model has a high abstraction level as it is the basis of the conception process. It consists of a schematic representation of phenomena and how they are related. The conceptual model not only provides a basis for schematizing, but is also a tool for discussion, and therefore a good conceptual model must be easily understandable. The model sharing may be done by using narrative language, but the transfer to the next stage is easier if a more formal language is used.[96] Until recently ER (entity relationship)[23] has been a popular tool for designing the conceptual data model. In the ER model, the world of interest is partitioned into entities (objects), which are characterized by attributes and interrelated relationships. Associated with the ER model is the ER diagram, which gives a graphic representation to the conceptual model. In the ER diagram entities are represented as boxes, attributes as ovals connected to the boxes, and relationships as diamond boxes. Recently, UML has become a standard for conceptual and logical model design. The UML class diagram is the counterpart of the ER diagram. UML is discussed in more detail in Section 5.5.

5.3 LOGICAL MODEL

In the phase of logical design the conceptual model is translated into a logical model. In this phase the conceptual schema is translated into the data model of a particular type of DBMS. Often the term *logical model* is associated with data structure, as in this phase the database structure is designed. Three types of database models are distinguished here (other examples are network models and hierarchical models): relational model, object-oriented model, and object relational model. These models are described in Sections 5.3.1, 5.3.2, and 5.3.3, respectively.

5.3.1 RELATIONAL MODEL

The relational model was introduced by Codd.[27] A relation is an organized assembly of data that meets certain conditions. A relational database is a collection of relations. A relation has a number of attributes or data items representing some property of an entity. Relational models have been widely adopted by the market and have been implemented in mainstream DBMSs.

A table in a relational database represents a relation, and each column of a table is called an attribute. An object type can be defined by one or more relationships. The relationships between tables are established by keys. A key is an attribute (or combination of attributes) that contains unique values for each row in the table. Certain constraints on the relational schema must be maintained to ensure the logical consistency of the data. Three kinds of constraints can be distinguished:

1. Key constraints: The key constraint specifies that every relation must have a primary key. There may be several keys in a relation. The one that is used to identify the entities is the primary key.
2. Entity integrity constraints: The entity integrity constraint states that no primary key can be null.
3. Referential integrity constraints: Logically consistent relationships between the different relations are maintained through the enforcement of referential integrity constraints. This constraint can be implemented using a foreign key. A foreign key is a set of attributes in a relation that is duplicated in another relation. The referential integrity constraint stipulates that the value of the attributes of a foreign key either must appear as a value in the primary key of another table or must be null. Thus, a relation refers to another relation if it contains foreign keys.

Data definition and data manipulation of relational models can be done with Structured Query Language (SQL). A short introduction of SQL follows.

SQL

SQL is the most widely implemented database language for relational models. SQL has two components: the DDL (Data Definition Language) and the DML (Data Manipulation Language). The schema of the database (containing definitions for tables and constraints) is specified with the DDL. The DDL is used to create, delete, and modify the definition of the tables in the database, while the actual queries are posed and rows are inserted, updated, and deleted in the DML. The basic principles of SQL[205] are described below to provide understanding of the SQL statements used in this book. Oracle SQL is used here as an example, although slight differences can be present between the SQL in different relational DBMSs. A table can be created using the DDL component of SQL:

```
CREATE TABLE subject (
subject_id                    number(12),
name                          varchar2(128),
street                        varchar2(24),
place                         varchar2(24),
PRIMARY KEY subject_id)
```

The name of the created table is "subject." The table has four attributes, and the name of each column and its corresponding data type is specified. Tables no longer in use can be removed from the database using the "drop table" command. After the table has been created, data can be inserted in the table ("populating the table"). This is done in the DML component of SQL. The following statement adds one row to the table "subject":

```
INSERT INTO subject VALUES (999, `Stoter', `Jaffalaan 9', `delft')
```

Adding another row with the same subject_id will be rejected by the DBMS because of the primary key constraint specified in the "create table" statement. The alternative to the insert command is "bulk loading," which can be used to save time when inserting high volumes of data. Once the database schema has been defined and the tables are populated, queries can be expressed in SQL to extract the subsets of interest. The return values of a select query can also be the result of operations on the resulting subset. The basic operations are union, intersection, and difference. All the rows of the table are scanned and the ones where the sought value is found are returned as results. The basic form of a select query (which is part of the DML) is:

```
SELECT column_names FROM relations WHERE row-constraint
```

Operations can be specified after both the SELECT and the WHERE keyword.

5.3.2 Object-Oriented Model

Although objects always have been the basis in the conceptual phase of data modeling, the existing technologies forced the data model to be implemented in other structures such as the relational structure. However, relational modeling is table and record oriented and not object oriented, which has proved to have its limitations when modeling the real world:[70]

- A restricted set of data types is available even for less complex data.
- The structure of relational databases (tables, rows, columns) does not accommodate complex data types easily.
- Complex data can be stored as BLOBs (binary large objetcs), which can be retrieved from relational databases, but not searched, indexed, or manipulated.
- Relational tables offer an inadequate model of real-world objects, because objects can only be modeled as a set of relationships, e.g., how to deal with behavior of objects.
- Relational tables offer poor support for integrity constraints.
- Operations are available only in a limited way.
- In relational DBMSs it is difficult to handle recursive queries.

The basic idea of object oriented modeling is to make a direct correspondence between real-world entities and their computer representation. In an object-oriented data structure, classification is the main principle. Classification is the mapping of objects or instances to a common type. The combination of classes, objects, and operations (methods), together with the inheritance principle, characterizes the object-oriented model, in contrast to the record-oriented relational model.[96]

Classes and objects: Classes are collections of objects with the same behavior. Instances are particular occurrences of objects for a given class. Within classes, subclasses can be defined, for example, the class trees can be divided into leaf trees and fir trees. The subclasses are specializations of the superclass, as was mentioned in Section 5.1.

Attributes: Objects have attributes associated with them with their data types (which can be user-defined data types). Attributes are the descriptive properties of the object. Instances of an object have all the attribute types of the class in common. Attribute values can be defined at either the class or the instance level.

Methods: Classes are characterized not only by attributes but also by methods. "Method" refers to an operation on objects: a procedure that can be applied to a class of objects. A method is a member function of the class.

Inheritance: In classification hierarchies, an object in a subclass (specialization) inherits all attributes of the corresponding higher-level superclass. For example, if we have a superclass LineString we can define subclasses LinearRing and Line, which both inherit the operation Length from LineString.

In object-oriented modeling the spatial and non-spatial attributes of spatial objects do not differ very much from each other. The attributes "area" and "geometry" of a land parcel are not treated differently as other alphanumerical attributes. According to Reference 224 there are some problems with object-oriented databases that cause performance to be a difficulty in object-oriented databases:

- Provision of query optimization is made difficult by the complexity of object types. Many operations are available compared with the few operations in relational DBMSs. It is therefore difficult to estimate the cost of execution and to choose between different strategies to execute a query.
- Indexing is difficult. The difficulty is that indexes rely on direct access to attribute values, while an object is only accessible via messages through its protocol and identified by the object-id.
- Transaction in object-oriented databases may be of a much higher level of complexity than simple transactions within a relational DBMS. As a consequence of the hierarchical nature of much object data, transactions may cascade downward and affect many other objects.

According to Reference 70 the problems of the object-oriented approach are that there is no standard data model, object orientation has no clear theoretical basis, and, most importantly, there is no standard query language, such as SQL in relational databases. Because of these problems, object-oriented modeling has been less adopted in mainstream DBMSs than has relational modeling.

5.3.3 OBJECT RELATIONAL MODEL

The object relational model[189] introduces the advantages of object-oriented models in relational models. In relational databases the set of data types is fixed. In object relational modeling this limitation is overcome because of the built-in support for user-defined data types: Abstract Data Types (ADTs). Like classes in object-oriented technology, a user-defined type consists of internal attributes and member functions to access the values of the attributes. Member functions are callable within SQL and can modify the values of the attributes in the data type. A user-defined type can appear as a column attribute type in a relational schema. The term *abstract* is used because the end user does not need to know the implementation details of the associated functions. The structure is hidden from the user, who can access it only through the operations defined on it. All that the end users need to know is the interface, i.e., the available functions and the data types of the input parameters and output results.[179] The ADTs appear at the same level as base data types, such as float or string.

Spatial Data and Abstract Data Types

Spatial database applications must handle complex data types such as points, lines, and polygons in 2D and 3D as well as 3D primitives such as polyhedrons. Traditional relational DBMSs support only a set of alphanumerical data types (date, string, number). It has been stated that the principal demand of spatial SQL is to provide a higher abstraction of spatial data by incorporating concepts in relational databases closer to our perception of space.[42] This can be accomplished by incorporating the object-oriented concept of user-defined ADTs. When a user-defined type "point" is created, we can define a column name "location," of type "point." The operations that can be performed on the data type are stated in the type definition. For the point type, for example, a function "distance" can be defined, which computes the distances between two points. Another example is a land parcel stored in the database. A useful ADT may be a combination of the type polygon and some associated function (method), say, "is_adjacent." The adjacent function may be applied to land parcels to determine if they share a common boundary.

The OpenGIS Consortium (see Section 5.7.1) has defined specifications for incorporating 2D spatial ADTs in SQL. These ADTs include topological and geometrical operations. How mainstream DBMSs have implemented these specifications is described in detail in Chapter 6.

5.4 PHYSICAL MODEL

In the phase of the physical design, the logical model is translated into hardware and software architecture. The physical model is hidden from the user. The design of the physical model is critical to ensure reasonable performance for various queries. Therefore, the physical design must enable the operations for manipulating the logical model in an efficient way. At the physical level the following tasks are handled by the DBMS:[173]

- Storage. The DBMS manages an efficient organization of the data on a persistent secondary storage unit (mostly one or many disks). The representation at this level might be completely different from that shown to the user according to the logical data model. A table might be stored in several files, possibly distributed over many disks. Data sets are often too large to fit in the primary memory of the computer and accessing secondary memory is much slower than accessing primary memory caused by moving the head of the disk reader. On the other hand transporting data between primary and secondary memory may also cause a performance bottleneck. The goal of good physical database design is therefore to keep the amount of data transfer between primary and secondary memory to an absolute minimum.

- Access paths and (primary) indexes. In response to a query, the spatial access method should only search through a relevant subset of objects to retrieve the query answer set. This can be achieved by primary and secondary indexes. Primary indexes are built with the table itself, while secondary indexes are additional structures. A DBMS provides data access methods or access paths that accelerate data retrieval. A typical data structure that accelerates data retrieval is the B-tree.[28]

- Query processing. Processing (evaluating) a query usually involves several operations. To evaluate the query efficiently, these operations must be properly combined. An important issue in query processing is the design of efficient join algorithms.

- Query optimization. Because most query languages are purely declarative, it is the responsibility of the system to find an acceptably efficient way to evaluate a query.

- Concurrency and recovery. The DBMS manages concurrent access to data and resources from several users and should guarantee the security and consistency of the database, as well as the recovery of the database to a consistent state after a system failure.

Another aspect can be added to this list:[179]

- Clustering. The goal of clustering is to reduce seek and latency time in answering queries that result in a range of data. For spatial data this implies that objects that are close to each other in the real world and are commonly requested jointly by queries should be stored physically together in secondary memory. The design of spatial clustering techniques is more difficult compared with traditional clustering, because a storage disk is a one-dimensional device. What is needed is a mapping from a higher-dimensional space to a one-dimensional space that is distance preserving. Several mappings to accomplish this are Z-order, Gray code, and Hilbert curve.[47]

5.5 UML

UML[206] has become a standard language for object-oriented software design at the conceptual level, as well as for many other applications. The language can be used to model the structural schema of a data model at the conceptual level. There are many types of UML diagrams: use-case diagram, class diagram, object diagram, sequence diagram, collaboration diagram, statechart diagram, activity diagram, component diagram, and deployment diagram. Apart from the diagrams the UML standard offers a

Parcel
+objectId:Number
firstLineId:Number
returnPolygon:Geometry

FIGURE 5.1 Example of UML notation of a class.

language to formally describe limitations and constraints in the diagrams: the Object Constraint Language (OCL). UML has two diagrams to describe the static structure of a system: class diagram and object diagram. Both diagrams show the elements of the system and the structural relationships. The class diagram contains the classes in the system with their attributes, operations, relationships (associations), and constraints. The class diagram is a model: it describes the structure and the limitations of the objects. The object diagram is a representation on a certain timestamp of the objects that have been created according to the structure of the class diagram. In most cases only the class diagram is used. In this book the class diagram of UML is used to describe the data models. UML notation for class diagrams is briefly described in this section (see also Reference 218).

Class: Class is the encapsulation of all objects that share common properties in the context of the application. The UML notation of a class is a rectangle with three parts. In the top of the rectangle the name of the class is stated, in the second part the attributes, and in the third part the operations (Figure 5.1).

Object: An object is denoted in UML with a rectangle containing underlined text, starting with a colon, followed by the name of a class (e.g., : Parcel).

Attributes: An attribute is information, maintained by an object (instance of a class). Every attribute has exactly one value for every instance of the class. These values represent the state of an object. Attributes are stated in the middle part of the representation of a class. The type of the attribute is reflected after the colon, after the name of the attribute. A "+" for the attribute indicates that the attribute is known outside the object (public attribute), a "−" indicates that the attribute is known only within the object (private attribute).

Operations: The collection of operations of an object represents the behavior of an object. Since all instances of a class have the same operations, the operations are described within the class. An operation can have arguments and a return value. Operations are stated in the bottom part of the representation of a class. Parameters are given with round brackets after the operation name. The type of a return value is given behind a colon after the parameters. The "+" or "−" can be added to indicate whether the operation is public or private.

Association: An association is a structural relationship between two classes. Structural means that an instance from one class is related to an instance from the other class during its existence. The relationship can change over time. Between two classes more than one association can be defined. For example, a person can work for a company, and a person can be a customer of a company. In UML an association is drawn with a line. The name of an association is typed along the line. The names of the relationship are drawn from left to right and from top to bottom. If this is different, an

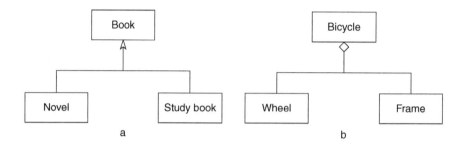

FIGURE 5.2 Examples of UML notations. (a) Generalization/specialization; (b) aggregation.

arrow indicates the direction of the relationship. The following types of associations can be distinguished:

- Generalization/specialization
- Aggregation
- Composition

Generalization/specialization: Generalization is the grouping of classes into new classes. A new class can be specified if there is more than one class with identical characteristics (operations or attributes). The original classes inherit these identical characteristics from the newly created class. The new class is called superclass or generalization; the classes with the identical characteristics are called subclasses or specializations. In UML a generalization/specialization is drawn with a large, open arrow. The arrow points to the superclass (Figure 5.2a). A superclass can represent an abstract class. An abstract class is a class of which no instances can exist. In UML this is denoted by giving the name of the class in italics, optionally followed by {abstract}, or by denoting it as a stereotype, using <<*name_stereotype*>>. A stereotype can be used to specify that the class or object belongs to a more general group of classes or objects, which give them specific characteristics, e.g., interface, enumeration, application, implementation, abstract, etc.

Aggregation: Aggregation is a special kind of association to show that one or more classes are part of another class. The parts can exist independently from the complex class. An example of an aggregation is a bicycle with wheels and a frame. The wheels and frame can exist individually and can be taken from the bicycle to be used for another bicycle. In UML an aggregation is denoted with a white diamond on the side of the complex (Figure 5.2b).

Composition: In a composition relationship, also a special kind of association, a part can belong to only one complex and there is a restriction that a part ceases to exist when the complex ceases to exist: a part cannot exist independently from the complex. This is called lifetime dependency. An example is a polygon existing of linear rings, when the polygon is removed, the linear rings defining the polygon also cease to exist. A composition is denoted with a black-filled diamond on the side of a complex.

Multiplicity: The multiplicity is the number of instances of the associated class with which one instance of the class can have a relationship. In UML the multiplicity

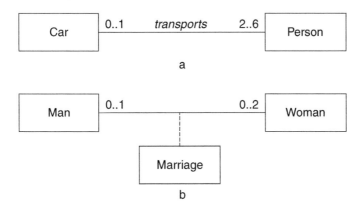

FIGURE 5.3 Examples of UML notations. (a) Multiplicity; (b) association class.

is drawn with an asterisk or a number. When nothing is defined, the multiplicity is one. The possible notations for multiplicity are as follows:

- 5: exactly 5
- *: zero or more
- 1..*: one or more
- 2..5: two to five
- 2,5: two or five

The multiplicity can be drawn on both sides of the association. The multiplicity in Figure 5.3a is read as "a car transports two to six passengers and a passenger is transported by zero or one car."

Association class: An association class is a class related to an association. This means that the class is identified with the association, which contains additional details (attributes, operations). As soon as there is a relationship between two instances, an instance of the association class exists. An example of an association class is a marriage, which is an association class between a man and a woman, and in some countries also between a man and a man or a woman and a woman (or between one woman and one or more men, or vice versa) (Figure 5.3b). An association class is used when the association has attributes, when the association has operations, or when the association itself has associations with other classes than the two on which this association is based. An association class is like a normal class and therefore it has the same characteristics as normal classes within UML. In UML notation an association class is drawn with the class symbol, which is linked with a dashed line to the association it belongs to.

Constraints: A constraint is a limitation on one or more elements in the class diagram. In UML constraints can be defined using OCL (Object Constraint Language). An OCL constraint is denoted with the notation {OCL-constraint} in a notebox linked to an object or class, e.g., {area of parcel > 0}. There are two predefined constraints in UML:

1. The ordered collection of objects with multiplicity greater than one is denoted with {ordered}, e.g., the ordered collection of linear rings in a polygon.
2. The symbol {XOR} with a dashed line to two or more associations indicates that only one of the associations can be instanced; e.g., a cadastral object can be an apartment unit or a parcel, but not both.

An example of an UML class diagram is shown in Figure 5.5, Section 5.7.1.

5.6 SPATIAL DATA MODELING AND DBMS

Spatial data are mostly part of a complete work and information process. Therefore, in many organizations there is a growing need for a central DBMS (at least at the conceptual level) in which spatial data and alphanumerical data are maintained in one integrated environment. Consequently, DBMSs are an essential part of the new-generation GIS architecture.

An extended description on how GISs have evolved with respect to DBMSs can be found elsewhere.[211] GISs used to be organized in a dual architecture consisting of (1) data management for administrative data in a (relational) DBMS and (2) data management for spatial data in a GIS. This was caused by the different nature of alphanumerical and spatial data and the inability of early DBMSs to handle spatial attributes. In the dual architecture (Figure 5.4, left) the two parts are connected to each other via links (unique ids). The spatial attributes are not stored in the DBMS and therefore they are unable to use the traditional database services (query, index). In the dual architecture the consistency of the data is hard to manage. For example, if a parcel is deleted in the spatial part, persons can no longer have a relationship with this parcel, which is maintained in the non-spatial part.

The solution to the problems of dual architecture was a layered architecture in which all data are maintained in a single (relational) DBMS. Since spatial data types were at that time not supported at the DBMS level, knowledge about spatial data types was maintained in middleware (Figure 5.4, middle). Spatial information was

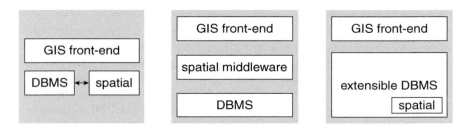

FIGURE 5.4 Evolving architectures of GIS. Left: dual architecture; middle: layered architecture; right: integrated architecture. (From T. Vijlbrief and P.J.M. van Oosterom, in *Proceedings of the 5th International Symposiom on Spatial Data Handling*, Charleston, South Carolina, August 1992. With permission.)

maintained in the DBMS by means of BLOBs. SQL cannot process data stored as BLOBs and therefore the data depends on the host application code, which handles the data in BLOB format. This solution requires data transport from the DBMS to middleware and consequently queries cannot be implemented optimally.

In recent times DBMSs have evolved toward an integrated architecture in which all data are maintained in one object relational DBMS (Figure 5.4, right). At present, most mainstream DBMSs support spatial data types and spatial functions by means of ADTs. This architecture ensures an integrated and consistent set of data. Chapter 6 describes the state of the art of geo-DBMSs in this new integrated GIS architecture.

5.7   STANDARDIZATION INITIATIVES

Since the same geo-information is used by more and more people and applications, interoperability of geo-information and geo-processes (together named geo-services) has become a major issue in geo-sciences. With respect to interoperability, two standardization initiatives are discussed and taken into account in this book, i.e., OpenGIS and ISO TC/211.

5.7.1   OPENGEOSPATIAL CONSORTIUM

The main mission of the OpenGeospatial Consortium (former Open GIS Consortium: OGC) (OGC), founded in 1994, is to enable interoperability of geo-services. Interoperability is the ability of digital systems to (1) freely exchange all kinds of spatial information and (2) cooperatively run software capable of manipulating such information over networks.[149] The OGC Specification and Interoperability Program provides an industry consensus process to plan, develop, review, and officially adopt OGC specifications for interfaces, encodings, and schemas that enable interoperable geo-services, data, and applications.[149] At the moment, about 280 public and private organizations participate in OGC. An important concept in the OGC model is a spatial (or geographical) feature, which is an abstraction of a real-world phenomenon associated with a location relative to the earth[144] and a geometry. The basic spatial class of the geometries is "GM_Object" (Figure 5.5).

OGC produces Abstract Specifications and Implementation Specifications.[142] The aim of the Abstract Specifications is to create and document a conceptual model sufficient to create the Implementation Specifications. The Implementation Specifications translate the Abstract Specifications into common distributed computing environments (e.g., Corba, DCOM, Java, HTTP). UML is (mainly) used as basic language for the formalism of models defined in the Abstract and Implementation Specifications. Examples of Implementation Specifications are as follows:[151]

- OpenGIS Location Services (OpenLS): Consist of the composite set of basic services comprising the OpenLS Platform for location-based services (mobile GIS).
- Catalog Interface: Defines a common interface that enables diverse but conformant applications to perform browse and query operations against distributed and potentially heterogeneous catalog servers.

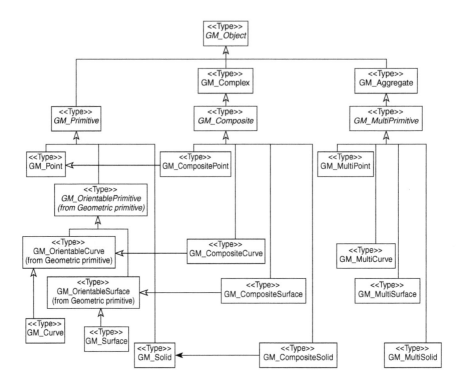

FIGURE 5.5 UML class diagram of geometry basic classes with specialization relations. (From OpenGIS Consortium, Project Document 01-101, Wayland, Massachusetts, 2001 copyright OGC.)

- Coordinate Transformation Services: Provide interfaces for general positioning, coordinate systems, and coordinate transformations.
- Grid Coverages: Designed to promote interoperability between software implementations by data vendors and software vendors providing grid analysis and processing capabilities.
- Simple Features—CORBA: Provide application programming interfaces (APIs) for publishing, storage, access, and simple operations on Simple Features (point, line, polygon, multipoint) using CORBA.
- Simple Features—SQL: Provide APIs for publishing, storage, access, and simple operations on Simple Features (point, line, polygon, multipoint) using SQL.
- Simple Features—OLE/COM: Provide APIs for publishing, storage, access, and simple operations on Simple Features (point, line, polygon, multipoint) using OLE/COM.
- Geography Markup Language (GML 3.0): The Geography Markup Language (GML) is an XML (eXtendible Markup Language; see Section 7.4) encoding for the transport and storage of geographic information, including both the spatial and non-spatial properties of geographic features.

Geography Markup Language and 3D

An example of GML code to describe a polygon in 3D space, is :

```
<gml:PolygonPatch>
 <gml:exterior>
   <gml:LinearRing>
     <gml:coordinates>
       105111.588,448909.588,9 105132.743,448884.341,9
       105137.45,448888.285,12 105116.295,448913.532,12
       105111.588,448909.588,9
     </gml:coordinates>
   </gml:LinearRing>
 </gml:exterior>
</gml:PolygonPatch>
```

The conceptual model underlying the representation of geometry and topology in GML[147] is that of Topic 1 of the OGC Abstract Specification,[144] which adopted the ISO 19107 standard (see next section and Chapter 6). The ISO model describes the correspondence of topological and geometrical relationships up to three dimensions. GML 3.0[147] includes the ability to handle complex properties, to describe coordinates with x, y, and z values (already possible in version 1 and 2), and to define 3D objects. A topological volume in GML is described using the TopoSolid type. A TopoSolid type is defined by faces, faces are defined by edges, and edges are defined by nodes. The user is free to choose where to explicitly store the geometry: at face, edge, or node level. However, the topology has to be defined fully to node level. The user is also free to choose whether to define co-boundary relationships as well, i.e., the face–solid relationships, the edge–face relationships, and the node–edge relationships.

OGC Web Services

OGC has defined several Implementation Specifications for Web Services for disseminating geo-information across the Internet: Web Map Services (WMS), Web Feature Services (WFS), Web Coverage Services (WCS), and Web Terrain Services (WTS). A service is a collection of operations, accessible to a user through an interface.[148] OGC compliant applications operating on user terminals (e.g., desktop, notebook, handset, etc.) can then "plug into" a server supporting the services to join the operational environment. Web Services are based on the general request–response rules used by Hypertext Transfer Protocol (HTTP). Support of GET and POST methods are available within this protocol. An example of a HTTP request to a WMS is:

```
http://www2.dmsolutions.ca/cgi-bin/
mswms_world?SERVICE=WMS&VeRsIoN=1.1.1&Request=GetMap&LAYERS=WorldGen_Outline
```

The OGC specifications for Web Services describe how to define a request string to be appended to the URL sent to the specific Web Service. They also define what requests are possible and what the output format of the responses should be.

Web Map Services

The Web Map Service Specification (WMS)[145] was the first OGC Implementation Specification to standardize the way a client requests maps. Clients communicate with a WMS by sending a URL request (using the HTTP protocol) to a WMS instance via general Web Server software like Microsoft Internet Information Server or Apache. The URL contains the name of the layer and other parameters such as the size of the returned map as well as the spatial reference system to be used when drawing the map. The WMS defines three operations:

1. GetCapabilities: The response to a GetCapabilities request is general information about the service itself and specific information about the available maps.
2. GetMap: Returns a map image with a defined spatial extent and spatial reference system.
3. GetFeatureInfo (optional): Returns information about features shown on the map based on the x,y position indicated by a click action of a user.

Web Feature Services

The next step was the Web Feature Service Specification (WFS),[146] which provides further extension of Web functionality, i.e., insert, update, delete, and query of geographic features. A WFS delivers GML (vector) representations of features in response to queries from HTTP clients instead of image representations in the case of a WMS. Clients access features through WFS by submitting a request for just those features that are needed for an application. A WFS can either be a basic WFS (read-only) or a transaction WFS. A basic WFS implements three operations:

1. GetCapabilities: Similar as in WMS.
2. DescribeFeatureType: Returns a schema of the data structure of the data set maintained at the data host on which the WFS has been implemented.
3. GetFeature: Returns a set of features in GML according to the query of the user based on spatial and non-spatial attributes of features.

With a transaction WFS it is, apart from querying features, also possible to insert, delete, and update data. Therefore, a transaction WFS implements, in addition to supporting all the operations of a basic WFS, the Transaction operation (and optionally the LockFeature operation).

Web Coverage Services

The Web Coverage Service Specification (WCS)[150] defines Web-based access to raster data. The raster data can be delivered in image format and can be further processed, e.g., rendered by visualization software at client side or used as input into scientific models. Operations in WCS are very similar to WMS operations, which work only on vector data.

Web Terrain Services

The Web Terrain Services Specification (WTS)[141] (not yet fully adopted as OGC specification) defines how to create views out of 3D data, like city models and digital

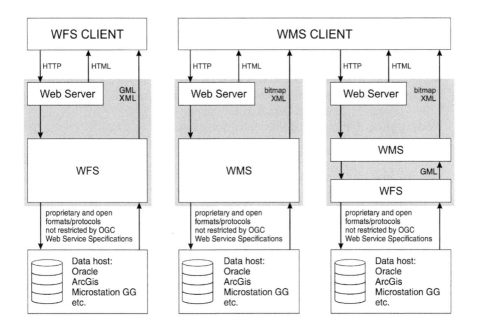

FIGURE 5.6 System architecture for disseminating geo-information using Web Map and Web Feature Services.

elevation models. The view, or 3D scene, is defined as a 2D projection of 3D features into a viewing plane. The view is created based on input parameters, such as point of interest and horizontal angle between the north direction and the horizontal projection. The service returns a rendered 2D image of the 3D view.

We illustrate the working of OGC Web Services by showing the system architecture for a WFS and WMS (Figure 5.6). A client sends a URL, defining a request, to a Web Server. The Web server sends the HTTP request to an OGC Web Service (WMS or WFS). The OGC Web Service translates the request and sends it to a data host. The data host sends the resulting data set to the OGC Web Service, whereupon the OGC Web Service translates the resulting data set into a format understandable to the client, as an image in case of a WMS or GML format in case of a WFS. The OGC Web Service sends the image or GML file back to the client. To be able to view the data, the client needs to be able to "understand" the image format file respectively GML document format.

5.7.2 ISO TC/211

The ISO Technical Committee 211 (TC/211): Geographic Information/Geomatics also defines standards related to GIS. TC/211 prepares geographic information standards in cooperation with other ISO technical committees working on related standards such as IT standards. Project 19107, Geographical Information: Spatial Schema defines a conceptual model of geometry and topology related to geographic features.

TC/211 is divided into several working groups. In total, nine working groups have been started, of which four have been disbanded since they achieved the goals of the specific working group.[82] Working groups that were disbanded are:

Working group 1: Framework and Reference Model
Working group 2: Geospatial Data Models and Operators
Working group 3: Geospatial Data Administration
Working group 5: Profiles and Functional Standards

Five working groups are still operating:

Working group 4: Geospatial Services
Working group 6: Imagery
Working group 7: Information Communities
Working group 8: Location Based Services
Working group 9: Information Management

Since 1997, ISO and OGC have worked together based on the large overlap of their area of interest. Today, OGC is working, via formal liaisons, with ISO TC/211 to harmonize abstract and implementation specifications. OGC members have access to key ISO documents and contribute (indirectly) to their evolution and in turn some of the future OGC specifications (geometry, meta-data) will essentially be ISO specifications repackaged under agreement. In the future, the same specifications will be published by both ISO and OGC (i.e., "double branding").

5.8 CONCLUSIONS

In this chapter the main topics of spatial data modeling were described. Applying these topics to the 3D cadastre research, a few concluding remarks can be made.

Object-Based or Field-Based: For the 3D cadastre model an object-based (vector) approach instead of a field-based approach (raster) has been chosen for the spatial modeling part. The character of cadastral data (parcels, property) favors an object-based approach (no continuous character, identifiable objects, human-made objects). However, the terrain elevation aspects could be treated with a field-based model.

Core Objects of a 3D Cadastre: In 2D, the core object for a cadastre is the real estate object (parcel) that is registered in the cadastral system. A parcel is not always easy to identify in the field. The parcel has a relationship with persons via rights and/or restrictions, as was seen in Section 2.1. For the 3D cadastre, the objects to be considered are:

- Representation of physical objects as they occur in the real world (tunnel, cable, pipeline).
- "Property objects," which are representations of 3D property units. Property objects are not always directly identifiable in the field, for example, if a right of superficies has been established while the actual construction has not yet been built or when a right of superficies has been established for a tunnel which also includes a safety zone.

Phases of Data Modeling: In Chapter 9 the conceptual model for a 3D cadastre that has been designed during this research is described. The next step is the translation of the conceptual model into the logical model (i.e., database structure of DBMS). As seen in Section 5.3.2 relational models have basic drawbacks when modeling the real world, especially when modeling topological and geometrical characteristics of spatial objects. An object-oriented approach overcomes these drawbacks. However, true object-oriented DBMSs have only been implemented and used in limited ways, and object-oriented technology still needs to be further developed and optimized, as seen in this chapter. Object relational models, which are the compromise between the relational and the object-oriented paradigm, are likely to be the leading DBMS technology for the next decades. In addition, object relational models offer sufficient functionality for the 3D cadastre domain. Therefore, an object relational DBMS was selected for the 3D cadastre prototypes. As this research does not aim at a complete operational application (although parts of a 3D cadastre have been developed in different prototypes), the logical model for a 3D cadastre is not completely designed in this book. Only the main parts of the data model are translated into a logical model and implemented in prototypes. In Chapter 10 principles of the DBMS model for a 3D cadastre are considered, as well as what issues should be taken into account when designing the logical model for a 3D cadastre.

The physical model for a 3D cadastre is beyond the scope of this book.

6 Geo-DBMSs

In Section 5.6 it was concluded that DBMSs play a central role in new-generation GIS architecture. Within this architecture spatial and non-spatial information on objects is maintained in one integrated DBMS environment, called a geo-DBMS. This chapter describes how spatial information on objects can be structured in DBMSs and how this information can be used, e.g., in spatial analyses (see also Reference 171).

The OGC adopted the ISO 19107 international standard[83] as Topic 1 of the Abstract Specifications: Feature Geometry.[144] These Abstract Specifications provide conceptual schemas for describing the spatial characteristics of spatial objects (geographic or spatial features, in OGC terms) with vector geometry and topology up to three dimensions embedded in 3D space. The Abstract Specifications also describe a set of spatial operations consistent with these schemas. According to the specifications, the spatial object is represented by two structures: (1) structure of geometrical primitives (i.e., simple feature) and (2) topological structure (i.e., complex feature). While the geometrical structure provides direct access to the coordinates of individual objects, the topological structure encapsulates information about their spatial relationships.

Geometrical primitives are a combination of geometry (coordinates) and a coordinate reference system. Topological primitives make use of id references to low-dimensional primitives; e.g., a polygon refers to its edges and nodes. The coordinates are stored only with the low-dimensional primitives. In principle, topological primitives are introduced to accelerate the computational geometry algorithms replacing them with combinatorial ones. Topological primitives have meaning only within a topological model. The OGC Abstract Specifications have been transformed into Implementation Specifications, of which the most relevant for this research is the OGC Simple Features Specification for SQL,[140] which supports spatial objects up to two dimensions in object relational DBMS environments. Mainstream DBMSs have adopted these Implementation Specifications.

This chapter describes how mainstream DBMSs can maintain spatial objects, using both a structure of geometrical primitives (Section 6.1) and a topological structure (Section 6.2). Section 6.3 describes spatial analyses that can be performed in DBMSs distinguishing between analyses that can be performed on geometrical primitives and analyses that can be performed on topological structure.

As will be seen in this chapter current support for 3D in DBMS is limited. Therefore, as part of this research a 3D primitive was implemented in a mainstream DBMS. The implementation is described in Section 6.4. When talking about 3D, the 2.5D representation of the terrain in a TIN (Triangular Irregular Network) structure should also be a topic of attention. The issue of TIN structures representing heights is elaborated in Chapter 8. This chapter ends with concluding remarks (Section 6.5) including a discussion on which spatial analyses should be DBMS built-in functionality and which spatial analyses should be reserved for front-end applications.

6.1 GEOMETRICAL PRIMITIVES IN DBMSs

Mainstream DBMSs (Oracle,[152] IBM DB2,[78] Informix,[79] and Ingres[80]) and also popular noncommercial DBMSs such as PostgreSQL[166] and MySQL[117] have implemented spatial data types and spatial operators (also called "spatial functions") more or less similar to the Simple Features Specification for SQL of OGC. The implementation described in the Specification for SQL consists of an SQL extension using ADTs that supports storage, retrieval, query, and updating of simple spatial features (points, lines, and polygons). The spatial features are stored in geometrical primitives. Topological relationships between geometries can be retrieved by the use of spatial operators (see Section 6.3). OGC Implementation Specification for SQL has so far been in 2D. Also the implementations of spatial data types in mainstream DBMSs are based on supporting 2D primitives in 2D and 3D space. With the implementations of the geometrical primitive, it is possible to store and query spatial features in a DBMS, but the relationships between neighboring spatial objects is not standardized and can be determined only with a geometrical query. In addition, the geometrical primitive causes redundancy in the case of a planar partition such as a cadastral map: shared edges and shared nodes are stored twice. In this section we distinguish between 2D (Section 6.1.1) and 3D geometrical primitives (Section 6.1.2). To illustrate how spatial objects can be maintained in DBMSs, Oracle Spatial 9i is used. Oracle Spatial 9i is not fully OGC compliant, since the spatial ADT as defined in Oracle differs slightly from the ADTs as defined by OGC. The OGC Implementation Specification for SQL defines separate data types for different types of geometry (points, linestrings, polygons, etc.) while Oracle Spatial has only one data type for all types of geometry. Oracle is OGC compliant at level 1 (relational encoding of geometry) and not at level 2 (types and functions). However, the experiments show generic aspects for supporting geometry in a DBMS.

6.1.1 2D GEOMETRICAL PRIMITIVES IN DBMSs

The supported spatial features in Oracle Spatial 9i are points, lines, and polygons (including arcs, boxes, and mixed geometry sets in 2D and 3D). The object relational model in Oracle defines the object type sdo_geometry as:

```
CREATE TYPE sdo_geometry
AS OBJECT (
SDO_GTYPE NUMBER,                              type of the geometry (e.g.
                                               point, linestring, polygon)
SDO_SRID NUMBER,                               reference to the spatial
                                               reference system
SDO_POINT SDO_POINT_TYPE,                      specific entry for pints
SDO_ELEM_INFO MDSYS.SDO_ELEM_INFO_ARRAY,       indicates how the coordinate
                                               array should be interpreted
SDO_ORDINATES MDSYS.SDO_ORDINATE_ARRAY);       list of coordinates
```

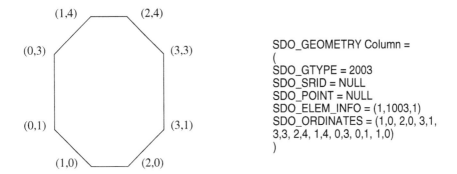

FIGURE 6.1 Example of storing a polygon using Oracle's spatial data type.

An example of using the Oracle object relational model to represent a polygon is shown in Figure 6.1. In sdo_gtype = 2003, the first position indicates the dimension (2D in this case), and the last position indicates the element type (3 indicates a polygon). In sdo_elem_info, the combination 1003,1 indicates that this is a polygon containing straight lines (1003 for polygon, and 1 for straight lines). The first position in 1,1003,1 (1 in this case) indicates that the first (and only) element starts at offset 1 in the coordinate list.

The next SQL statements illustrate how a box with 0,0 as lower-left and 100,100 as upper-right coordinates is stored in Oracle (sdo_geometry type) in the "geom2d" table. Another way to represent a box is with a special element type in the sdo_elem_info array by which only the lower-left and upper-right coordinates are needed (which is not illustrated in this example).

```
/* creation of the table */
CREATE TABLE geom2d (shape mdsys.sdo_geometry not null, ID number(11) not null);

/* inserting data (2D box) */
INSERT INTO geom2d (shape,id)
VALUES (
mdsys.SDO_GEOMETRY(2003, NULL, NULL,
mdsys.SDO_ELEM_INFO_ARRAY(1, 1003, 1),
mdsys.SDO_ORDINATE_ARRAY(0,0, 100,0, 100,100, 0,100, 0,0)
), 8);
```

Besides the tables representing the geometries of the objects, meta-data can be maintained describing the dimension, lower and upper bounds, and tolerances in each dimension. In the following statements the information on the table geom2d is inserted in the metadata table. Finally, a spatial index is created on the table to speed spatial queries. In this case an R-tree index is built, but a Quad-tree spatial index is also possible in Oracle Spatial. A spatial index can only be built when meta-data have been inserted for the specific table:

```
/* inserting metadata, 2D table*/
INSERT INTO user_sdo_geom_metadata VALUES
(`GEOM2D', `SHAPE', mdsys.sdo_dim_array(
mdsys.sdo_dim_element(`X', 0, 500, 0.5),
mdsys.sdo_dim_element(`Y', 0, 500, 0.5) ), NULL);

/* creating index */
CREATE INDEX geom2d_i ON geom2d(shape) INDEXTYPE IS mdsys.spatial_index;
ANALYZE TABLE geom2d COMPUTE STATISTICS;
```

6.1.2 3D GEOMETRICAL PRIMITIVES IN DBMSS

2D primitives are also supported in 3D space, for example, a "geom3d" table can be created by the following query in Oracle:

```
/* creation of the table */
CREATE TABLE geom3d (
shape mdsys.sdo_geometry not null,
ID number(11) not null);
```

Note that the commands to create a 2D table and a 3D table are the same. The following query inserts the box as used in the 2D example with a height of 50:

```
/* inserting data, a 3D box *
INSERT INTO geom3d (shape, id) VALUES (
mdsys.SDO_GEOMETRY(3003, NULL, NULL,
mdsys.SDO_ELEM_INFO_ARRAY(1, 1003, 1),
mdsys.SDO_ORDINATE_ARRAY(0,0,50, 100,0,50, 100,100,50, 0,100,50, 0,0,50)
), 9);
```

Meta-data can be inserted after which a spatial index (R-tree in 3D) can be created on the "geom3d" table:

```
/* inserting metadata, 3D table*/
INSERT INTO user_sdo_geom_metadata VALUES
(`GEOM3D', `SHAPE', mdsys.sdo_dim_array(
mdsys.sdo_dim_element(`X', 0, 500, 0.5),
mdsys.sdo_dim_element(`Y', 0, 500, 0.5),
mdsys.sdo_dim_element(`Z', 0, 300, 0.5)
), NULL);

/* creating index */
CREATE INDEX geom3d_i ON geom3d(shape)
INDEXTYPE IS mdsys.spatial_index parameters(`sdo_indx_dims=3');
ANALYZE TABLE geom3d COMPUTE STATISTICS;
```

Most DBMSs (including Postgres, IBM DB2, Ingres, and Informix) support the storage of points (0D), lines (1D), and polygons (2D) in 3D space as illustrated by this example, but not of 3D volumetric data types. However, volumetric objects can be stored in a geometrical primitive within current techniques using 3D polygons. 3D objects can be represented as polyhedra (body with flat faces) in two ways: as a set of

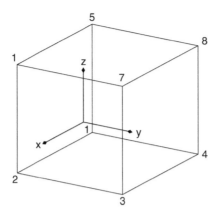

FIGURE 6.2 Cube to be stored in the DBMS.

polygons or as multipolygon (one object consisting of several polygons). To illustrate this, the cube in Figure 6.2 has been used.

In the first option (defining a 3D object as a set of 3D polygons) two tables are used: a table "BODY" and a table "FACE." In the table "BODY," the 3D spatial object is defined by a set of records representing a polyhedron with references to the (flat) faces it consists of. In the table "FACE," the actual geometries of faces are stored as polygons in 3D space (sdo_gtype: 3003, sdo_elem_info: (1,1003,1)). This structure is partly a topological structure, since the body is defined by references to the faces and the faces can be shared by neighbor bodies. However, shared edges and nodes are stored with coordinates with every face they belong to, which leads to many redundant coordinates. The generated tables for the cube are shown in Table 6.1 (x1, y1, z1 refers to the x, y, and z-coordinate of point 1 in Figure 6.2).

In the second representation (defining a 3D object as a multipolygon), a body is stored as one record instead of a set of records. The multipolygon, which is also supported in Oracle Spatial, is used for this representation (sdo_gtype: 3007, sdo_elem_info: (starting offset,1003,1)). This has also been implemented. The

TABLE 6.1
Tables Representing a 3D Cube Using a Set of 3D Faces

BODY		FACE	
BID	FID	FID	sdo_ordinate array
1	1	1 (lower face)	x4,y4,z4, x3,y3,z3, x2,y2,z2, x1,y1,z1, x4,y4,z4
1	2	2 (side 1)	x3,y3,z3, x4,y4,z4, x8,y8,z8, x7,y7,z7, x3,y3,z3
1	3	3 (side 2)	x4,y4,z4, x1,y1,z1, x5,y5,z5, x8,y8,z8, x4,y4,z4
1	4	4 (side 3)	x1,y1,z1, x2,y2,z2, x6,y6,z6, z5,y5,z5, x1,y1,z1
1	5	5 (side 4)	x3,y3,z3, x2,y2,z2, x6,y6,z6, z7,y7,z7, x3,y3,z3
1	6	6 (upper face)	x5,y5,z5, x6,y6,z6, x7,y7,z7, z8,y8,z8, x5,y5,z5

TABLE 6.2

Table Representing a 3D Cube Using a 3D Multipolygon

BODY table

Bodyid	Geometry
1	SDO_GEOMETRY(3007, – *3007 indicates a 3D multipolygon*
	NULL, NULL, SDO_ELEM_INFO_ARRAY(– *offset of polygons is specified*
	1, 1003, 1,
	16, 1003, 1,
	31, 1003, 1,
	46, 1003, 1,
	61, 1003, 1,
	76, 1003, 1
	),
	SDO_ORDINATE_ARRAY(
	x4,y4,z4, ,x3,y3,z3, x2,y2,z2, x1,y1,z1, x4,y4,z4, –*end of 1st (lower) polygon*
	x3,y3,z3, ,x4,y4,z4, x8,y8,z8, x7,y7,z7, x3,y3,z3,
	x4,y4,z4, ,x1,y1,z1, x5,y5,z5, x8,y8,z8, x4,y4,z4,
	x1,y1,z1, ,x2,y2,z2, x6,y6,z6, z5,y5,z5, x1,y1,z1,
	x3,y3,z3, ,x2,y2,z2, x6,y6,z6, z7,y7,z7, x3,y3,z3,
	x5,y5,z5, ,x6,y6,z6, x7,y7,z7, z8,y8,z8, x5,y5,z5 – *end of last (upper) polygon*
	))

resulting table "BODY," in which the cube of the example is stored, is shown in Table 6.2.

An advantage of 3D multipolygons (compared to a set of polygons) is that they are identifiable as one object by front-end applications (GIS, CAD) that can access objects stored in the DBMS. Another advantage of the 3D multipolygon approach is the one-to-one correspondence between a record and an object. A disadvantage of both representations is that the topological structure between objects cannot be used, which implies risks for consistency as well as redundant storage of coordinates (and in the 3D multipolygon solution also of faces). In addition, topology within one object is not maintained. However, the main disadvantage of these implementations is that no true 3D geometrical primitive (as volumetric data type) is supported by the DBMS and therefore it is not recognized as such by the DBMS. In addition, functions on 0D, 1D, and 2D primitives that are defined in 3D space project the primitives on a 2D plane (as illustrated in Section 6.3.2).

These disadvantages can be overcome with the implementation of real 3D (volumetric) data types. An extension of Oracle Spatial 9i has been proposed with support of a true 3D data type: the polyhedron primitive.[192] This primitive has been implemented in this research including the data model, validation functions, and spatial functions in 3D[4] (see Section 6.4).

6.2 TOPOLOGICAL STRUCTURE IN DBMSs

Topological structures are generally used to represent planar or space partitions without redundancy and to represent (linear) networks. In this book, the focus is on planar and space partition. Therefore, linear networks are not further considered. In planar partitions (2D topological structures) and space partitions (3D topological structures) spatial objects are defined on the basis of non-overlapping objects.

A large number of 2D topological structures are already available in literature, of which some have been implemented in commercial[94] and user-defined systems[129] and populated with data. Many 3D topological structures are also reported but only a few of them have been tested for large data sets, e.g., Reference 233.

In general, many questions related to topological structures in relation to DBMSs still have to be resolved. How many and which primitives to store persistently? How many and which relationships to store explicitly? Is it sufficient to maintain the relationships to only low-dimensional objects (edges and nodes in the case of polygons) or does the relationships to high-dimensional objects (co-boundary relationships, e.g., edges that refer to their left and right polygon) also need to be maintained? In this respect, it is likely that a data model appropriate for a certain application may fail to serve another application. Thus, a simultaneous maintenance of several topological structures in the DBMS might be needed. Organization of many topological structures in the DBMS has been suggested by using a detailed description in a meta-data table.[136]

An extensive argumentation for the need to organize the topology support at the DBMS level is provided elsewhere.[136] As specified there, a topological structure at DBMS level has many advantages:

- It avoids redundant storage (more compact than a full geometrical model).
- It is easier to maintain the consistency of the data after editing.
- It is more efficient during the visualization in some types of front ends, because less data have to be read from disk and transferred to clients.
- It is the natural data model for certain applications; e.g., during surveying an edge is collected (together with attributes to a boundary).
- It is more efficient for certain query operations (e.g., find neighbors).

An Implementation Specification for topological structures (complex features in OGC terms) is currently being developed by OGC in cooperation with ISO. A request for a proposal on this topic was issued in 2001 (and not updated since then).[143] The request aimed at extending the interfaces in the OGC Simple Features Implementation Specification. The new interfaces will build on the OGC Simple Features Specification to address feature collections and more complex objects and concepts including curves and surfaces in 2D and 3D, compound geometries, arcs and circle interpolations, and topology. Note that GML is able to model complex objects and 3D objects as defined in the OGC Abstract Specifications Topic 1.

In the current Implementation Specifications for Simple Features topological relationships can be derived by spatial operations on geometrical primitives (see also Section 6.3.1).

Relational DBMS has proved that it can efficiently store the topological references: a face left and right of an edge, boundary to boundary references, treatment of islands, etc., i.e., the modeling aspect of topology. The problem with a standard relational DBMS, however, is that the declarative language SQL cannot handle the "navigational access" needed to obtain the geometry of a topological primitive. In SQL it is not possible to express the statement: "follow the next references of the boundary until we are back at the beginning." This functionality has to be provided by embedded queries using programming languages (able to "loop" the data), e.g., PL/SQL (procedure language of Oracle) or Java. This functionality is, however, already available in every object-oriented DBMS implemented within methods associated with classes. Currently, few user-defined and commercial implementations of topological structure in DBMSs exist using object relational technology. Also the latest version of Oracle Spatial (10g) has some support for topological structure.

To illustrate possibilities of topological structure in current DBMSs, this section describes a user-defined implementation of 2D topological structure (Section 6.2.2) and two commercial solutions (Laser-Scan Radius Topology[94] and Oracle Spatial 10g;[153] Section 6.2.3). All these implementations represent a planar partition structure. First, a description of planar partition topology according to both OGC and ISO is given (Section 6.2.1). Like the 3D geometrical primitive, 3D topological structure has not (yet) been implemented as part of a DBMS. In Section 6.1.2 a data structure was described in which the faces of a 3D body are geometrically described in a face table. The body table in this data structure contains references to the faces where the body consists of, but the data structure does not contain references to edges and nodes. In this data structure, bodies can share faces. Section 6.2.4 describes user-defined implementations of a full 3D topological structure.

6.2.1   OGC, ISO and Planar Partition Topology

Spatial models defined by planar partitions are based on faces, edges and nodes. Polygon is the geometrical equivalent of the topological primitive "face." This section describes how ISO and OGC define the feature "face" and the geometrical equivalent "polygon" in a planar partition topological structure.

ISO/TC 211

The ISO standard 19107 "Geographic information—Spatial schema" defines geometrical primitives for which the code starts with GM, and related topological primitives, for which the code starts with TP. A TP_FaceBoundary consists of one or more TP_Rings. One of these rings is distinguished as being exterior of the boundary. Each ring is oriented so that the face is on its left, which means an anticlockwise orientation for outer rings and a clockwise orientation for inner rings. A TP_Ring is used to represent a single component of a TP_FaceBoundary. It consists of a number of TP_DirectedEdges in a cycle. The endNode of a TP_DirectedEdge is the startNode of the next TP_DirectedEdge. Since TP_Rings are used in TP_FaceBoundary objects, the ring will be oriented so that the face is on its left.

According to the ISO/TC 211 standard a face is defined by edges and those edges are anticlockwise oriented in case of outer rings (and clockwise in case of inner rings):

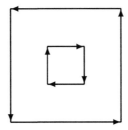

Every edge has a reference to the preceding and succeeding edge. The associated geometrical primitive of a face is "polygon." From the specifications it is not clear whether the outer boundary of a polygon is allowed to touch itself, nor is it clear if inner rings can touch the outer boundaries or other inner rings.[133] However, since only one outer boundary is allowed, a polygon with two outer boundaries (defining potentially disconnected areas) is certainly invalid.

OGC Specifications for SQL

The ISO definition of the topology of a face is at the abstract level. As was stated before, the OGC Consortium adopted the ISO Spatial Schema as Abstract Specifications and transformed these to the implementation level in the OGC Simple Feature Specification for SQL.

Since the OGC Specification for SQL does not define topology, we will have a look at the geometrical primitive of a polygon according to this Implementation Specification. A polygon is defined as a simple surface that is planar. A very precise definition of the polygon is given in the OGC specifications. The main characteristics from this definition relevant for the topology implementations described below is that rings may touch each other in at most a point. However, since polygons are built of LinearRings and since LinearRings are simple geometries, self-intersection of outer and inner rings is not allowed.[133] Inner rings, which divide the polygon into disconnected parts, are also not allowed. Note that the Simple Feature Specification does not say anything concerning the orientation of polygons.

6.2.2 User-Defined DBMS Implementation of 2D Topological Structure*

To explore the possibilities of using topology in spatial DBMS, a data set of cadastral parcels was selected, provided by the Netherlands Kadaster. This data set is modeled topologically in a relational DBMS; i.e., the geometry of the parcels is not stored explicitly, but can be inferred from the cadastral boundaries that are stored.[129] The most important tables are "boundary" (cadastral boundaries) and "parcel" (parcel identifiers). There is no need for the geometric data type "polygon," because the

---

* This section is based on Reference 167.

area features (parcels) are stored topologically in the "parcel" and "boundary" table using the winged edge structure.[8] The edges in the boundary table contain references to other edges according to the winged edge structure, which are used to form the complete boundary chains (parcels). The edges in the winged edge structure also contain a reference to the left and right parcel.

According to Reference 129 there are a number of reasons the Netherlands Kadaster has chosen to maintain parcels in a topology structure:

- The approach allows calculations on correctness of topology after updates.
- It makes possible to relate attributes to the boundaries between parcels, e.g., date of survey, name of person locating the boundary, etc.
- If each parcel were represented in the DBMS by a closed polygon, it would be complicated to represent the basic object of cadastral surveying: one boundary between two neighbor parcels.
- Closed polygon representation would lead to double (or triple or even more) storage of all coordinates (except the territorial boundary), which complicates data management in a substantial way.
- Closed polygon representation can result in the introduction of gaps and overlaps between parcels, which is not related to reality.

A parcel has exactly one reference to one of the surrounding boundaries and one reference to a boundary of each enclave. The structure of the topological references and the relationship between parcels and boundaries are visualized in Figure 6.3.

The apparent disadvantage of storing spatial objects in a user-defined topological structure in the DBMS is that the DBMS is not aware of the geometry of spatial objects. Because there is no geometry attribute in the parcel table, it is, for example, not possible to calculate the area of a parcel or use the geometry of a parcel in overlap functions. By extending the DBMS with a function that materializes (*realization* in OCG terms) the geometry from the topological relationships, it is possible to store data topologically and still use the spatial operations offered by the DBMS built on the geometrical model.

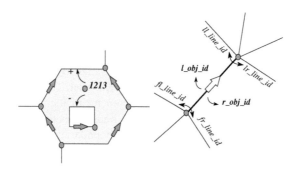

FIGURE 6.3 Topological structure in the spatial DBMS of the Netherlands Kadaster. (From P.J.M. van Oosterom and C.H.J. Lemmen, *CEUS*, 25(4–5): 509–528, 2001. With permission.)

Therefore, a function "return_polygon" has been implemented, which realizes the geometry of a polygon. The implementation is done in Oracle Spatial 9i. To achieve high performance and to avoid unnecessary conversions and data communication between DBMS and client, the return_polygon function must be performed within the geo-DBMS itself. In Oracle Spatial 9i, this can be done by stored procedures or functions that work within the database. The stored procedures and functions can be written in PL/SQL and/or Java, both of them using SQL to access the data. With the help of the spatial index, spatial clustering, and an index on the ids of objects, this should lead to good performance. The return_polygon function can be used in an SQL statement, e.g., in a query to compute the area of a parcel:

```
SELECT sdo_geom.sdo_area(return_polygon(object_id), 1) FROM parcel;
```

The function to realize the geometry of polygons has been implemented in two ways. The first solution uses only the information on the relationships between the preceding and succeeding edges. The second solution is based on the left–right information of edges. Both implementations are described and compared in the next paragraphs of this section.

A function-based spatial index is created on the face of the parcels in order to optimize the performance. Since version 9i, Oracle has offered function-based indexes, i.e., an index' that is created on the return value of a function in addition to a normal index created directly on the value of an attribute. A function-based spatial index facilitates queries that use locational information of type sdo_geometry returned by a function. The spatial index is created based on the pre-computed values returned by the function. This is implemented in Oracle 9i in two steps. First, the user_sdo_geom_metadata table was updated (defining the lower and upper bounds and tolerance in each dimension) to specify the function name:

```
INSERT INTO user_sdo_geom_metadata VALUES(
`PARCEL', `return_polygon(object_id)', mdsys.sdo_dim_array (
mdsys.sdo_dim_element(`X', 82291, 84261, 0.0005),
mdsys.sdo_dim_element(`Y', 453039, 455632, 0.0005)), NULL);
```

The next step is to create a spatial index by specifying the function name and parameters. For example, creating an R-tree index, is done with the following SQL statement:

```
CREATE INDEX parcel_idx ON parcel(return_polygon(object_id))
INDEXTYPE IS mdsys.spatial_index;
```

Without a function-based spatial index it would not have been possible to index the faces properly. During an overlap query or any other query using the spatial index, objects are filtered by means of this index; that is, using the pre-computed bounding boxes that are stored in the R-tree. Then the return_polygon function is executed to obtain the complete geometry of filtered objects to be used in the exact overlap test. The return_polygon function depends on the values in other tables. Therefore, when the index is built, it contains the results of evaluating the function as at the time of index build. If the function does not produce the same results next time it is evaluated, the index search algorithm will give the wrong results. Therefore, the index needs to be rebuilt each time an update is done that affects the bounding box of any parcel. A trigger on the edge updates could probably do the job to also update the appropriate index entries in the R-tree. However, this was not tested.

Realizing Geometry of Polygons Based on Relationships between Edges

The function return_polygon based on the relationships between edges has been implemented in PL/SQL. The function starts with the table "parcels" and uses the "boundary" table. The function creates a polygon geometry, of which the orientation is valid according to the Oracle Spatial (and OGC) rules: the coordinates of the outer ring are listed in anticlockwise order and the coordinates of the enclaves are listed in clockwise order. In the data set the winged edge structure is defined in both directions, since every boundary contains a reference to its four connecting boundaries (which is dissimilar to the ISO definition that define references only to the succeeding and preceding edge).

The relevant attributes in the "parcel" table used in the construction of polygons are:

- object_id: the unique identifier of parcels
- line_id1: reference to one of the surrounding boundaries (stored in the boundary table)
- line_id2: reference to one of the boundaries of the first enclave (also stored in the boundary table)

If there is more than one enclave, the "parcelover" table is used. The relevant attributes in this table are:

- object_id: the unique identifier of parcels
- line_id1: reference to one of the boundaries of the 2nd enclave
- line_id2: reference to one of the boundaries of the 3rd enclave
- ....
- line_id10: reference to one of the boundaries of the 11th enclave

These line_id's contain also references to a line in the boundary table. If a parcel has more than 11 enclaves, the parcelover table has more than one entry for that object_id. Consequently the attribute "line_id1" in the parcelover table may refer to the 2nd, the 12th, the 22nd, etc. enclave, the attribute "line_id2" may refer to the 3rd, the 13th, the 23rd, etc. enclave, etc.

The relevant attributes in the "boundary" table are:

- object_id: unique identifier of boundaries
- geo_polyline: geometry of the line
- fl_line_id: reference to the first line on the left, seen from the middle point of the line, looking back to the beginning
- ll_line_id: reference to last line on the left, seen from the middle point of the line, looking toward the end
- fr_line_id: reference to the first line on the right, seen from the middle point of the line, looking back to the beginning

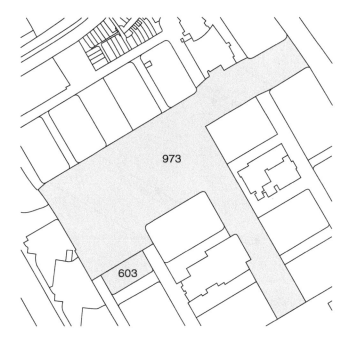

FIGURE 6.4 Parcel 603 and 973 are used in the examples.

- lr_line_id: reference to last line on the right, seen from the middle point of the line, looking toward the end
- l_parcel: parcel that is located at the left-hand side from the directed boundary (when looking from the beginning to the end of the boundary)
- r_parcel: parcel that is located at the right-hand side from the directed boundary (when looking from the beginning to the end of the boundary)

Note that these references are different from those in Figure 6.3. In Figure 6.3 the references at the start of the edge are "left" or "right" seen from the starting point of the edge. In contrast, in the data set these references are "left" or "right" seen from the middle point of the edge and therefore they are reversed (which is the Dutch interpretation of the winged edge structure).

How the function works, is illustrated with an example in which the polygon of parcel 603 is realized (Figure 6.4). The attributes of line_id1 and line_id2 in the parcel table are:

```
SELECT object_id, parcel, ·line_id1, line_id2 FROM parcel WHERE parcel=603;

OBJECT_ID  PARCEL  LINE_ID1    LINE_ID2
---------- ------  ----------  ----------
310148953  603     310439663   0
```

The parcel has one reference to its outer boundary (i.e., line_id1) and no enclaves (because line_id2 = 0). The polygon of the parcel can now be constructed by starting

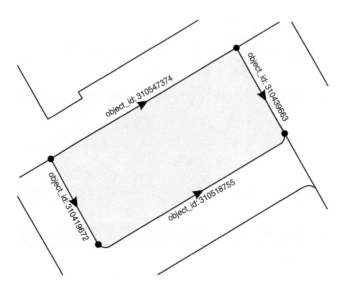

FIGURE 6.5 Ids and direction of boundaries of parcel 603.

with the first boundary, with object_id = 310439663. The boundary table is queried to look for the coordinates of this boundary and to look for the boundaries that are connected to it in anticlockwise direction. This step is repeated until the first boundary is found again. To avoid performing select statement for every next boundary, all boundaries together with the relevant attributes, which have parcel 603 on their right-hand or left-hand side, could have been selected first (see Figure 6.5 and the query below). However, the implementation of the function described here only uses the "connect" information, while the implementation as described in the next session only uses the left–right information.

```
SELECT object_id, fl_line_id, fr_line_id, ll_line_id, lr_line_id, l_parcel, r_parcel
FROM boundary WHERE l_parcel=603 OR r_parcel=603;
```

OBJECT_ID	FL_LINE_ID	FR_LINE_ID	LL_LINE_ID	LR_LINE_ID	L_PAR	R_PAR
310547374	310419672	-310419673	310594168	310439663	973	603
310419672	-310419673	310547374	310518755	310419671	603	960
310518755	310419671	-310419672	-310439663	310439732	603	605
310439663	-310547374	310594168	310439732	-310518755	960	603

After having followed all edges of the outer ring, the polygon can be constructed by connecting all line strings of the resulting boundaries. In this process the line strings, which are oriented in clockwise order (referred to with a minus), need to be reversed. The polygon geometry is realized in such a way that the coordinates at connection points are stored only once, and polygons are closed (first and last point is repeated). The collected geometry information is returned as a spatial data type of Oracle (polygon).

Now we will look at a polygon with enclaves: parcel 973 (see Figure 6.4). As can be seen from line_id2, parcel 973 has at least one enclave, starting with the boundary with object_id 310376490 (line_id2):

```
SELECT object_id, parcel, line_id1, line_id2 FROM parcel WHERE parcel=973;

OBJECT_ID    PARCEL    LINE_ID1      LINE_ID2
----------   -----     ----------    ----------
310152502    973       -310419676    310376490
```

The realization of the outer boundary of the polygon is performed in the same way as in the first example and will not be explained here. Parcel 973 contains one or more enclaves (line_id2 > 0). Therefore, the rings of the enclaves need to be constructed in clockwise direction according to ISO and Oracle rules. The first enclave starts with the boundary with object_id 310376490 (line_id2). In principle, we can follow the same procedure as in the case of the outer boundary: create a list with all connecting arcs (this time in clockwise order) to realize the geometry of the enclave.

The geometry of enclaves is constructed in the same way as the geometry of outer boundaries: linestrings are connected, duplicate coordinates are removed, linestrings in anticlockwise direction are reversed, and the polygon is closed. To see if this parcel has more than one enclave the "parcelover" table is checked:

```
SELECT * FROM parcelover WHERE object_id IN (SELECT object_id
FROM parcel WHERE parcel=973);

LINE_ID1    LINE_ID2    LINE_ID3   LINE_ID4   LINE_ID5 ......... LINE_ID10
--------    --------    --------   --------   --------           ----------
310379237   -310205718     0          0          0     ......... 0
```

The result is two more enclaves. The enclaves are generated in the same way as the first one. Again, the collected geometry information of enclaves together with the geometry of the outer boundary is inserted in the spatial data type of Oracle to create the polygon geometry of the parcel in Oracle.

Realizing Geometry of Polygons Based on Left–Right Information

The alternative version of the "return_polygon" function uses only the left–right information stored with every parcel boundary and a geometrical comparison to find and join connected boundaries in a ring. Here the boundaries that have the given parcel to the left or right are selected. By repeatedly joining boundaries that end in the same end point, we end up with the boundary of the complete parcel. Enclaves are realized in the same way. At the end of the procedure, which of the rings define the outer boundary and which of the rings define enclaves must be detected.

The attributes in the "boundary" table that are used by the algorithm are:

- geo_polyline: geometry of the line
- l_parcel: parcel, located at the left-hand side from the directed boundary
- r_parcel: parcel, located at the right-hand side from the directed boundary

The function has been implemented in the Java programming language and is integrated in the database server. The function accesses the database tables via an internal JDBC connection.

The first step is to retrieve all boundary lines that are part of the parcel:

```
SELECT geo_polyline FROM boundary WHERE l_parcel = 973 OR r_parcel = 973;
```

This query results in a collection of LineStrings. What needs to be done now is to glue these LineStrings together in such a way that they form an ordered collection of rings. This is done using two data structures:

1. Rings: In this variable we collect the completed LinearRings (LineStrings that form a loop) that are formed during the algorithm.
2. Graph: The graph structure contains all LineStrings that still need to be combined to form loops. The graph contains vertices (nodes) and edges. The end points of the LineStrings form the nodes of the graph. The edges in the graph are formed by the LineStrings and run between two nodes, being the start point and the end point of the LineString.

The algorithm first fills the graph structure and then tries to move all LineStrings from the graph into the ring structure from which the result is constructed.

```
// 1. Initialization.
for (all LineStrings that belong to the parcel boundary)
{
  Insert the LineString into the graph.
}

// 2. Main Loop.
while (graph contains a node with two edges)
{
    Delete the node and the two edges from the graph.
    if (the two edges at the node are the same edge)
    {
        we have found a loop and add the edge to the rings.
    }
    else
    {
        glue the two LineStrings together to form one big LineString.
        Insert the new LineString into the graph.
    }
}

// Now the graph should be empty. If this is not the case, the input
// data was incorrect.

// 3. Construct Polygon from rings.

Find the ring which encloses the largest area.
This is the outer boundary assuming that the input data is correct.
The rest of the rings are enclaves.
Construct a polygon using the boundary and the enclaves.
Calculate the orientation and return the polygon as Oracle's spatial data type.
```

Discussion on Self-Implemented return_polygon Function

The performance of both implementations is of course dependent on the complexity of the data: the more points in a boundary, the worse the performance; also the more boundaries in a polygon, the worse performance. In addition, following pointers as in the implementation based on the relationships between edges is not very compatible with the relational model, as it leads to "row at a time" processing. This also causes response time issues with increasing boundaries in a polygon. To test these statements, we did some tests; one of the results is shown in Figure 6.6.

On the x-axis of this figure the number of points in the resulting polygon is shown; on the y-axis the construction time per polygon in seconds is shown. For both implementations the trend of increasing construction time when the number of points in the resulting polygon increases is visible. This trend is more apparent in the left–right implementation. Probably this is because in the left–right implementation the boundaries are connected by finding common points. In this process computational costs increase with the number of points. Because the left–right method has been implemented in Java and the method based on relationships between edges in PL/SQL, the performance of both methods cannot be compared. Apart from performance, the implementations differ in the underlying geometrical primitive. In the relationships-between-edges implementation, the outer ring of a face can touch itself on the outer

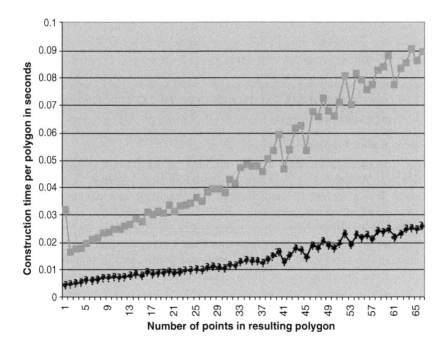

FIGURE 6.6 Construction time per polygon for different number of points in the resulting polygon. The black line represents the implementation based on the relationships between edges and the gray line represents the left–right implementation.

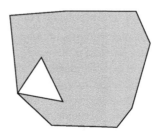

FIGURE 6.7 A polygon with a hole that touches the boundary.

boundary at exactly one point, and in the left–right implementation this is not possible. This difference can be illustrated by the polygon shown in Figure 6.7: a polygon that has an island that touches the boundary at exactly one point. The relationships-between-edges algorithm will generate a polygon with one self-touching outer ring, while the left–right algorithm will return a polygon with a boundary and an island. As described in Section 6.2.1, a self-touching boundary is not allowed according to the OGC Specification for SQL and not valid according to Oracle (rings may only touch other rings). Therefore, the relationships-between-edges method returns a non-valid geometry according to OGC and Oracle rules. Post-processing invalid polygons is possible, but requires so much geometrical and topological calculation that it is easier to use the left–right topology. From this it can be concluded that the winged edge structure as implemented in the cadastral data set is not OGC compliant, but also the OGC standard requires refinement.

6.2.3 COMMERCIAL DBMS IMPLEMENTATIONS OF 2D TOPOLOGICAL STRUCTURE

Laser-Scan Radius Topology

Compared to user-implemented models, the implementation of topology structure in Laser-Scan Radius Topology,[94] which is based on Oracle Spatial, is much more developed. It is a "complete" implementation of topology with support for linear networks and planar topology, including updates, insertions, and deletions.

To retrieve geometry from a topologically structured data set, Radius offers a function "get_geom" that is equivalent to the "return_polygon" function of our own implementations. Most users, however, choose not to use this function, but instead store a copy of the geometry explicitly. This increases the storage requirements, but it means that there is no performance penalty when accessing geometries (e.g., for display or geometric queries) since the geometry is instantly available and does not have to be computed. The use of database triggers in the Radius Topology architecture ensures that the geometries and their topological representation are always synchronized.

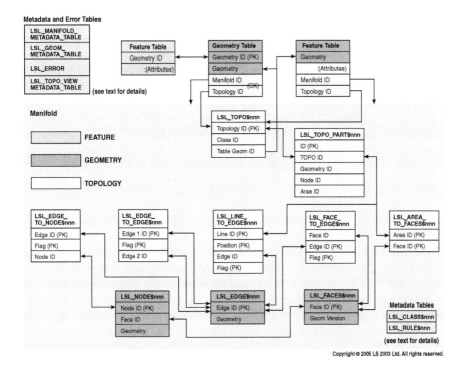

Metadata and Error Tables

| LSL_MANIFOLD_METADATA_TABLE |
| LSL_GEOM_METADATA_TABLE |
| LSL_ERROR |
| LSL_TOPO_VIEW_METADATA_TABLE |

(see text for details)

Feature Table

| Geometry ID |
| :(Attributes) |

Geometry Table

| Geometry ID (PK) |
| Geometry |
| Manifold ID |
| Topology ID |

(CK)

Feature Table

| Geometry |
| :(Attributes) |
| Manifold ID |
| Topology ID |

Manifold

	FEATURE
	GEOMETRY
	TOPOLOGY

LSL_TOPO$nnn

| Topology ID (PK) |
| Class ID |
| Table Geom ID |

LSL_TOPO_PART$nnn

| ID (PK) |
| TOPO ID |
| Geometry ID |
| Node ID |
| Area ID |

LSL_EDGE_TO_NODE$nnn

| Edge ID (PK) |
| Flag (PK) |
| Node ID |

LSL_EDGE_TO_EDGE$nnn

| Edge 1 ID (PK) |
| Flag (PK) |
| Edge 2 ID |

LSL_LINE_TO_EDGE$nnn

| Line ID (PK) |
| Position (PK) |
| Edge ID |
| Flag (PK) |

LSL_FACE_TO_EDGE$nnn

| Face ID |
| Edge ID (PK) |
| Flag (PK) |

LSL_AREA_TO_FACES$nnn

| Area ID (PK) |
| Face ID (PK) |

LSL_NODE$nnn

| Node ID (PK) |
| Face ID |
| Geometry |

LSL_EDGE$nnn

| Edge ID (PK) |
| Geometry |

LSL_FACES$nnn

| Face ID (PK) |
| Geom Version |

Metadata Tables

| LSL_CLASS$nnn |
| LSL_RULE$nnn |

(see text for details)

FIGURE 6.8 Radius Topology database tables (version 1.0). (Courtesy of Laser-Scan Limited.)

Additionally, support for topological querying (containment, adjacency, connectivity, overlap) is available in Radius Topology by means of a topo_relate operator.

All required topological references are stored explicitly: the winged edge representation (in the edge-to-edge table) makes up just a small part of the complete system (Figure 6.8). Topological primitives are stored in the NODE, EDGE, and FACE tables while faces are only stored by references to edges. A number of reference tables are used to store various types of topological references. The TOPO table is the link between the features and the topological structures. Topology is organized in "manifolds." Associated with each manifold and with the system as a whole are some meta-data and error tables. Before topologically structuring data in Radius Topology, the user can specify rules in order to control the way the structuring works (snap tolerances, which features/primitives are moved and which stay while snapping, etc.)

A performance test has been described in which the topological structure of Laser-Scan Radius Topology (version 1.0) was compared to the geometrical primitive of Oracle Spatial 9i.[105] In the topology case, fewer points are stored (by avoiding storing "common" boundaries twice). However, disk space requirements are much larger in the topological case due to the increased number of topological primitives and the references between them compared to the number of area features (and the way geometry is implemented in Oracle Spatial: small objects have relatively large overhead).

The total storage requirement for topology is intended for references, ids, and associated indexes that are required for the Radius Topology structure. The storage requirement will probably be more favorable for topology in the case of smaller-scale data and data with a relatively high number of intermediate points in the boundaries.

From the tests described in Reference 105 it can be concluded that performance of geometrical querying on a data set structured with Radius Topology is slower. This is due to the cost of computing the geometries on-the-fly from the topological information. This occurs when geometries are not stored explicitly alongside the topology. For this reason users often store the geometries explicitly, as described above.

Topology in Oracle Spatial 10g

Oracle Spatial 10g also supports 2D topology.[153] The basic topology elements in an Oracle topological structure are nodes, edges, and faces. A node is represented by a point and can be used to model an isolated point feature or edges. Every node has a coordinate pair associated with it to describe the location of the node. An edge is bounded by a start node and an end node and has a coordinate string describing the geometric representation. Each edge can consist of multiple vertices, represented by linear as well as circular arc strings. As each edge is directed, it is possible to determine which faces are located at the left- and right-hand side of the edge. A face corresponds to a polygon (that can be reconstructed from several edge strings) and has references to a directed edge on its outer and inner boundaries, if any. Each topology has a universal face that contains all other nodes, edges and faces in the topology.

In the results of a study that tested the support of topology with two data sets from the Netherland Kadaster have been reported.[161] The tests have shown that the current implementation does not completely avoid redundant data storage. The geometry is stored both in node and edge tables. However, as long as the user uses the supplied tools for data editing instead of directly updating the node, edge, and face tables, data consistency can be efficiently maintained. More experiments are needed to explore the offered functionality.

6.2.4   USER-DEFINED DBMS IMPLEMENTATION OF 3D TOPOLOGICAL STRUCTURE

In 3D, there is as yet no consensus on a single topological structure. Different topological structures can be defined depending on the number of primitives to maintain, and also the number and nature of relationships to explicitly store. The problems of defining 3D topological structures are relatively many compared to 2D. As a result of the large amounts of data and higher complexity, one data structure representing a specific topological structure, which is appropriate for a certain application, may not be easy to serve another application. Unfortunately, 2D topological structures are not

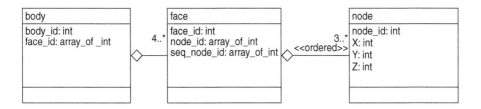

FIGURE 6.9 UML class diagram of Simplified Spatial Model.[229]

directly extendable to 3D. 2D structures are mostly built around the properties of an edge. One edge has exactly two neighboring nodes (begin and end) and exactly two neighboring faces (left and right). This property is not true in 3D space. An edge can have more than two neighboring faces; i.e., the order of the faces has to be specified.

Because the 3D topological structure of Zlatanova[229] is one of the few implementations of a topological structure defining volumetric objects, and because the implementations showed good results,[30,234] we took a closer look at this model. The Simplified Spatial Model (SSM) is a typical boundary representation. The role of the edge (= boundary) in 2D is now the role of the face (= boundary) in 3D. Nodes describe faces; faces describe bodies. The 1D primitive as part of a body (edge) is not explicitly stored in the model (Figure 6.9). Shared faces and nodes are only stored once. This 3D topological structure is described in detail elsewhere.[229]

This 3D topological structure can be implemented in several ways in an object relational DBMS. The first approach is the relational implementation. The conceptual model can be converted directly into a relational data model. For each object (node, face, and body) a separate relational table is created. The NODE table contains the id of the node and the three coordinates of the points. The FACE table contains the id of the face, a column denoting the order (anticlockwise) of the nodes in a face and the ids of nodes that the face consists of. A BODY table contains references to the ids of faces it consists of. Because the relationship between a face and constituting nodes (and between a body and constituting faces) is one-to-many, multiple rows (or columns) represent one face (and one body) in a traditional relational implementation using only plane relational tables and traditional data types. In the multiple-column representation the number of columns is fixed and a high number of columns must be chosen to be able to represent also faces with a large number of nodes and bodies with a large number of faces. This leads to a table with a large number of zero fields and consequently to overhead of information. Multiple-row representation (of faces and bodies) is therefore preferred in order to avoid the large number of zero fields in a single row presentation (with enough columns for the maximum number of references).

Another possibility is the object relational implementation. The list of ids referring to lower-dimensional objects (faces, nodes) is stored in a single column. This means that the number of rows in the object table is reduced to the actual number of the higher-dimensional object (body, face). Object relational implementation is a two-step procedure, i.e., creating objects (ADTs) and creating tables. The object relational implementation of 3D topological structure is illustrated with Oracle Spatial 9i. Two extended Oracle data types are used, which are intended for representing the one-to-many relationship, i.e., varrays (variable arrays) and nested tables. The syntax of the commands to create a data type of type varray is:

```
CREATE TYPE NodeArray AS varray (10000) OF number (5);
```

Utilizing the newly created data type NodeArray, the FACE object can be stored in the database in the following way:

```
CREATE TABLE face
(fid NUMBER(11) NOT NULL, num NUMBER(11) NOT NULL,
 nids NodeArray NOT NULL);
```

The other way to represent one-to-many relationships using only one column is nested tables. The commands to create a data type of type table and to use this new data type in the FACE object are:

```
CREATE TYPE NodeTable AS table OF number(5);
CREATE TABLE FACE (
FID number(11) not null,
NUM number(11) not null,
NIDS NodeTable not null);
```

As can be concluded from Reference 234, the nested table shows slower performance than the tables with varrays. This is probably because nested tables are less efficient than varrays because more overhead is produced during the implementation.

To be able to use the spatial operations of the DBMS on topologically structured data, a realization function was written. This function realizes the geometry of the 3D spatial objects, based on the topological tables. The function is based on the relational implementation. In the function, the nodes of one 3D spatial object are retrieved by the following query:

```
/* for the body bid=1 */
SELECT body.bid,face.fid, face.seqn, node.nid, node.x, node.y, node.z
FROM body, face, node
WHERE body.fid=face.fid AND face.nid=node.nid AND body.bid=1;
```

After this, the obtained nodes are translated to a complex geometrical object of 3D polygons, or a 3D multipolygon (see Section 6.1.2), or a polyhedron primitive (see Section 6.4).

6.3   SPATIAL ANALYSES IN DBMSs

Spatial analyses in the context of a DBMS are related to operations that are performed on spatial objects (in vector format) in which often no distinction is made between the spatial and thematic components of spatial objects. In this section we concentrate on the part of spatial analyses that is related only to the spatial component.

The Abstract Specifications of OGC distinguish between two sets of operations (also called operators or functions) defined for both geometrical and topological primitives while some of them are identical. The operations can be classified as unary (performed on one object) and binary (performed on two objects). For example, 15 unary (mbRegion, representativePoint, boundary, closure, isSimple, isCycle, distance, dimension, coordinateDimension, maximalComplex, transform, envelope, centroid, convex hull, buffer) and 7 binary relations (contains, intersects, equals, union, intersection, difference, symmetricDifference) are suggested within the geometry schema. Within the topology schema, the unary operations are 7 (dimension, boundary, coBoundary, interior, exterior, closure, maximalComplex). The binary operations for the topology schema can be a different number depending on the formalism used for detecting relationships. Three frameworks are accepted as fundamental: Boolean set of operations (considering intersections between closure and exterior), Egenhofer operations (taking into account exterior, interior, and boundary of objects),[44] and Clementini operations using the same topological primitives as Egenhofer but considering the dimension of the intersection.[24] It should be noticed that the Abstract Specifications do not discuss implementation environments. The current Implementation Specification for SQL[140] specifies eight relationships based on the Egenhofer framework, i.e., equals, disjoint, intersects, touches, crosses, within, contains, and overlaps, which are only defined for Simple Features, i.e., geometry.

In this section spatial analyses in DBMS are considered, distinguishing between spatial analyses on geometrical primitives (2D in Section 6.3.1 and 3D in Section 6.3.2) and spatial analyses on a topological structure (Section 6.3.3). In Section 6.3.4 a case study is described that compares the same spatial analysis (using the same test area) performed on geometrical primitives on the one hand and on a topological structure on the other hand.

6.3.1   2D SPATIAL ANALYSES USING GEOMETRICAL PRIMITIVES

The OGC Simple Feature Specification for SQL[140] describes geometrical and topological functions that should be supported at DBMS level as part of the implementation of the geometrical primitive. The defined operations to obtain the topological relationships do not give the dimensionality of the relationship as a result. For example, the query "Find all adjacent parcels to a query parcel" (using the touch relationship) gives all parcels that touch the query parcel as a result, regardless the dimensionality (touch at edge or point). To restrict the result data set to only parcels that touch at an edge, the query should be extended with the condition that boundaries of two parcels should also overlap. Overlap results "true" if the intersection results in geometry of the same dimension as the input geometries.

In Ingres, support for topological relationships is minimal. Oracle, IBM DB2, Informix, and PostGIS support geometrical and topological functions defined by OGC and often more functions than these.[132]

Oracle Spatial 9i is used to illustrate the possibilities of spatial analysis using the geometrical primitive in DBMSs. Currently, Oracle Spatial supports three groups of selection operations, i.e., topological relationship operations, metric operations, and specialization operations.

TABLE 6.3

Topological, Metric, and Specialization Operations in the DBMS According to Implementation Specifications of OGC and the Oracle Spatial Implementations.

Topological Operations		Metric and Specialization Operations	
OGC	**Oracle**	**OGC**	**Oracle**
		Unary Metric Operations	
Equals	Equal	Area	sdo_area
disjoint	disjoint	Length	sdo_length
intersects	anyinteract	**Unary Specialization Operations**	
touches	touch	Buffer	sdo_buffer
crosses	overlapbdydisjoint	Convexhull	sdo_convexhull
within	inside	Centroid	sdo_geomcentroid
contains	contains	**Binary Metric Operations**	
overlaps	overlapbdyintersect	Distance	sdo_distance
	coveredby	**Binary Specialization Operations**	
	covers	Intersection	sdo_intersection
	on	Union	sdo_union
		Difference	sdo_difference
		Symdifference	sdo_xor

Topological relationship operators between two geometries are implemented with respect to the nine-intersection model of Egenhofer and Herring.[44] The names of the operations slightly differ from the ones suggested by OGC. In Oracle Spatial 9i all these topological relationships are implemented using one function (sdo_geom.relate) or operator (sdo_relate), where the type of relationship is passed as a text string (Table 6.3, left). The spatial operator requires and utilizes a spatial index and is therefore faster than the spatial function, which also works without a spatial index.

In the Egenhofer model each spatial object has an interior, a boundary, and an exterior. The boundary consists of points or lines that separate the interior from the exterior. The boundary of a line consists of its end points. The boundary of a polygon is the line that describes its perimeter. The interior consists of points that are in the object but not on its boundary, and the exterior consists of those points that are not in the object. Some of the topological relationships of the nine-intersection model have names associated with them that specify the type of relationship, e.g., INSIDE and COVEREDBY. INSIDE returns true if the first object is entirely within the second object and the object boundaries do not touch; otherwise, it returns false. COVEREDBY returns true if the first object is entirely within the second object and the object boundaries touch at one or more points; otherwise, it returns false.

TABLE 6.4
Examples of Aggregate Functions in Oracle Spatial 9i

SDO_AGGR_CENTROID	Returns a geometry object that is the centroid ("center of gravity") of the specified geometry objects
SDO_AGGR_CONVEXHULL	Returns a geometry object that is the convex hull of the specified geometry objects
SDO_AGGR_MBR	Returns the minimum bounding rectangle of the specified geometry objects
SDO_AGGR_UNION	Returns a geometry object that is the topological union (OR operation) of the specified geometry objects

Besides the relationship operations, many metric and specialization operations are proposed by OGC that can take one (unary operations) or two geometries (binary operations) or other parameters (e.g., buffer size) and calculate some values or new geometries. The most important of them together with their Oracle equivalents are given in Table 6.3, right. An example is when one wants to obtain a new geometry that is the intersection between the geometry of parcels and the geometry of the extent of a tunnel. The query to create these new geometries is the following (to speed this query a "where" clause could be added using "anyinteract"):

```
CREATE TABLE new_geometry AS
SELECT t.object_id, p.parcel_number,
  sdo_geom.sdo_intersection(t.shape, p.shape,1) shape
FROM parcel p, tunnel t;
```

Another class of spatial operations in Oracle Spatial returns an aggregate of a collection of geometries. These are not defined within OGC (Table 6.4).

6.3.2 3D Spatial Analyses Using Geometrical Primitives

Our experiments showed that it is possible to maintain objects with 3D coordinates in Oracle Spatial 9i (see Section 6.1.1). However, the current implementations of geometry operators (e.g., compute area of 3D polygon) in Oracle Spatial 9i omit the z-value.

In the following example, a table (geom) is created in Oracle Spatial 9i in which a 2D polygon and a polygon defined in 3D space are inserted. After that, the geometrical operators area and length (perimeter) are performed on both polygons. The operator "validate" is performed to show that the polygons are both valid. As can be seen in the results of the queries, sdo_area and sdo_length (both spatial operators in Oracle), return the same value for both polygons, although the 3D polygon actually has a greater area and length (perimeter). In these calculations, the 3D polygon is projected on the surface.

```
/* 66: a 2D polygon */
INSERT INTO geom (shape,tag) VALUES (mdsys.sdo_geometry(2003, NULL, NULL,
  mdsys.sdo_elem_info_array(1, 1003, 1),
  mdsys.sdo_ordinate_array(12,15, 15,15, 15,24, 12,24, 12,15)), 66);

/* 88: a 3D polygon */
INSERT INTO geom (shape,TAG) VALUES (mdsys.sdo_geometry(3003, NULL, NULL,
  mdsys.sdo_elem_info_array(1, 1003, 1),
  mdsys.sdo_ordinate_array(12,15,0, 15,15,0, 15,24,999, 12,24,999, 12,15,0)), 88);

SELECT tag,
sdo_geom.sdo_area(shape, 1) area,
sdo_geom.sdo_length(shape, 1) length
sdo_geom.validate_geometry(shape, 1) geom_validate
FROM geom;

TAG AREA    LENGTH    GEOM_VALIDATE
--- ----    ------    -------------
66   27     24        TRUE
88   27     24        TRUE
```

Many other DBMSs support a similar set of geometry operators as most of them also skip the z-coordinate. Some exceptions are PostGIS (PostgreSQL)[165] and the MapInfo Spatialware Datablade[110] (based on Informix) that do have limited support for geometry calculation in 3D, such as length and perimeter in 3D. This is illustrated in the next PostGIS example.

First, four tables are created: line2D, line3D, polygon2D, and polygon3D in which, respectively, a 2D line, a 3D line, a 2D polygon, and a 3D polygon are inserted (\g in PostGIS is used to end a command):

```
/* a table with a 2D line */
CREATE TABLE line2d (id int4)\g
SELECT addgeometrycolumn(`test',`line2d',`shape',0,`LINESTRING',2)\g
INSERT INTO line2d (id, shape) VALUES(1,
geometryfromtext(`LINESTRING(1 1,2 2)',0))\g

/* a table with a 3D line */
CREATE TABLE line3d (id int4)\g
SELECT addgeometrycolumn(`test',`line3d',`shape',0,`LINESTRING',2)\g
INSERT INTO line3d (id, shape) VALUES(1,
geometryfromtext(`LINESTRING(1 1 0,2 2 50)',0))\g

/* a table with a 2D polygon */
CREATE TABLE polygon2d (id int4)\g
SELECT addgeometrycolumn(`test',`polygon2d',`shape',0,`POLYGON',2)\g
INSERT INTO polygon2d (id, shape) VALUES(1,
geometryfromtext(`POLYGON((0 0, 1 0, 1 1, 0 1, 0 0))',0))\g

/* a table with a 3D polygon */
CREATE TABLE polygon3d (id int4)\g
SELECT addgeometrycolumn(`test',`polygon3d',`shape', 0,`POLYGON',3)\g
INSERT INTO polygon3d (id, shape) VALUES(1,
geometryfromtext(`POLYGON((0 0 0, 1 0 0, 1 1 100, 0 1 100, 0 0 0))',0))\g
```

In the next step the following queries are executed:

```
SELECT length(shape) FROM line2d\g
SELECT length3d(shape) FROM line3d\g
SELECT perimeter(shape) FROM polygon2d\g
SELECT perimeter3d(shape) FROM polygon3d\g
```

As the results show, length and perimeter do work in 3D:

```
length       1.4142135623731 (1 row)
length3d     50.0199960015992 (1 row)
perimeter    4 (1 row)
perimeter3d  202.009999750012 (1 row)
```

The other functions (overlap, area, distance) in PostGIS (and also in the SpatialWare Datablade of MapInfo) are performed in 2D. PostGIS also has a box3D function that gives the maximum extents in 3D as a result.

6.3.3 SPATIAL ANALYSES USING THE TOPOLOGICAL STRUCTURE

Some spatial operations are specific to topological structure, for example, validation functions on topological structure (e.g., is loop closed?) and network computation (e.g., find shortest path). Another spatial operation specific to topological structure is realization of geometry, which is the basis for nearly all metric operations and needed for visualization of the objects. The complexity of the realization functions varies considerably with respect to the different implementations of the topological structure. For example, the geometry (coordinates) of a body can be extracted by only one SQL statement (in case of relational implementation) if the geometry is maintained explicitly, but a PL/SQL in Oracle script is required if the body is represented as a variable array of ids of faces and the coordinates are only stored at node level.

Although not yet very common, spatial analysis on topological structure is available in some DBMS software (e.g., Laser-Scan Radius, Oracle Spatial 10g) but the support is still limited. A lack of native topology support of DBMS is compensated by many user-defined implementations of topological structure. Each implementation has its own set of topological operations available with the model. It depends on the topological structure and thus on the relationships defined in the topological structure, which topological operations are available. For example, if a topological structure of planar partition is implemented with information only on connecting edges, without information on the left and right face of an edge, an adjacent analysis (give all polygons adjacent to this polygon) cannot be performed on the topologically structured data. Therefore, first the geometries of polygons have to be realized and the analysis has to be performed on the geometrical primitive.

We have developed several functions to realize geometry (PL/SQL and Java) related to two different topological structures, i.e., winged edge (in 2D, Section 6.2.2), and SSM (in 3D, Section 6.2.4). Both topological structures were user-defined implementations in a object relational DBMS. As can be concluded from the experiments with the realization functions, the required realization of geometry requires traverse of all the relational tables, which may result in poor performance of metric analyses for large data sets.

6.3.4 CASE STUDY: TOPOLOGICAL STRUCTURE OR GEOMETRICAL PRIMITIVES?

As has already been mentioned a number of times before, it can be expected that spatial queries relying only on topological references perform very well on the topological structure compared to the geometrical primitive, e.g., to find all features that are adjacent to a certain feature. In contrast, the performance of metric and specialization operations will be slower on the topological structure. These last operations need the coordinates of the objects, which, if performed on the topological structure, will initiate a join of all the relational tables (dependent on the type of implementation). It has also been concluded[73] that some operations (compute area, distance, etc.) on topological structured data will be slower than on geometrical primitives because it also requires querying and joining different relational tables. Another explanation for the better performance of these spatial operations on the geometrical primitives is the internal optimizations provided by the DBMSs and the possibility of applying spatial indexes.

To illustrate the power of topological structure in performing relationship operations, an experiment was carried out in Oracle Spatial 9i on a data set that is a selection of the cadastral database of the Netherlands. The test data set contains 1,788,019 parcels and 5,599,089 boundaries. The first query that we use in this experiment is to find all adjacent parcels to the parcel with object identifier 6862 (Figure 6.10). The query was performed on both a topologically structured data set and a geometrically

FIGURE 6.10 Query parcels (6862 and 7142) used in test queries.

structured data set. The geometries of the parcels were therefore stored explicitly in a table with the geometrical primitive of Oracle Spatial and populated with the return_polygon function (Section 6.2.2). A spatial index was built on the geometry column to speed spatial analyses. The topological structured data set was described in Section 6.2.2. Note that performance depends also on spatial clustering, which was not taken into account in this test. For the data set described by geometrical primitives, the query to find all adjacent parcels is given below (using a "subselect" structure), in which the polygons of parcels are stored in the table "parcels_geom" in the column named "shape." The query finds all parcels that have a "touch" relationship with parcel "6862" using the spatial operator "sdo_relate," which is implemented on geometrical primitives.

```
SELECT object_id FROM parcels_geom
  WHERE sdo_relate(shape,(SELECT shape
FROM parcels_geom WHERE object_id=6862),
  `MASK=TOUCH, QUERYTYPE=WINDOW') = `TRUE';
```

The query returned the following result:

```
OBJECT_ID
----------
7142
2067
2066
7141
2065
6862
6861
Elapsed: 00:00:22.05
```

For this query we used Oracle's spatial operator since spatial operators use the spatial index in contrast to the spatial functions in Oracle (see Section 6.3.1). The query plan of the query was checked to verify that the query indeed used the spatial index. As shown in the result, the time needed to perform the geometrical query is about 22 seconds.

In the topologically structured data set, all adjacent parcels to parcel 6862 can be found when all the boundaries are selected that have the specific parcel on the left or right side. The next step is to find the parcel that is located on the other side of the selected boundaries. The result, in 0.01 seconds, is:

```
SELECT l_obj_id, r_obj_id FROM boundary
  WHERE r_obj_id=6862 OR l_obj_id=6862;

L_OBJ_ID    R_OBJ_ID
--------    --------
2066        6862
6862        7141
6861        6862
6862        7142
Elapsed: 00:00:00.01
```

The same test was performed for parcel 7142 (with 28 adjacent parcels). The processing time for this second query was 22.56 seconds for the geometrical query and 0.01 seconds for the topological query. The queries were repeated a number of times, which resulted in processing times of the same order, every time. These examples show that this topological query is indeed faster on a topologically structured data set than on the data set described with geometrical primitives.

There is another conclusion that can be drawn from the first query: the results differ.

The topological query does not give parcels 2067 and 2065 as a result because these parcels touch parcel 6862 only at a point and are therefore not seen as adjacent parcels from the topological point of view as defined in the winged edge structure. The result set in spatial analyses using topological structure depends therefore on the topological structure implemented.

The geometrical query does find parcels 2067 (neighbor on the right of parcel 2066) and 2065 as adjacent parcels because they do touch parcel 6862, even if it is at a point. The geometrical query could be further specified by adding the condition that boundaries of two parcels should also overlap (see Section 6.3.1). It is a moot point which of these results is "correct." Some applications require the "corner contact" parcels to be returned as well, and other applications do not.

6.4 IMPLEMENTATION OF A 3D GEOMETRICAL PRIMITIVE IN A DBMS*

Present geo-DBMSs do not support 3D geometrical primitives, although 3D objects can be modeled within current techniques, as seen in Section 6.1.2. The absence of a real 3D primitive in geo-DBMSs results in two main problems:

1. Geo-DBMSs do not recognize 3D spatial objects, because they do not have a 3D primitive to model the 3D object. This results in DBMS functions not working properly (e.g., there is no validation for the 3D object as a whole and functions only work with the projection of these objects, because the third dimension is ignored).

2. Where 3D objects are stored as one multipolygon or a set of polygons, no relationship exists between the different 2D polygons defining the object. Besides the facts that no validation can be performed and that any set of polygons can be inserted, the main disadvantage is that the same coordinates are listed multiple times (causing risks of inconsistencies) and there is no information about outer or inner boundaries of the polyhedron. Where 2D polygons that bound a 3D object are stored in multiple records, a 1:n relationship exists between the object and the number of records; a clearer and more efficient administration of large data sets requires a 1:1 relationship between objects in reality and objects in the database.

---

* This section is based on References 4 and 5.

ISO/TC211 spatial schema[83] adopted by OGC defines 3D geometry primitives in an abstract mathematical manner. However, 3D geometrical primitives are not yet included in the OGC Implementation Specification for SQL. To fill this gap we worked on a solution in the form of a design and an implementation of a real 3D primitive within a DBMS context. This section presents this solution and describes how 3D spatial objects can be modeled, i.e., stored, validated, and queried in a geo-DBMS using a 3D geometrical primitive (also using 3D spatial functions). Many concepts have been developed in the area of 3D modeling.[30,90,114,162,163,177,229] In the research presented here the developed concepts have been translated into a prototype implementation of a true 3D primitive in a DBMS environment. The implementation has been based on a proposal for extending the spatial model of Oracle Spatial 9i with support for a 3D primitive.[192]

6.4.1 Definition of 3D Primitive

Definition

There are a number of 3D geometrical primitives possible to model 3D spatial objects:

- **A set of tetrahedra:** This is the simplest 3D primitive and consists of four triangles that form a closed object in 3D coordinate space. The tetrahedron is well defined, because the three points of the four triangles always lie in the same plane. It is relatively easy to create functions that work on this primitive. The disadvantage is that it could take many tetrahedra to construct one factual object; this does not solve the disadvantage of not having a 1:1 relationship between the factual object and the object's representation in the database.

- **Polyhedron:** This is the equivalent of a polygon, but in 3D. It is made up by several flat faces that enclose a volume. An advantage is that one polyhedron equals one factual object. Because a polyhedron can have holes in the exterior and interior boundary, it can model many types of objects. A disadvantage is that the buffer operation results into a non-polyhedral object, because it will contain spherical or cylindrical patches, which cannot be represented by the polyhedron primitive. The solution is to approximate the result of the buffer operation.[214]

- **Polyhedron combined with spherical and cylindrical patches:** This is the equivalent of the current 2D geometry data model of most geo-DBMSs (i.e., straight lines and circular arcs). This solution makes it possible to model 3D objects more realistically (although it is also not closed under the buffer operation). However, modeling with this primitive is complex.

- **CAD objects:** There are many possibilities,[115] such as Constructive Solid Geometry, cell decomposition, octree,[20] and objects with curved faces. These objects either do not fit with the present OpenGIS/ISO 2D geometry data model or are complex to model without an advanced graphical user interface.

To choose a suitable 3D primitive, a number of criteria were evaluated.[1] The implementation should lead to valid objects. It should be easy to specify instances and to create and enable efficient algorithms. Furthermore, the size and redundancy of storage (conciseness) should be taken in consideration.

The tetrahedron was not selected, because there are several primitives necessary to model one object. CAD objects with curved faces can model a spatial object very realistically, but are complex to model without an advanced graphical user interface, and also 2D CAD objects do not always fit within the present 2D geometry data model. That leaves the polyhedron option with and without the cylindrical/spherical patches. The one with spherical and cylindrical patches would fit better to the present 2D geometry data model (in which geometry is defined not only by straight lines but also by circular arcs), but ease of creation and implementation favor the polyhedron without spherical and cylindrical patches at first. Therefore, the polyhedron is chosen as the 3D primitive in this research to start with. If needed, spherical and cylindrical patches are approximated by several flat faces. It was also expected that choosing a relatively simple primitive will give more insight into the problems that occur when implementing more complex primitives in the future.

Implementation

The 3D primitive has been implemented in a geometrical model with internal topology (i.e., topology is maintained within one instance of the object and not between objects). Managing topological structures between objects (e.g., sharing common faces) is not within the scope of the polyhedron primitive. The polyhedron is defined by storing the vertices explicitly (x,y,z) and describing the arrangement of these vertices in the faces of the polyhedron. Internal topology within one object is maintained because only the vertices are stored (no polygons or lines). Faces are defined by internal references to nodes and nodes are shared between faces (Figure 6.11). This yields a hierarchical boundary representation.[1,229] Note that edges are not stored explicitly in this model. The Oracle Spatial geometry type has been extended to support the polyhedron primitive. The vertices and arrangement of faces are all stored in the sdo_ordinate array.

The interpretation code of the faces (Figure 6.11) describes if the list of node references refers to an outer or inner boundary of a polyhedron (face) or if the list of node references refers to an outer or inner ring of a face, either outer or inner. Most polyhedra will just have an outer boundary, but an inner boundary can for example be used to create a hollow object: the inner boundary will then describe this hollow space. Most faces will have just an outer ring, but inner rings can be used to create cavities in the polyhedron. With these elements it is possible to model complex objects, e.g., objects with cavities or objects that are hollow inside. This set of elements is enough for the implemented functions in the next sections to understand what the 3D spatial objects look like.

In the field of computer graphics (see, for example, Reference 123) it is a custom to order all the vertices of outer boundaries (rings) anticlockwise, seen from the outside of an object, and the vertices of inner boundaries (rings) clockwise. That is, the normal

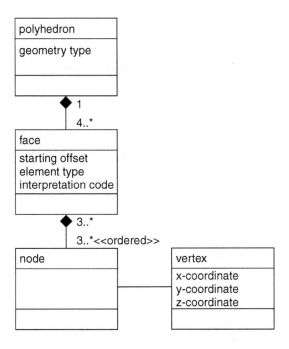

FIGURE 6.11 UML class diagram describing storage of polyhedron primitive.

vector of the face points to the outside of the object. This practice is followed in the implementation (details and examples in Reference 3). A table can now be created to hold polyhedra:

```
CREATE TABLE polyhedron_table
   (id NUMBER, geometry MDSYS.SDO_GEOMETRY);
```

Then the meta-data table can be updated:

```
INSERT INTO user_sdo_geom_metadata VALUES (
`POLYHEDRON_TABLE',`GEOMETRY',
mdsys.sdo_dim_array(
mdsys.sdo_dim_element(`X',-100,100,0.001),
mdsys.sdo_dim_element(`Y',-100,100,0.001),
mdsys.sdo_dim_element(`Z',-100,100,0.001)),
NULL);
```

To be able to use the 3D R-tree index of Oracle, the polyhedron primitive is defined as an existing sdo_gtype: "3002." This corresponds to a fictive 3D polyline going through all the coordinates of the defined polyhedron. When creating a 3D spatial index, a bounding box is created around this line. This bounding box equals the bounding volume around the polyhedron. Oracle Spatial ignores all elements with sdo_gtype or e_type = 0. If the sdo_gtype = 0, the object is also ignored by the spatial index. These values are therefore used for the remainder of the elements of the polyhedron (flat faces).

Summarizing, the following parameters are used for storing a cube as a polyhedron primitive defined as an extension of the sdo_geometry type:

- sdo_gtype = 3002 (3D line)
- sdo_srid = NULL (no spatial reference system)
- sdo_point = NULL (no point data)
- sdo_elem_info = 1,2,1 (line consisting of straight segments), and x,0,1006 (6 times an exterior polyhedron boundary, x is the starting offset in the array with ordinates)
- sdo_ordinates: contains eight coordinate triplets and six face descriptions

The query to insert a cube in the table is:

```
INSERT INTO polyhedron_table (id, geometry) VALUES (1,
mdsys.sdo_geometry(3002, -- geometry type: 3D polyline
NULL, NULL,
mdsys.sdo_elem_info_array(1,2,1, 25,0,1006, 29,0,1006, 33,0,1006,
  37,0,1006, 41,0,1006, 45,0,1006),
-- starting offset, e_type, interpretation code,
-- first triplet is fictive polyline, followed by 6 faces
mdsys.sdo_ordinate_array(
1,1,0, 1,3,0, 3,3,0, 3,1,0, -- vertices
1,1,2, 1,3,2, 3,3,2, 3,1,2,
-- bottom, top, front face, defined by references to nodes:
1,2,3,4, 8,7,6,5, 1,4,8,5,
-- back, left, right face, defined by references to nodes:
2,6,7,3, 1,5,6,2, 4,3,7,8
)));
```

Note that in a full implementation of the 3D primitive (that starts from scratch) two arrays would be used: one for the coordinates and one for the references. However, because the sdo_geometry framework was used, both arrays were combined in the sdo_ordinate array. A 3D R-tree index can be created by the following SQL statement:

```
CREATE INDEX polyhedron_table_index ON polyhedron_table(geometry)
INDEXTYPE IS mdsys.spatial_index parameters(`sdo_indx_dims=3');
```

6.4.2 VALIDATION

It is important that the spatial data are checked (validated) when they are inserted in the DBMS or when they are updated. Valid objects are necessary to make sure that the objects can be manipulated correctly; e.g., it is impossible to compute the volume of a cube when the top face is omitted; this would be an open box without a volume. Validating may seem quite easy for humans, but a computer needs an explicit set of rules to check the spatial data. To allow for checking the spatial data, it is important

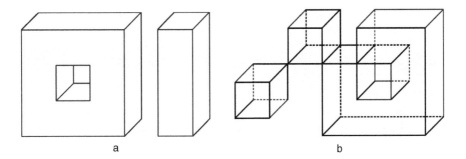

FIGURE 6.12 Examples of polyhedra. (a) Two polyhedra; (b) pseudo-polyhedron.

to give an accurate definition of the 3D primitive. In Reference 1 definitions of both a polyhedron and a pseudo-polyhedron are given:

- **Polyhedron:** A polyhedron (Figure 6.12a) is a bounded subset of 3D coordinate space enclosed by a finite set of flat polygons such that every edge of a polygon is shared by exactly one other polygon (adjacent polygons). The vertices and edges of the polygons are the vertices and edges of the polyhedron; the polygons are the faces of the polyhedron. The edges and faces are two manifold (see below).
- **Pseudo-polyhedron:** A pseudo-polyhedron (Figure 6.12b) is a bounded subset of 3D coordinate space enclosed by a finite set of planar faces such that (1) every edge has at least two adjacent faces, and (2) if any two faces meet, they meet at an edge.

Polyhedra are, therefore, a subset of pseudo-polyhedra. Edges and vertices, as boundary elements for polyhedra and pseudo-polyhedra, may be either two manifold (in case of polyhedra) or non-manifold (in case of pseudo-polyhedra) elements.

In the case of edges, they are two(non)-manifold elements when every point of it is also a two(non)-manifold point, except that either or both of its ending vertices might be a point of the opposite type. A two-manifold edge is adjacent to exactly two faces, and a two-manifold vertex is the apex of only one cone of faces.

In our implementation we used the definition of a polyhedron of Reference 1, which is a two-manifold element. Consequently, a valid polyhedron bounds a single volume, which means that from every point (also on the boundary), every other point (also on the boundary) can be reached via the interior. Based on this definition, a validation function has been implemented.

Tolerance

The validation function and some of the 3D functions have a tolerance value as input parameter. The points that make up the polygon can be slightly out of the flat plane, because of the geodetic measuring methods[201] and the finite representation of coordinates in a digital computer. To solve this problem a tolerance value has been introduced. The faces of a polyhedron are flat within this tolerance. This tolerance value should not be too large; otherwise invalid objects will be accepted as valid. A good value for the tolerance is the standard deviation of the geodetic measurements.

Implementation

The definition of the polyhedron primitive is the basis for a set of validation rules that have been implemented to evaluate the validity of stored objects. All the rules together enforce the correctness of the stored polyhedra. According to the implemented validation rules, a polyhedron is valid when (see below):

- It has been stored correctly.
- It has flat faces.
- It is two manifold (it bounds a single volume).
- Its faces are simplicit.
- It is orientable.

Correct storage   First, a check is needed on the storage of the data. It is important for the validation function to work properly that the spatial objects are stored as described in Section 6.4.1. This means that valid interpretation codes need to be used and that node references in the faces should correspond with an existing vertex. If the spatial object is correctly stored, the next test can be carried out.

Flatness characteristics   The next test evaluates the flatness of the faces. At the same time whether an inner boundary of a face is in the same plane as its corresponding outer boundary is tested. All faces should be flat within a given tolerance. This is tested by estimating a least squares plane through the average coordinate of all vertices:

$$x_c = \frac{1}{n} \sum_{i=1}^{n} x_i \qquad y_c = \frac{1}{n} \sum_{i=1}^{n} y_i \qquad z_c = \frac{1}{n} \sum_{i=1}^{n} z_i$$

A least-squares plane minimizes:

$$\sum_{i=1}^{n} (Ax_i + By_i + Cz_i - D)^2$$

where A, B, and C are the components of the normal vector, D is the distance to the origin, x_i, y_i, and z_i are the vertices, and n is the number of vertices. If the average coordinate is substracted from the vertices, the plane goes through the origin, which results in $D = 0$. The components of the normal vector are now the unknowns and the eqations can be solved. To retrieve the plane equation, D can be computed by:

$$Ax_c + By_c + Cz_c + D = 0$$

where x_c, y_c, and z_c are the average coordinates of all vertices. The derived plane equation is used to compute the distance from each vertex to this least-squares plane. If all distances are smaller than the tolerance value, the face is planar.

Two-manifold characteristics   The next step is to test if the polyhedron bounds a single volume in 3D space (two-manifold polyhedron). To test if a polyhedron is a two-manifold polyhedron, a set of rules has been constructed and implemented to enforce the two-manifold characteristic of a polyhedron:

- All edges (defined by two vertices) occur exactly twice in opposite order.
- Inner or outer faces should not intersect (touch is allowed).
- A polyhedron can contain only one connected volume containing one or more holes.
- Vertices related to one shell structure should be two manifold.

Simplicity characteristics   The faces should be simplicit. Therefore, a test was implemented to check that faces have an area, they are not self-intersecting, and they are not built of disconnected parts. Also an inner boundary of a face should not intersect (touch is allowed) with its outer boundary.

Orientation characteristics   The final test of the validation is to check if the vertices in the faces are orientated correctly, i.e., anticlockwise (looking from the outside) for outer boundaries and clockwise for inner boundaries. Only one face of the polyhedron has to be tested, because if the edges are two manifold, the whole object is either orientated correctly or incorrectly. It is important which face to test. From the bottom face we know that the normal vector should be pointing toward the negative z-direction. The cross product of two following edges of a convex part of this bottom face gives the normal vector. The z-component of this normal vector should be negative.[201]

If all the criteria in the validation are met, then the spatial object is valid. The following SQL statement tries to validate the two objects shown in Figure 6.13. How the validation function has been implemented is described in Section 6.4.4.

```
SELECT validate_polyhedron(geom,0.05) VALID FROM table;

VALID
------------------------
Not a 2-manifold object
Not a 2-manifold object
```

Both objects are detected to be invalid within a tolerance value of 0.05. Note that the coordinates of these objects are measured in meters. A tolerance value of 0.05 then corresponds to a maximum error of 5 centimeters.

Critical Objects

The statement that 3D data structures are very complex compared to 2D data structures and that therefore a correct and finite definition of a polyhedron is not easy to give

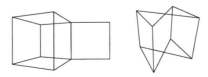

FIGURE 6.13 Invalid objects, because of dangling face (left) and intersecting faces (right).

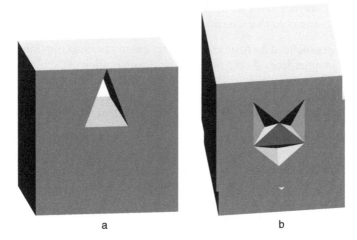

FIGURE 6.14 Valid polyhedra that are determined invalid by the implementation. (a) Edge in upper face is used four times, (b) front face contains disconnected parts due to three tetrahedral dents.

is underlined by the fact that still some valid polyhedra are determined as invalid by our validation test. The definitions above exclude the valid polyhedron as shown in Figure 6.14a, which is a cube with a triangle-shaped hole that touches the upper face. The upper face is divided into two parts, by which the middle edge of this face occurs four times in the definition of the polyhedron. According to the definition (and our implementation) this polyhedron is not valid, since the edge in the upper face is used more than twice. However, in this case this does not cause division of the polyhedron into disconnected parts. Therefore, this polyhedron should have been determined as valid. If the same polyhedron is modeled by not dividing the upper face and only defining a hole (that touches the upper face), the object is determined as valid.

Another valid polyhedron, which is not valid according to the implemented rules, is shown in Figure 6.14b. This is a polyhedron with three pyramid-shaped dents in the front face, by which the front face contains disconnected parts. The front face is not simplicit, because the inner rings of the face divide the face into disconnected parts. However, because the face does not divide the polyhedron into disconnected parts, the polyhedron should have been determined valid.

6.4.3 Spatial Indexing in 3D

Oracle Spatial 9i supports R-trees[64] up to four dimensions and the 2D quadtree (no support for octree). Therefore, the Oracle R-tree indexing can be used for the 3D primitive. Storing the 3D objects in a special way makes using the Oracle spatial index possible, as was mentioned before. A 3D polyline going through all the coordinates of the defined polyhedron can be imagined. When creating a 3D R-tree in Oracle, a bounding volume is created around this line, which equals the bounding volume around the polyhedron.

2D or 3D Spatial Index?

In many spatial applications the extent of the domain in the x,y plane is larger than in the z-direction. For example, a city plan typically covers an area of 5×5 kilometers with buildings up to 50 meters tall. This, plus the fact that queries usually try to find all the objects in a specific x,y region (with possibly objects that are on top of each other), may make a 3D spatial index hardly any more useful than a 2D index (on x,y coordinates only). In these kinds of queries the x- and y-coordinates are more selective than the z-coordinates. This means a 2D spatial index might work just as well as or better than a 3D spatial index.

A test was executed to see if one might just as well use a 2D spatial index instead of a 3D spatial index.[3] The test data set consisted of 1348 3D objects that are stored with the 3D primitive. In the test (retrieving 3D objects that intersect with a 3D box) the efficiency of the spatial index was measured by determining the number of candidates that were selected by the spatial index compared to the actual number of intersections. Sdo_filter is the Oracle Spatial function that uses the spatial index to select candidates for spatial queries. It is the only Oracle Spatial function that works in 3D (in connection with the 3D R-tree).

The following SQL statement shows how to use this filter to retrieve the number of candidates:

```
SELECT COUNT(id) FROM buildings_table WHERE SDO_FILTER(geometry,
(SELECT geometry FROM querywindow WHERE id=1), `querytype = WINDOW')='TRUE';
```

To retrieve the number of actual intersections, a 3D Boolean intersection function is used that also was implemented (see Section 6.4.4). The function can be used in an SQL statement as follows:

```
SELECT COUNT(id) FROM buildings_table WHERE intersection(geometry,
(SELECT geometry FROM querywindow WHERE id=1), 0.05)=1;
```

To use the spatial index in the implemented function, the spatial filter has to be combined with the intersection function like this:

```
SELECT COUNT(id) FROM buildings_table WHERE SDO_FILTER(geometry,
(SELECT geometry FROM querywindow WHERE id=1), `querytype = WINDOW')=`TRUE'
AND
INTERSECTION(geometry, (SELECT geometry FROM querywindow WHERE id=1), 0.05)=1;
```

From the results of the test it can be concluded that a 2D index works as well as a 3D spatial index when the query window contains the ground level:[3]

Query box	Result	No Spatial Index		2D R-Tree		3D R-Tree	
		# cand.	eff.	# cand.	eff.	# cand.	eff.
0--50 m	509	1348	37.76%	510	99.80%	510	99.80%
20--50 m	59	1348	0.04%	510	11.57%	59	100%

However, if the ground level is not included in a 3D query window, then the 3D R-tree is significantly faster and more efficient, because most objects can be skipped.

With the knowledge that the overhead of a 2D R-tree and a 3D R-tree are both relatively small, there may be no reason to build a 2D R-tree on the data set instead of a 3D R-tree. The 3D R-tree performs equally well as the 2D R-tree when the query window contains the ground level height, but it performs a lot better when this query window does not contain the ground level height.

6.4.4   3D FUNCTIONS

As mentioned in Section 6.3.2, the standard functions in Oracle, just as in most geo-DBMSs, work only with the projection of 3D spatial objects onto 2D coordinate space, because the third dimension is ignored. To offer realistic functionality, some of the most common functions have been implemented in 3D (for 0D up to 3D primitives):

- Function to insert data: creating data from 3D multipolygons and VRML
- Function to validate polyhedron: validation function
- Functions that return a Boolean: point-in-polyhedron query and intersection test (polyhedron-polyhedron)
- Unary functions that return a scalar: area, perimeter, and volume
- Binary functions that return a scalar: distance between centroids
- Unary functions that return a simple geometry: bounding box, centroid, 2D footprint, and transformation functions
- Binary functions that return a simple geometry: line segment representing the distance between centroids

Note that this set of implemented functions is just a small sample of all possible functions. Obvious further implementations would be topology relationship operations in 3D (in the category "binary functions that return a Boolean") according to the nine-intersection framework[41] or to the dimension-extended framework.[24] As explained in Section 6.4.3, the 3D polyhedron data type can be indexed with a 3D R-tree. This is very important if large tables with polyhedrons are used in a query based on one of the above topological relationship functions: the index enables the DBMS to avoid evaluating pairs of polyhedrons, that are not relevant. Functions that return a complex geometry such as tetrahedrization and skeletonization are not implemented yet, but are also interesting, because of their analogy with 2D triangulation and generalization. The functions are implemented in Java, which has the advantage that the functions are available outside Oracle as well (although implementation in PL/SQL would probably show better performance).

It is clear that functions in 3D require more complex algorithms than 2D functions. This also has a big influence on the computational complexity. To maintain good performance, the algorithms have been implemented as efficiently as possible. Spatial data sets can contain many objects, so a slightly more efficient algorithm already will yield noticeably better performance when querying all these objects.

The next example shows how to compute the area, volume, and perimeter (length of edges) of the objects in Figure 6.15. The figure shows a tetrahedron (1), a cube (2), a cube with a dent in one of the faces (3), a hollow cube (4), and a cube with hole that runs through the whole cube (5).

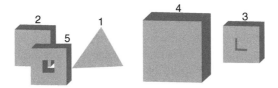

FIGURE 6.15 Set of five polyhedra used to show some 3D unary functions. Note that object 4 is hollow.

```
SELECT id, area3d(geom), volume(geom), perimeter(geom)
  FROM testobjects;

ID  AREA3D(GEOM)  VOLUME(GEOM)   PERIMETER(GEOM)
--  -----------   ------------   ---------------
1   22.9530689    5.5            22.0723224
2   54            27             36
3   58            26             48
4   204           98             96
5   64            24             56
```

6.5 CONCLUSIONS

This chapter showed that DBMSs are becoming increasingly mature in the maintenance of spatial objects.

Geometrical Primitive in DBMSs

Mainstream and popular noncommercial DBMSs offer support, maintenance, and some operations that allow spatial analysis of objects defined in geometrical primitives. However, the implementation of geometrical primitives is still not complete. Real 3D volumetric data types are lacking. A solution for 3D representation in the DBMS was designed as part of the research presented in this book as described in this chapter. The implemented geometrical primitive (polyhedron without spherical and cylindrical patches) showed that it is possible to support a true 3D primitive in the DBMS (including validation functions and geometrical functions in 3D) although the 3D primitive needs further development to be able to model more complex geometries.

Topological Structure in DBMSs

Support for topological structure management is a relatively new issue in DBMSs (recently available in Laser-Scan Radius Topology and Oracle Spatial 10g). The lack of topological structure has led to a variety of topological structures in front end applications missing uniformity as a result. Managing topology in front ends undermines data consistency and integrity at DBMS level. In addition, the conversion between the stored geometry in DBMS and the application dependent topological layer has its influence on performance. An issue that needs attention is the type of topology and dimensionality of the models. Current efforts aim toward providing 2D topological structure (planar partition, linear networks) that most probably will restrict the topological operators to 2D. Maintenance of several different types of topological structures appears unavoidable.

In this chapter two user-defined implementations of a topological structure in a DBMS were described: one in 2D (winged edge) and one in 3D (Simplified Spatial Model). The experiments with these structures, including realization functions, show potential. However, at the moment user-defined solutions of topological structure focus on organization of data. The consistency checks and updates still need to be performed outside the DBMS. In addition, performance of metrical operations on the topological structure might become critical in case of large data sets. Further, two commercial solutions of a 2D topological structure in the DBMS were described in this chapter (i.e., Radius Topology and Oracle Spatial 10g).

Topological Structure or Geometrical Primitive in DBMSs

Geometrical primitives are already supported by DBMS and, as first implemented, are considered a basic model. However, to improve data quality and data consistency topological structure offers better possibilities. As was illustrated in the example (Section 6.3.4), spatial queries relying only on topological references perform well in the topological structure compared to topological analyses on geometrical primitives. On the other hand, experiments with Radius Topology 1.0 with large-scale spatial data[105] showed that storage requirements and performance of the plain geometry approach are still superior in many cases. At the moment topological structure is therefore mainly appropriate for representing relationship operations and for checking the quality of data during updates.

To make geometrical analyses on topologically structured data possible, a function is needed to derive the geometry from the topology. On the other hand, for full support of topology in DBMSs, a function to derive the topology from geometry is also needed. Many spatial operations give geometries as a result, and it should be possible to convert these geometries into topological layers in order to get a topologically structured data set (also in case of topological results with redundant information such a function is needed). Consequently, geometry-based operators will always be necessary to build the topology: to find all the topological relationships in a new layer, functions based on geometrical primitives are required.

Spatial Functions in DBMS or in Front Ends

DBMS plays an important role in the new-generation GIS architecture. Mainstream DBMSs have implemented support for spatial data types and they are still improving support for geometrical primitives and topological structures. Does it mean that a DBMS will and should include all spatial analyses, including complex spatial analyses that have been optimized in GISs during recent decades? Does it mean that traditional GIS software (or extended with attribute maintenance CAD software) has to convert to a tool for import, visualization, editing, and exploration of spatial data?

Many spatial functionalities are, and probably will be, available only at the front end and not at DBMS level (e.g., spatial analyses that are specific for certain domains and applications, tools for inserting new data, interaction tools for starting spatial analyses, visualization tools). In addition, too many operations performed at a DBMS level may lead to overloading of the server and affecting the performance of the DBMS. On the other hand, too few operations provided by DBMS will result in development of many functionalities by the front end, i.e., duplication of development efforts and resources. The question now is: Which spatial operations should DBMSs take over?

The balance depends very much on the scope and constraints of spatial analysis: What is spatial analysis and what is spatial analysis in a DBMS context. DBMSs are essential in applications in which large amounts of large-scale geo-data in vector format need to be maintained and managed, such as cadastral data, or spatial data used in municipalities. In principle, generic spatial functionalities that are not specific to a certain application belong in the DBMS and not in front-end applications. Examples are the spatial functions that examine the topological relationships between spatial objects. Arguments for this are logical consistency of the data, better performance, and better maintenance of the quality of the data. Unnecessary transport and conversions of data between DBMSs and GIS front ends prone to errors can be avoided. In contrast to the group of selection operations, specialization and navigation in the spatial domain can be very complex and time-consuming. If they are performed at a DBMS level on the server, the performance can decrease drastically. Furthermore, such complex operations may not be needed for all kind of applications. Therefore, complex operations falling in the group of specialization and navigation operations can be considered to be left for implementation by the front end.

A relevant question in this whole discussion is whether spatial functionalities implemented in the DBMS will replace spatial functions that were originally built in GISs. GIS has become an important instrument in work processes of companies and governmental offices. A lot of money and effort has been invested by GIS vendors for selling their software and for giving support and by organizations to develop specific GIS applications. The future will prove if GIS vendors are willing to give up spatial analyses (which always have been an important part of GISs) by which GISs will be converted into visualization/interaction tools (including editing) built on top of geo-DBMSs and if organizations will move from spatial analyses in GIS applications to spatial analyses in DBMSs.

COLOR FIGURE 1.1 Example of complex property situations. Business district La Defense in Paris; a road and a metro in the subsurface intersect buildings and plazas.

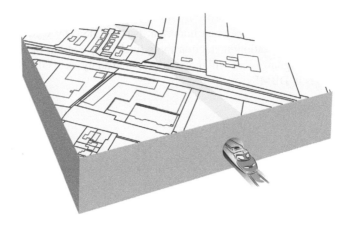

COLOR FIGURE 1.4 How to register 3D situations in a 2D cadastral registration.

COLOR FIGURE 3.1 Cellars below roads in Utrecht.

COLOR FIGURE 3.3 Building over a road.

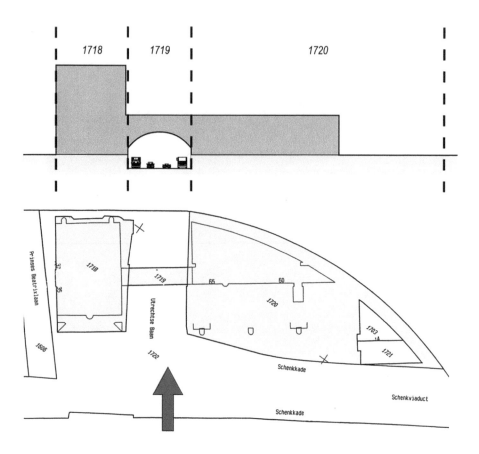

COLOR FIGURE 3.4 Cadastral map of the building in Figure 3.3. The arrow indicates the position of the camera.

COLOR FIGURE 3.5 The Hague Central Station, combination of a business center, a railway station, and a bus/tram station.

COLOR FIGURE 3.9 Rijswijk railway station (left) and kiosk (right).

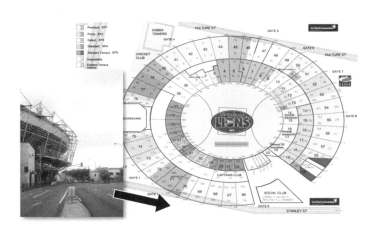

COLOR FIGURE 3.13 Overview of Gabba Stadium overhanging Stanley Street in the south and Vulture Street in the north, Brisbane, Australia.

COLOR FIGURE 3.17 Pan Am building above Grand Central terminal.

a

b

COLOR FIGURE 3.18 6th Street, double skyways connect 2nd and 4th floor. (From http://www.cgstock.com.)

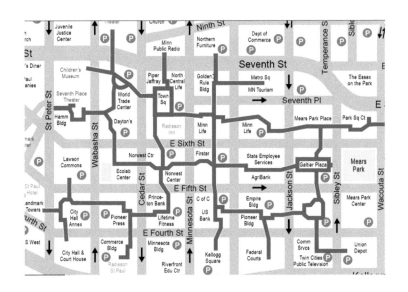

COLOR FIGURE 3.19 Map of the St. Paul skywalk system.

COLOR FIGURE 3.20 Registered Land Survey 554.

COLOR FIGURE 3.21 Arial photograph of some buildings connected by the skywalk system (Figure 3.20 shows the RLS 554, documenting the second level of the lower left tower of twin towers).

COLOR FIGURE 3.22 Base map fragment (showing parcels and structures) of the example used in this section.

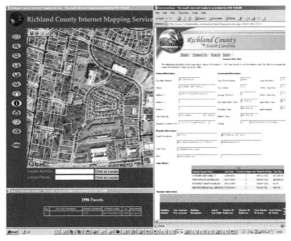

COLOR FIGURE 3.25 The Richland County Internet mapping service for locating parcels and obtaining related online taxation information (including sale history of the property); note the tall building, a little above the center of the map on the aerial photograph (Middleborough).

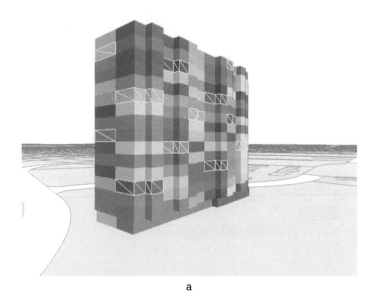

a

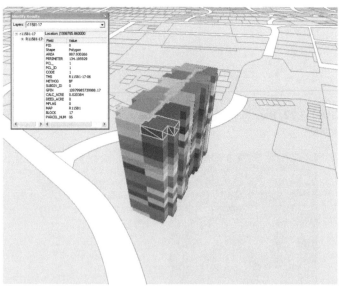

b

COLOR FIGURE 3.26 The 3D interface (ArcScene) of the 3D representation of the condominiums. (a) Selection of a number of units; (b) selection of a single unit and the display of a number of relevant attributes.

COLOR FIGURE 4.2 To avoid damage to cables, digging first by hand is necessary (*De Volkskrant* Newspaper, July 2000).

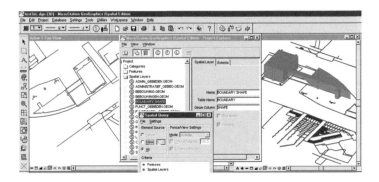

COLOR FIGURE 7.1 Querying spatial objects organized in Oracle Spatial, using MS GG.

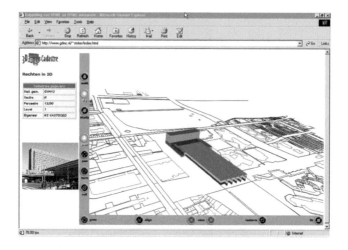

COLOR FIGURE 7.6 Screen dump of first prototype (ASP and VRML).

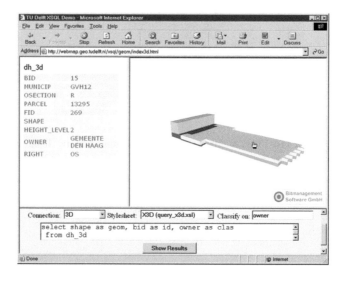

COLOR FIGURE 7.7 Screen dump of second prototype (based on XSQL and X3D).

COLOR FIGURE 8.1 Data sets used in this case study. AHN (above).

a

b

COLOR FIGURE 8.11 Conforming TIN in which point heights and 2D planar partition of parcels are integrated, (a) before and (b) after filtering.

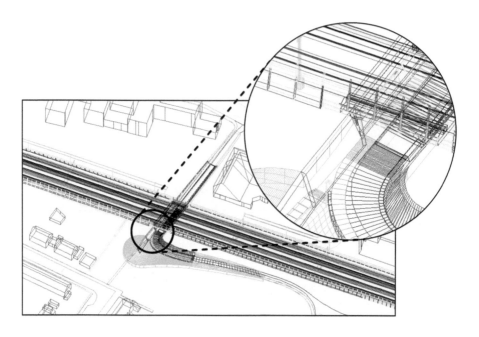

COLOR FIGURE 10.2 The CAD model designed for a cycle tunnel in Houten, the Netherlands. (Courtesy of Holland Railconsult.)

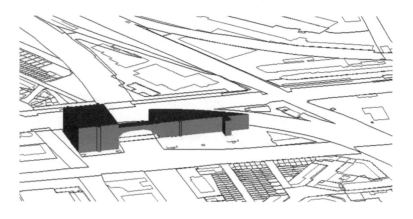

COLOR FIGURE 11.2 Registration of a 3D physical object.

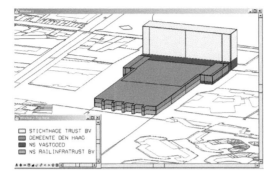

COLOR FIGURE 11.3 Cadastral representation for The Hague Central Station.

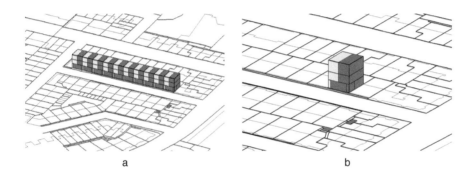

a   b

COLOR FIGURE 11.5 Apartments as right-volumes. The horizontal lines between the first and second floor are for visualization purposes. (a) All apartments in the street; (b) the apartment complex of the case study, which is the second complex from right.

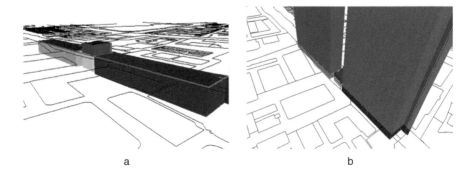

a   b

COLOR FIGURE 11.6 The first two alternatives for unrestricted right of superficies (third option is not displayed). Note that the lowest right-volumes (for the railway tunnel) are located below the surface (below the $z = 0$ plane).

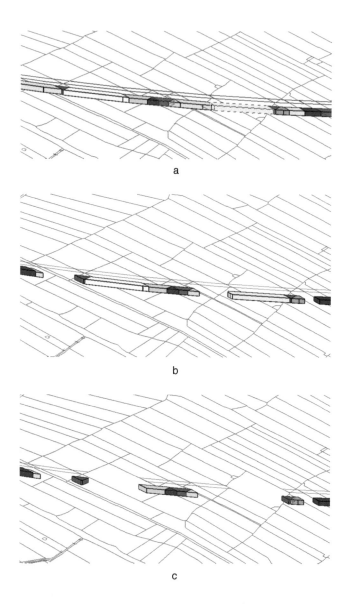

COLOR FIGURE 11.7 Three possible recordings of right-volumes in the case of a railway
tunnel. (a) All the parcels are encumbered by right of superficies; new parcels are created for
all intersecting parcels. (b) As a, but now three newly created parcels are in full ownership.
(c) Three newly created parcels are in full ownership; two parcels that are not subdivided are
in full ownership. All the other new parcels are encumbered by a right of superficies.

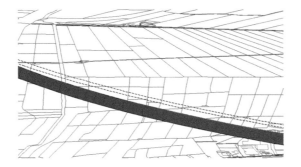

COLOR FIGURE 11.8 Registration of the 3D physical object in the case of the HSL tunnel. The dashed line is the projection of the tunnel on the surface. Note that the parcels are not divided into smaller parcels.

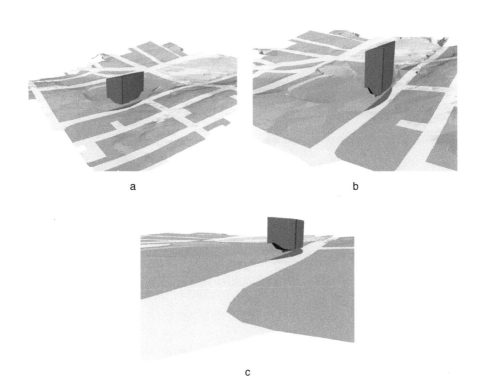

a

b

c

COLOR FIGURE 11.11 Visualization of 3D geometries of volumetric parcels together with the 2.5D cadastral base map, seen from different viewpoints.

7 3D GIS and Accessing a 3D Geo-DBMS with Front Ends

In Chapter 6 the possibilities of maintaining and analyzing 2D and 3D spatial objects in a geo-DBMS were described. A geo-DBMS is part of the new-generation GIS architecture, as seen in Section 5.6. For this research on 3D cadastre, 3D aspects of GIS other than geo-DBMS functionality are also important. The implementation of a 3D cadastre addresses the issues of inserting, maintaining, querying, editing, and visualizing 3D geo-objects in general. These are core topics of 3D GIS. Therefore, the state of the art of 3D GIS is described in Section 7.1.

Once 3D spatial objects have been stored in a DBMS, these objects can be optimally maintained, together with 2D spatial objects and non-spatial objects in an integrated DBMS environment. In object relational DBMSs, geo-information can be accessed only with SQL commands, of which the output is a sequence of characters and numbers (or binary output). To query and edit the spatial objects in a visual environment, spatial information maintained in DBMSs should be accessible in front ends having visualization utilities.

The aim of the second part of this chapter is to show the state of the art of the technology to access spatial objects, and 3D spatial objects in particular, which are stored in a DBMS with different front ends.

In this chapter three front ends are examined to access 2D and 3D spatial information organized in the geometrical model of Oracle: a CAD-oriented system (Section 7.2), a GIS (Section 7.3), and a self-developed Web-based front end (Section 7.4). The chapter ends with concluding remarks.

7.1 3D GIS*

2D GIS analysis has shown its limitations in certain applications, as discussed in Section 4.4. Therefore, the need for 3D geo-information is rapidly increasing.

Developments in the area of 3D GIS are motivated by a growing need for 3D information from one side and new technologies on the other side, e.g., improving techniques of 3D data collection, of 3D object reconstruction but also of computer hardware. Processors, memory, and disk space devices have become more efficient in processing large data sets (especially graphical cards also used by the games industry). Furthermore, elaborate tools to display and interact with 3D data are evolving.

---

* This section is based on Reference 195.

163

This section provides an extensive overview on the status of 3D GIS by considering the core topics of 3D GIS:

- Organization of 3D data (Section 7.1.1)
- 3D data collection and object reconstruction (Section 7.1.2)
- Visualization and navigation in 3D environments (Section 7.1.3)
- 3D analyzing and 3D editing (see Section 7.1.4)

7.1.1 Organization of 3D Data

3D Representations

Several approaches may appear very appropriate for 3D GIS models: Constructive Solid Geometry (CSG), voxel representation (regular space subdivision), irregular space subdivision (Tetrahedron Networks), and boundary representation. All approaches show advantages and disadvantages considering different criteria and depending on the specific application. The advantage of CSG is that it is very appropriate for computer-aided manufacturing: a brick with a hole drilled through is represented as "just that." The disadvantage for real-world modeling is that the objects and their relationships might become very complex. Voxels are appropriate in modeling continuous phenomena such as geology, soil, etc. Voxels are regular in modeling: the basic unit of the model is the same. A disadvantage of voxels is that high-resolution data require a large volume of computer space. Another disadvantage is that surface is not regular by nature: it is always somehow "rough." The tetrahedron object is well defined, because the three points of each triangle always lie in the same plane.[21,210] A disadvantage is that it could take many tetrahedra to construct one factual object. The main advantage of boundary representation is that it is optimal for representing real-world objects. The boundary of real-world objects can be observed, measured, and surveyed from properties that are visible (i.e., "boundaries"). Furthermore, most of the rendering engines are based on boundary representations (i.e., triangles). Unfortunately, boundary representations are not unique, and constraints (rules for modeling) may become very complex to implement (e.g., how to determine neighbors in 3D, how to ensure planarity of faces in 3D, etc.).

Logical Models of 3D Data

DBMS vendors still have not taken the step to implement 3D data types in their geometrical model, as seen in Chapter 6. Reasons for this may be that the OpenGIS Consortium (OGC) is still working on extension of the Simple Feature Specification to support 3D features and consensus on a 3D topological structure has not yet been achieved. Another limiting factor is the relatively low (but growing) market demand for 3D support in DBMS. The new-generation GIS architecture for 3D has not yet been adopted by GIS users. The current trend is to develop specific *ad hoc* solutions when using 3D geo-information instead of building a database for maintaining spatial objects. User-defined implementations of 3D GIS models can be found elsewhere.[20,139,174,217]

At present, 3D implementations defined by ISO/TC 211 and OGC focus on boundary representation. However, CSG may appear appropriate for designed large-scale real-world objects (trees, traffic signs, building ornaments, statues) and voxel representation for continuous phenomena.

7.1.2 3D DATA COLLECTION AND OBJECT RECONSTRUCTION

3D GIS requires 3D representations of distinct objects. Traditionally, 2D GIS makes use of data collection techniques such as surveying and measurements of the real world, while creating 3D models used to be done separately from GIS, either using CAD software or photogrammetric methods and modeling software. This subsection describes if and how CAD designs and 3D object reconstruction techniques can be used for 3D GIS models.

GIS and CAD

In the late 1980s and early 1990s many publications were written on GIS vs. CAD and how GIS and CAD could be effectively combined.[32,71,102,120,180] The tendency of these papers is "how to use CAD systems for certain GIS tasks." The typical tasks range from geographic data entry to automated map production (including some cartographic aspects). This was motivated by the fact that two decades ago, CAD systems were more generally available on the commercial (and research) market than were GISs. However, one could hardly observe the desire for true integration of the different data models and functionalities offered by CAD and GIS. About a decade ago the attention indeed shifted to the integration of CAD and GIS functionality driven by application domains such as urban and landscape architecture and planning.[75,116,178,187,197] The presented solutions are often very *ad hoc* in nature (capturing and transferring simple 3D models between the different systems) or require custom-made software solutions. Often these papers end with the remark that the off-the-shelf CAD/GIS functionality still needs to be integrated for better support of their applications. However, seldom is a clue given on how this could be achieved or what could be the fundamental issue causing the integration problems. More recent sources seem to be commercial and/or development notices,[108] where the emphasis is on providing data exchange mechanisms either through shared files, translators, or inter-APIs, but until now, there has been little care for the fundamental issues that need to be addressed, such as integrated geometrical data structures concerning 3D and topological support (see Reference 97 for an overview), harmonized semantics of the concepts used, and integrated data management (in contrast to independent and inconsistent information islands with data conversions and transfer).[135] The issue of a fundamental integration of GIS and CAD is further discussed in Section 10.2.4.

Object Reconstruction

In the last several years a lot of research has been conducted toward automation of 3D object reconstruction (especially human-made objects). There are a variety of approaches based on different data sources and aiming at different resolution and accuracy. For constructing 3D models, four general approaches can be considered:

1. **Bottom-up:** Using footprints from existing 2D maps and extruding the footprints with a given height using laserscan data, surveying, GPS, or photogrammetric data. The problem with this approach is that the details of roofs cannot be modeled. Since one value is used for every footprint, the buildings appear as blocks in the model. The approach, however, is very fast and sufficient for applications that do not need high accuracy (do not need roofs) and many details.

2. **Top-down:** Using the roof obtained from aerial stereophotographs, airborne laserscan data, and some height information from the ground (one or more height points near the buildings, DTMs). These approaches emphasize the modeling of roofs.[12,63] Obviously, the accuracy of the obtained 3D models is dependent on the resolution of the source data.

3. **Detailed reconstructing of all details:** The most common approach is to fit predefined shapes (building primitives) to the 3D point clouds obtained from laserscan data[212] or 3D edges extracted from aerial photographs.[54,106] The advantage of this approach is the full automation and the major disadvantage is that it is very time-consuming since the algorithms used are very complex.

4. **Combination of all approaches:** For example, laserscan data and topographic data,[74] aerial photographs, and maps,[65,198] etc. This approach contains some risks because many data sources are used and combined, all with different scale and quality. Using only a few data sources will introduce fewer inconsistencies to be solved during processing.

There is no universal automatic 3D data reconstruction approach. At the moment, the manual approach is still needed to reconstruct large-scale detailed 3D models, which is a bottleneck for modeling urban areas in 3D. More research is needed to make the process of 3D reconstruction automatic or semiautomatic. A tighter connection between 3D object reconstruction and GIS will support development in 3D GIS.

Important for 3D object reconstruction is to derive terrain elevation itself (Digital Terrain Models and Digital Elevation Models). Laser altimetry can be used to automatically derive terrain and elevation models with high accuracy, e.g., the AHN (*Actueel Hoogtebestand Nederland*), which is a DTM covering the whole area of the Netherlands with a density of one point per 16 square meters and in forest areas a density of 36 square meters (see Chapter 8).

7.1.3   VISUALIZATION AND NAVIGATION IN 3D ENVIRONMENTS

3D models usually deal with large data sets, requiring efficient hardware and software for visualization. Several techniques are being developed to improve efficiency of navigating through a 3D model, such as different levels of detail,[90,155] low-resolution graphics, and imposters (image of object instead of geometry of object).[156] All these techniques aim at visualizing high detail when objects are close by and low detail when objects are farther away. Different representations of objects can be either stored in the DBMS or created on-the-fly. The main problems of storing multirepresentations are fitting highly detailed data to data that are represented at a low level of detail and the redundant storage of representations.

A specific problem that comes with visualizing 3D geo-data compared to 2D geo-data is readability of the data (approaching realism). To make a view realistic one can add illumination, shade, fog, textures, shadow, and material to the geometry (apart from traditional characteristics such as color). Apart from visualizing 3D models, interacting in 3D environments (exploring 3D models) also requires specific techniques. These issues touch the fundamental difference between the Digital Landscape Model (DLM) and the Digital Cartographic Model (DCM), which is a well-known issue in traditional map production. The stored data set of a specific study area is called the Digital Landscape Model. This model has to be converted into a Digital Cartographic Model to make the (spatial) data set suitable for communication to other persons. The DCM consists of series of instructions to the plotter, printer, screen, etc. to produce dots, dashes, or patches, in different sizes, colors, and textures to make the content of the data set readable.[92] In 3D this means that apart from spatial and non-spatial information of spatial objects (DLM), characteristics for visualizing and interacting with the objects (DCM) also need to be maintained. Although visualizing in 2D also requires organizing cartographic aspects apart from the content of the data, the DCM aspects to be considered in 3D are much more, such as physical properties of objects (texture and material), behavior (e.g., on-click-open), and different levels of detail representations. This requires several new elements be organized in the database compared to 2D data.

Virtual Reality and Augmented Reality

Virtual Reality (VR) and Augmented Reality (AR) are supporting techniques for improving visualizations of and interaction with 3D geo-data,[209] e.g., putting textures on objects and facilitating navigation through the 3D environment.[62] VR is a realistic representation of data (2D, 2.5D, and 3D), which means that details and physical properties are represented highly realistically even together with sounds and behaviors of the objects. Manipulation and interaction in the views can take place by mouse click, animations, navigation, and exploration. In AR a user explores and navigates in the real world augmented by computer-generated data. Several research efforts have already addressed the issue to link 3D GIS with VR.[209]

All kinds of devices are today available to support visualization in VR/AR environments,[230] such as elaborate 3D display (Head Mounted Device, workbench, panorama, CAVE, Cockpit), wire and wireless devices for positioning (gyros, accelerators, GPS, GSM, WLAN), sensor devices to track the movements of the user (Power Glove, indoor outdoor tracking systems), and various acceleration hardware units.

3D GIS and the Internet

3D Web visualization is also progressing. The research on spatial querying and 3D visualization using VRML (Virtual Reality Modeling Language), X3D (eXtensible 3D), and/or GML (Geographic Markup Language) has resulted in several prototype systems.[14,31,35,93,100,229] It should be noted that research on spatial querying and visualization of geo-objects organized in a DBMS using Web-based techniques is not yet available (see Section 7.4).

7.1.4 3D ANALYSES AND 3D EDITING

GIS software tools have also made a significant movement toward 3D GIS. A survey on mainstream GIS software has been presented[232] including: ArcScene,[46] Imagine VirtualGIS,[45] PAMAP GIS Topographer,[158] and GeoMedia Terrain.[81] In Reference 232 it is concluded that major progress in 3D GIS has been made on improving 3D visualization and animation. However, 3D functionality is still lacking, such as generating and editing 3D geo-objects, 3D structuring, 3D manipulation, and 3D analyses (3D overlay, 3D buffering, 3D shortest route on polyhedral or TIN surface). An example of the implementation of a specific 3D analysis, 3D buffering, is described elsewhere.[214] Concerning editing of 3D geo-data in the new-generation GIS architecture, CAD and GIS front ends should be able to read 3D output and write it to both topological and geometrical structures in DBMSs in which the front end has to be able to preserve the topology of the 3D object. This topic has not yet been addressed.

7.2 ACCESSING A GEO-DBMS WITH A CAD FRONT END

This section describes how to access spatial objects that are stored in Oracle Spatial using a CAD-oriented front end. Bentley's MicroStation GeoGraphics[11] is an extension of the CAD software MicroStation containing functions specific for geo-information and for connection to Oracle Spatial. The organization of data within MicroStation GeoGraphics (MS GG) is defined in a project hierarchical structure. Project represents the data for the entire study area. The second level is the category, which groups features with a similar theme (e.g., buildings, rivers). One project can have many categories but a category may belong to only one project. Feature is at the third level and represents one or more spatial objects with the same thematic attributes (e.g., the bank building, the school building). A category may have many features but a feature may belong to only one category. Feature is the basic structural unit in MS GG.

With MS GG there are two tools delivered to query and post data from Oracle Spatial: a MS GG tool and a Java applet, "Spatial Viewer." Here we focus on the MS GG tool. The Spatial Viewer is described at the end of Section 7.2.3.

Visualization of spatial data from Oracle Spatial 9i using the MS GG tool is relatively simple and straightforward. The user has to create a project and connect to Oracle Spatial. MS GG checks the Oracle meta-data table for the name of the table or tables and corresponding columns that contain spatial data. These are supplied to the user for display. In this case, the geometries will be visualized, but will not be available for querying and editing. More steps have to be taken to distinguish between different spatial objects stored in Oracle Spatial (e.g., "identify" or "query") and also to edit objects maintained in Oracle Spatial. In general, each spatial object in the Oracle database has to be assigned to a predefined feature in MS GG, but depending on the original source of the data (Oracle or MS GG), different steps have to be followed. We completed a number of case studies with MS GG (version 8.1.0.7) following the two different approaches of representing 3D objects (as a set of 3D polygons and as a multipolygon defined in 3D), having the data initially organized either in Oracle (user-defined tables) or in MicroStation (graphics in a design file [DGN]) (see also Reference 232).

7.2.1 GEOMETRICAL DATA INITIALLY ORGANIZED IN ORACLE SPATIAL 9I

The required steps to assess and query the objects that are originally organized in Oracle Spatial 9i are as follows:[232]

- Create semantics, i.e., project (by specifying the Oracle connection and the Oracle database), categories, and features. This step is enough for only visualizing spatial layers from Oracle Spatial.
- Register the spatial table, stored in Oracle Spatial as MS GG layer by creating a new MS GG layer and referencing this new layer to the corresponding Oracle table and column that contain the geometries.
- Link features (the code of the feature) to the corresponding spatial objects (id of spatial object). Running an appropriate script within Oracle is one of the easiest ways to complete this operation in the case of many objects. Both the features and spatial objects are maintained in the DBMS.

To illustrate the steps to visualize and query spatial objects maintained in Oracle Spatial with the MS GG tool, we use a table with buildings, represented as a set of faces (3D polygons). The data set with buildings is organized in a relational table (BODY) that originally consisted of only three columns (BODY_ID, FACE_ID, and SHAPE). The column SHAPE contains the geometries of the objects as mdsys.sdo_geometry type, i.e., the polygons. The links between FACE_ID and SHAPE is 1:1 and the link between FACE_ID and BODY_ID is m:1.

1. Creating project, category, and features   Bearing in mind the basic conceptual structure of MS GG we created a project (cadastre), a category (buildings), and several features (build1, build2, build3, and build4) in MS GG. The four buildings (polyhedrons) are instances of the type (category) buildings. This operation resulted in 12 new relational tables in Oracle Spatial. The names of the tables created by MS GG and us (in bold) are: **BODY**, CATEGORY, FEATURE, MAPS, MSCATALOG, UGCATEGORY, UGCOMMAND, UGFEATURE, UGJOIN_CAT, UGLAYER, UGMAP, UGMAPINDEX, and UGTABLE_CAT. UG refers to MicroStation GeoGraphics. Among all these tables, MSCATALOG and FEATURE are of practical interest. The first table maintains reference to all the tables used in the project. The second contains information (names, codes, unique identifiers, etc.) related to all the features created by the user. The spatial data (BODY table in our case) become visible after this step in the Query tool in MS GG; i.e., it is possible to query and display the entire layer. To be able to post data in the database and to query individual objects, the table has to be linked to a spatial layer and the objects to features.

TABLE 7.1

Columns in the Spatial Data Table after Creating the Spatial Layer in MS GG

Column-name	Type
BODY_ID	NUMBER(10)
FACE_ID	NUMBER(10)
SHAPE	MDSYS.SDO_GEOMETRY
BODY_DFLAG	NUMBER(10)
BODY_UDL	RAW(200)
BODY_LOCK	NUMBER(10)
BODY_FID	FCODE_LIST
BODY_CREATED	DATE
BODY_REVD	DATE
BODY_RETIRED	DATE
BODY_XML	XMLTYPE (or BLOB or VARCHAR2)
BODY_TXT	VARCHAR2(1024)
MSLINK	NUMBER(10)

2. Creating spatial layer   Table 7.1, with the geometry (i.e., BODY) with geometry column SHAPE, has to be referred to as a spatial layer in MS GG. Further, all the features that are to be associated with objects in this layer need to be assigned to the layer (again, in MS GG). This operation extended our table BODY with nine new columns, all starting with BODY_. We also added a mslink column (as primary key), since MS GG requires a column, named mslink with unique values to be able to query attributes:

3. Linking features to spatial objects   First, one should make sure that the table with the spatial data (i.e., BODY) is declared in the table MSCATALOG. The project tables CATALOG and FEATURE are automatically registered in the CATALOG table by MS GG under entity numbers 1 and 2. Second, the column BODY_FID (in the BODY table) has to be populated. The FID (feature id) column contains all database linkages that are related to elements in a DGN-file. The column is an object of type Fcode_list, which is an array of Fcode_item objects. *Fcode_item (p1, p2, p3, p4)* provides the link between a feature (from FEATURE) and a particular spatial object (from BODY). The first of (in this case) two *fcode_item*s is related to the feature as it is described in the FEATURE table and the second to the spatial object from the BODY table. Parameter *p1* is the number of the table in the MSCATALOG (as it appears under the column ENTITY). Parameter *p2* is the number of the feature in the FEATURE table (given in the mslink column) in the first *fcode_item* and the identifier of the object (i.e., FACE_ID) in the second *fcode_item*. The third parameter gives indication of whether the description is for feature (informational object, i.e., 1) or spatial (non-informational) object (i.e., 0). Cases in which more than one feature refers to the same object are resolved by introducing a new *fcode_item* in the *fcode_list* description. The operation to fill the FID column can be performed either in MS GG or Oracle Spatial. Last, all the values in the column BODY_LOCK (giving information about the owner of the data) have to be set to zero (i.e., belong to the owner of the table). A PL/SQL script completes these two operations within Oracle:

```
... FOR i in n..m LOOP
        UPDATE body SET body_fid =
                        fcode_list (fcode_item (2,4,1,0),
                        fcode_item (5,i,0,0))
          WHERE face_id=i;
        UPDATE body SET body_lock = 0 WHERE body_id=i;
  END LOOP; ...
```

Note that in this case, one feature (i.e., number 4) is assigned to several objects, e.g., to attach the set of polygons to one 3D object.

7.2.2 GEOMETRICAL DATA INITIALLY ORGANIZED IN MICROSTATION DESIGN FILES

3D objects that are initially stored in MicroStation dgn (design) files can be imported in Oracle Spatial directly by the three following major steps: (1) create project, features, categories, and spatial layers, (2) select the entire geometry (polygons or groups of polygons) per spatial object in MS GG and attach a feature to it, (3) post the spatial objects to the database.

In both approaches, after completing all the required steps, it was possible to query, visualize, edit, and post the spatial objects as they are defined in Oracle Spatial (Figure 7.1). Evidently, only spatial objects can be posted that are described with geometrical primitives that are supported in Oracle Spatial. A query can be specified either per layer or per feature and can be performed on the basis of the semantic characteristics of the objects as defined in MS GG. For example, query on feature "buildings" will result in visualizing all the buildings. Apparently, such a possibility brings advantages for editing and updating large 3D models. Instead of working with the entire model, the user can query and work with only one object. Thus, rendering thousands of polygons can be easily avoided.

MS GG was able to visualize and edit both 3D geometrical representations (i.e., set of 3D polygons and 3D multipolygons) by following the steps described above. It should be noted that MS GG interprets the two representations in a different manner. In the first case the building is visually one object, but in the Oracle Spatial table, it is a set of individual polygons (Figure 7.2). The entire building can be selected only by placing a fence around all the polygons. In the second case, the building is one "group"; i.e., a single click of the mouse will highlight the entire building. To edit the object, however, the group has to be divided ("dropped" in MS GG terms) into individual polygons; i.e., the 3D object cannot be edited as a whole. To send the changes back to the database, grouping of the objects will be required again. Otherwise, the object will be considered a set of several new polygons. It would be more efficient and less sensitive to errors to be able to edit the 3D object as it is defined, without dropping the element into 3D individual elements.

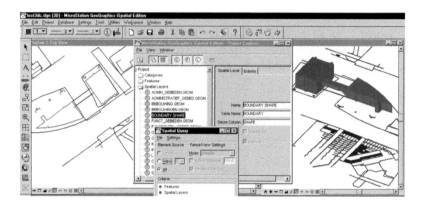

FIGURE 7.1 (Color figure follows page 176.) Querying spatial objects organized in Oracle Spatial, using MS GG.

7.2.3 SPATIAL VIEWER

The Spatial Viewer is an example to show the possibilities of the MS GG API, in this case: an implementation of how to handle spatial information without a GG project. The Spatial Viewer is a Java applet and is delivered together with MS GG. The Spatial Viewer is especially meant to show the possibilities for implementing one's own data model.

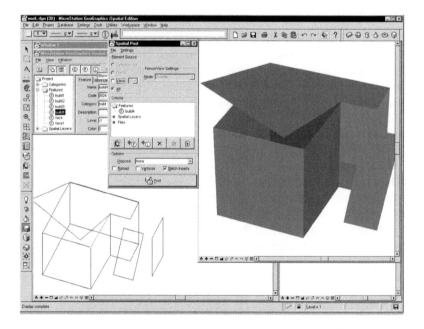

FIGURE 7.2 Editing and posting a 3D object as set of polygons using MS GG.

Using the Spatial Viewer one can visualize, query, and post (= update) elements. Also here, an mslink column containing unique numbers needs to be added and populated with unique values for accessing attribute data. The table name needs to be added in the MSCATALOG table to be able to post data. The Spatial Viewer reads the Oracle meta-data table for available tables with geometries. In our example the relationship between mslink and face_id is 1:1. In the case one would like to use another column as the key (e.g., body_id for a reference to the whole 3D object), this can be achieved by using the available API. Using the Spatial Viewer, a MS GG project is not required; only the table MSCATALOG is needed and therefore the Spatial Viewer requires less customization and less work for querying and posting data from Oracle. In addition, the functionality can be adjusted to meet the user requirements. The main disadvantage of the Spatial Viewer is that it is not directly available in the menu of the MicroStation environment.

7.3 ACCESSING A GEO-DBMS WITH A GIS FRONT END

ESRI software (ArcMap for 2D and ArcScene for 3D, both part of the complete package ArcGIS),[46] is able to access data that is stored as a sdo_geometry type in Oracle with ArcSDE. ArcSDE is middleware that facilitates managing spatial data in a DBMS (IBM DB2, IBM Informix, Microsoft SQL Server, and Oracle). Originally ArcSDE was developed for the SDE binary format, which is a format for spatial data types in the DBMS (stored as BLOBs) developed by ESRI. Since spatial data types have become available in DBMSs, ArcSDE now also supports spatial data types. We did experiments to see if and how 2D and 3D geo-objects stored in Oracle Spatial can be accessed with ArcGIS version 8.3.

There are two methods to access data stored in Oracle Spatial via ArcSDE:

1. Using SDE client/server software
2. Using "direct connect," which does not use the SDE server, but only the SDE client software, which is part of ArcGIS; the required SDE server functionality is included in the client software

For the user who visualizes the data, both connections work similarly. The difference is the way the connection is defined and how the connection works behind the GUI. The "direct connect" makes direct connection to the DBMS without using the ArcSDE application server. On the other hand, for the "direct connect" one needs Oracle client software on the client platform (PC), which is not needed for the ArcSDE connection. According to the manual, the "direct connect" is easier to install and maintain. However, for this connection one still needs the tables of the user "sde" in Oracle (the ArcSDE system tables) in combination with the Oracle Spatial meta-data tables. An advantage of the "direct connect" is that one does not need an ArcSDE licence to visualize data, in contrast with the SDE client/server connection. However, also in the "direct connect" case one needs a license when one wants to edit the data.

There are four steps to visualize data organized in Oracle in ArcMap and ArcScene:

1. Insert meta-data in the Oracle Spatial metadata table
2. Register the table that contains the geometry in the SDE system tables
3. Define the DBMS connection in ArcCatalog (the ArcGIS program for management of GIS-layers)
4. Obtain the data in ArcMap or ArcScene

Step 1: Insert meta-data in the Oracle Spatial meta-data table The insertion of meta-data in the Oracle meta-data table was shown in Section 6.1.

Step 2: Registering the table containing geometries For both connections the table containing geometries needs to be registered as sde layer. The registration of a table "test2d" containing line features in a geometry column "shape" and a primary key in the column "ID" is registered as follows:

```
sdelayer -o register -l test2d,shape -k SDOGEOMETRY -e 1
-u stoter -p password@database_name -i sde:oracle9i -c id -C USER
```

For 3D information, the element type should be polygons in 3D. Also multipolygons need to be supported. This is handled by the -e a3+ option (instead of -e l); a for area, 3 for 3D, and + for multipolygons. Furthermore a keyword is needed that is available in the sde.dbtune table to describe the dimension and tolerance of the spatial layer, for example, -k TEST3D (instead of -k SDO_GEOMETRY, which is the default). The registration of an sde-layer also works without an ArcSDE license and can be performed on both the client and the server. The registration edits only the sde tables stored in Oracle on the server. In these tables ArcSDE maintains the information of the geometry tables (dimension, spatial data types, geometry column). When the layer is "unregistered" (with the sdelayer and the sdetable command) sometimes the Oracle meta-data table is updated (i.e., entity for the table with geometry is deleted). This is not the way it should be since ArcSDE should edit only tables belonging to ArcSDE, and not tables that also exist without ArcSDE. The influence of a spatial index is also not clear. Without a spatial index a layer can be registered without any problems; however, layers without indexes cannot be visualized. It is also not clear why ArcSDE does not use only the Oracle Spatial meta-data table for the dimensions and the tolerances of the layers, but needs its own meta-data table.

Step 3: Define the DBMS connection in ArcCatalog In ArcCatalog both DBMS connections (ArcSDE connection and "direct connect") need to be added: "add spatial database connection." The connections are defined with the name of the machine on which the database is stored, the name of the database, the user, and the password of the user. The difference between the "direct connect" and the SDE client/server connection is the service that specifies the type of connection: "sde:oracle" for the "direct connect" and the port number for the SDE client/server connection (usually 5151).

Step 4: Obtain the data in ArcMap or ArcScene The spatial data stored in Oracle can now be visualized and queried in ArcMap or ArcScene by means of the defined database connections. With the "add data" option, all tables that are accessible to the user are checked, as well as tables that are not registered with sde and tables that do not contain geometry columns. This means that tables owned by other users who have granted select privileges are also checked, although ArcMap and ArcScene cannot do anything with the geometries in those tables as long they are not registered with ArcSDE. This is not optimal because for this query different sde tables need to be queried, which is time-consuming. A better option could be to just query the Oracle Spatial meta-data table of the specific user or only layers that are registered with ArcSDE. This is only one table and in this way the user can choose which (geometry) layers should be available for ArcGIS, although tables with no geometry are in this case not available. The available layers are shown in the "add data" dialog window. The icons preceding the table names show if the layer has been registered in ArcSDE and therefore has a geometry column that can be visualized in ArcScene or ArcMap (Figure 7.3). In addition, the data type (point, line, or polygon) is indicated by this icon (as it has been defined during registration).

Findings of Accessing Spatial Data Stored in Oracle with ArcGIS

2D data that are stored as an sdo_geometry type can be visualized, queried, and updated in ArcGIS with both connections, without modifying the geometry tables (e.g., adding columns), although in the sde registration process the Oracle meta-data table is sometimes updated. There are some minor problems. A table cannot contain more than one geometry column (for example, when a bounding box is stored together with the geometries in one table or when a label point is stored for polygons). A solution to this might be to create different views for one table. Also only one geometry type per column is possible (so lines and polygons cannot be stored in the same sdo_geometry column of a table). Views that contain only geometries of a

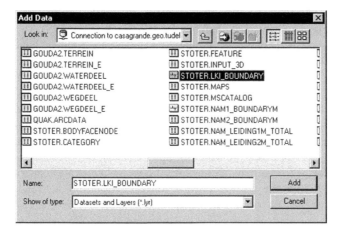

FIGURE 7.3 Dialog window in ArcMAP showing the tables stored in Oracle (note that the tables of users stoter, quak, and gouda2 are all shown).

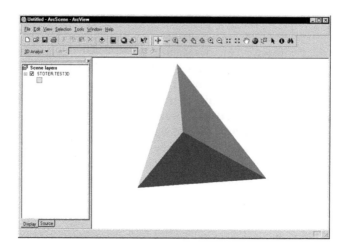

FIGURE 7.4 A tetrahedron defined by four separate polygons in Oracle, visualized in ArcScene.

specific "gtype" might be the solution. When there is a geometry stored in the table that is not valid according to Oracle, it might happen that none of the geometries in the table can be visualized. A last remark is that the primary key should be of type number(38) to avoid visualization problems. This solution—adding primary key of type number(38)—is not straightforward and for a less-experienced user difficult to find in the manual.

Concerning 3D data, experiences showed that the z-values that are stored in Oracle are recognized by ArcScene (which is the 3D module of ArcGIS). Therefore, it is possible to visualize 3D polygons (Figure 7.4). However, a problem in the visualization is that "vertical" polygons (polygons that are perpendicular with the x,y-plane) are not supported. This is because ArcSDE first performs a validation on the geometry in the DBMS. Vertical polygons are not valid according to Oracle Spatial 9i because in 2D their area equals zero. Consequently, vertical polygons cannot be accessed by ArcSDE. This is a major problem in urban modeling as vertical polygons are basic elements in models for buildings. It should be noted that ArcScene does support vertical polygons when they are stored in other formats, e.g., as multipatches or as ArcView 3D shape format. Therefore, when the spatial tables are converted into, for example, a 3D shape file, vertical polygons can be visualized in ArcScene. It also should be noted that 3D (or, actually, 2.5D) information can only be visualized and queried, and not edited with ArcGIS. It is possible to edit 2D spatial objects organized in a DBMS in ArcGIS; however, ESRI, as other GIS software, does not have a graphical user interface to edit 3D data (although it is possible to individually change z-coordinates per vertex in a special dialog).

7.4 ACCESSING A GEO-DBMS USING WEB TECHNOLOGY*

Outside the GIS domain, Web-based tools have been developed both to access data that is organized in a DBMS and to visualize 3D objects via the Internet. To explore whether these techniques can be combined to be used in the 3D cadastre prototypes, we built two prototypes using different Web technologies. The aim of the prototypes is to show the possibilities and constraints of accessing 3D geo-information organized in a geo-DBMS by means of Web-based open standards and open source software. The advantages of using Web-based technology are that its use is free (no licenses are needed) and that it is usually easier to use for end users. Therefore, a larger public has access to the data. Related research can be found elsewhere.[9,107,118,232]

For the prototypes we used VRML and X3D.

7.4.1 VRML AND X3D

In 1994, the Web3D Consortium launched VRML (Virtual Modeling Language), which became an international ISO standard in 1997 (ISO/IEC 14772–1:1997). The basis for the development of VRML was to have a simple exchange format for 3D information. This format is based on the most-used semantics of modern 3D computer graphics applications: hierarchical transformations, illumination models, viewpoints, geometry, fog, animation, material characteristics, and texture. VRML is a language to describe 3D models and to make them accessible on the Internet. Interaction and visualization are done by plug-ins for Web browsers (e.g., Cosmoplayer, Cortona[124]).

The development of VRML has stopped since the Web3D Consortium started to work on a XML version of VRML in order to integrate with other Web technologies and tools: X3D (eXtensible 3D). The specifications of X3D became available in May 2003. In our research we use both X3D and VRML to visualize 3D geo-information.

The data structure of a X3D document is very comparable to the data structure of a VRML file. As far as the underlying data model is concerned, X3D contains similar functionalities as VRML.[219] The difference lies in the notation (the syntax) used. While VRML is text, with accolades for structuring, X3D is coded in XML, with "tags" for structuring. This is a major advantage for on-the-fly retrieval, because of the ease of use of XML in Internet applications.

For the prototypes we examined how 3D spatial objects with their non-spatial information can be displayed using VRML/X3D. For modeling the 3D polyhedral object we used the VRML/X3D geometry type IndexedFaceSet. In this type, first all coordinates of the 3D object are listed, then the faces of the 3D object are defined by references to the coordinates. Every face definition ends with -1.

For example, the VRML code for a cube, defined with an IndexedFacedSet is as follows:

---

* Part of this section is based on References 196 and 215.

```
#VRML V2.0 utf8
Shape
{
        appearance Appearance {
                                                material Material { }
                                                }
        geometry IndexedFaceSet {
                coord Coordinate {
                        point [
                                        0 5 5,
                                        5 5 5,
                                        5 5 0,
                                        0 5 0,
                                        0 0 5,
                                        5 0 5,
                                        5 0 0,
                                        0 0 0
                                        ]
                                }
                coordIndex [
                        0, 1, 2, 3, -1,
                        7, 6, 5, 4, -1,
                        1, 0, 4, 5, -1,
                        1, 5, 6, 2, -1,
                        3, 2, 6, 7, -1,
                        0, 3, 7, 4, -1
                        ]
                        }
}
```

There are basically two methods to display non-spatial information linked to 3D objects using VRML/X3D:

1. Using VRML/X3D for both spatial and non-spatial information
2. Using VRML/X3D for spatial information in combination with HTML for non-spatial information

Both methods will be explained using VRML. Since X3D is the XML version of VRML and has the same functionalities, X3D supports similar functionality.

Using VRML for Both Spatial and Non-Spatial Information

In the first case, only one VRML file is created, which contains both spatial and non-spatial information for objects. Non-spatial information becomes visible as text in the VRML browser on "mouse-click" or "mouse-on" the object of interest. Figure 7.5 shows an example of viewing the attributes of an object (a building, represented as a cube). The text becomes visible when the user places the cursor on the building.

Because a VRML browser is not a complete GUI (the point-and-click operation is not a responsibility of the browser), the interaction has to be explicitly described in the VRML. This can be organized by two additional VRML nodes. First, a particular sensor (e.g., TouchSensor) has to be attached to the object (a Shape), which will monitor whether the cursor interacts with the object. Second, a billboard node has to be introduced to visualize the attributes in text format. In our example, we have

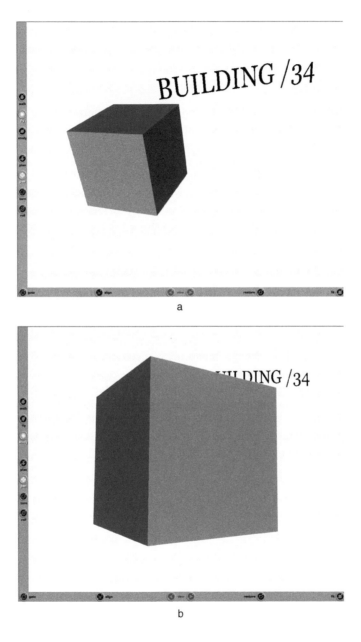

FIGURE 7.5 VRML containing spatial and non-spatial information. (a) Attribute information attached to a billboard; (b) text is not readable when billboard is behind object.

designed a new "proto" node. The node is basically a TouchSensor extended with a Javascript code (included in the VRML file), which controls the text that is visualized (in our case attribute information). The code provides a link between the attributes and the geometry. This link needs to be defined for every object using the specific code. The VRML code for the example of Figure 7.5 (see also Reference 229) is presented in Appendix A.

The major drawbacks of this approach are related to the size of the file and the visualization of the text.

Depending on the size of attribute information for visualization and the number of spatial objects with attributes, the VRML file can become between five and ten times larger compared to the VRML containing only the geometries. In case of large VRML models this can result in long time needed to create the VRML file (which is an important issue when creating VRML files on-the-fly), to transfer the file, and to display it on the screen. Because the attributes are visualized on a billboard (i.e., another 3D shape in VRML), they may be occluded by the object (Figure 7.5b) or even invisible (if a user observes the billboard shape from a direction perpendicular to the axis of rotation of the billboard).

VRML for Spatial Information and HTML for Non-Spatial Information

The problem of getting very large (and consequently slow-working) VRML files when including the attribute data in the VRML files can be overcome by using HTML files for non-spatial information. Again, the 3D object is represented as an IndexedFaceSet but now an anchor node is attached to every object. An anchor can be used to link an URL to an object. The anchor contains fields specifying the anchor. The complex object is defined within the anchor.

For every object a single HTML file is generated containing the attributes of the object. When one clicks on the object the corresponding URL (which indicates the specific HTML file) is opened in a frame defined in the parameter field of the anchor with the keyword "target=<frame>." For one object the VRML fragment looks as follows (with the attributes stored in t_1.htm):

```
Anchor {
parameter "target=leftframe"
description "Test 3D Cadastre"
url "domain-name/attributes/t1.htm"
children [
   Shape {
      appearance Appearance {
         --------------------
       }
    }
      geometry IndexedFaceSet {
        coord Coordinate {
          point [
            ----------------------------
          ]
```

```
        }
            coordIndex [

                ----------------------------

            ]
        }
    ]
}
```

Using Web-based technology to access spatial and non-spatial information stored in a DBMS, the information must be converted on-the-fly into a format accessible for Internet clients. Therefore, the VRML/X3D combined with HTML solution was selected for our prototypes because this solution promises better performance.

7.4.2 PROTOTYPES

The basic idea of the prototypes is to organize 3D geo-objects in a DBMS and to query them via an Internet browser. Geo-objects contain both spatial and non-spatial information. The spatial information can be visualized after conversion into (dynamic) VRML or X3D, and the non-spatial information can be presented in (dynamic) HTML pages. When using dynamic files, both the VRML/X3D files and the HTML files are generated on demand, which means they are not present on the Web server or on the DBMS server. On a client's request a connection is made to the DBMS and the spatial information of interest is selected from the DBMS and converted into X3D/VRML. A browser plug-in at the client side makes it possible to view the VRML or X3D output. VRML and X3D provide the possibility to start a script when a user clicks on an object. This functionality is used to retrieve the non-spatial information that is linked to a 3D geo-object. Via the VRML/X3D plug-in a URL request is sent to a Web server. The Web server receives and interprets the incoming information, sends the request to the DBMS, and sends an HTML with the required information back to the browser. For retrieving (and posting) the spatial and the non-spatial information from and to the DBMS a technique is needed to communicate between a client and a Web server and between a Web server and a DBMS server. For this communication, several techniques are available, such as ColdFusion and ASP (Active Server Pages). The choice of the technique used is dependent on the Web server used.

To show the possibilities to query 3D geo-objects via an Internet client, first a simple prototype was built, based on Microsoft technology (MS Access, Microsoft Internet Information Server). The aim of this first prototype was to study the functionalities of accessing non-spatial information stored in a DBMS linked to spatial objects using common Web technology. After good results of the first prototype, a second, more advanced prototype was built. An implementation of the same principles based on MySQL have been described[229] in combination with CGI (Common Gateway Interface) for the communication with the database and Apache as Web Server.

Prototype I: ASP, VRML and MS Access

In the first prototype (see also Reference 122) only the non-spatial data can be dynamically retrieved from the DBMS. For the spatial information, a (static) VRML-file is created beforehand containing all the 3D geo-objects in the data set. This is not the optimal way to do it, since the VRML file may become very large in case of large data sets. A Java program was written that converts the spatial objects stored in the Simplified Spatial Model (see Section 6.2.4) in the DBMS (Oracle) to a VRML file. For the prototypes we used a data set of the building complex of The Hague Central Station (see Chapter 3), which is divided into 15 property units, stored as polyhedral objects (see Figure 7.6 and Figure 7.7 and Section 11.1.2). For every object non-spatial information (such as ownership) is stored in MS Access. In this prototype Microsoft Internet Information Server 5.0 (IIS) is used as Web server software and ASP (Active Server Pages, part of the Web server software) for the communication between the Internet client and the DBMS server. For the communication between the MS Access database and ASP an ODBC connection is set up. The operations to obtain the information in the correct format are performed at the server and not at the client side (Figure 7.8). For the prototype we use an interface of an HTML page consisting of two frames, one frame to display the VRML data and the second frame to show the attribute information. The user opens the VRML file in the browser and when the user clicks on an object, a URL string containing the unique key for the object is sent to the server. The ASP page connects to the DBMS, retrieves the requested attribute information in HTML, and sends it to the left frame as dynamic HTML.

The URL string for object with id 5 looks as follows:

```
http://domain-name/searchresults.asp?id=5&submit=SEARCH
```

The URLs are linked to the objects by storing the URL string with the relevant parameters with every object in the VRML file using the anchor construction. ASP contains a method Request.QueryString. This method filters the needed parameters from the URL string that is sent to the server, in this case the unique id of the object. The ASP page also contains a SELECT statement to obtain the requested information

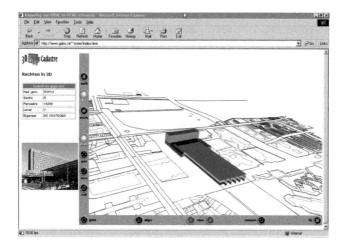

FIGURE 7.6 (Color figure follows page 176.) Screen dump of first prototype (ASP and VRML).

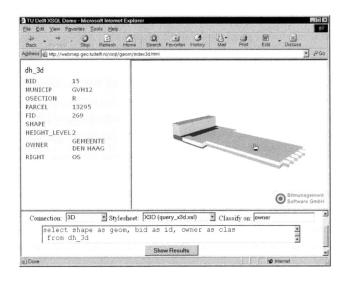

FIGURE 7.7 (Color figure follows page 176.) Screen dump of second prototype (based on XSQL and X3D).

from the database. In this case we defined the following (hardcoded) select statement (which could vary for different types of geographical objects):

```
SELECT object_id, section, parcel, level, owner, type_of_right
FROM 3D_rights
WHERE object_id = `varName'
```

A screen dump of the prototype, using Cortona as VRML browser (see link in Reference 124), is shown in Figure 7.6. This first prototype has three main disadvantages:

1. The transformation of spatial information into the 3D format is not performed on-the-fly.
2. The non-spatial data and the spatial data are stored in different DBMS, MS Access was selected for this prototype, to explore possible techniques.

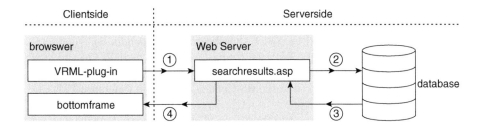

FIGURE 7.8 Model of the architecture of the prototype.

3. MS Access does not support spatial data types, although 3D objects could be stored in a topological model in MS Access.

To overcome these disadvantages a second, more advanced prototype was built.

Prototype II: XSQL, X3D, and Oracle

The second prototype is based on the Oracle XSQL Servlet and on X3D. The XSQL Servlet is part of Oracle's XML Developers Kit (or XDK), aimed at "XML enabling" the Oracle DBMS technology.[202] In this prototype both the spatial and non-spatial information are stored in Oracle. XSQL operates in combination with an XML Parser, an XSL processor (for the processing of XSLT stylesheets), and the XML SQL Utility (or XSU), also part of XDK. Servlets are Java classes that operate in a Web server environment. They process requests that are passed to them by the Web server when HTTP requests are received from a client. In this respect XSQL performs the same role as ASP in the prototype we described in the previous section (server-side processing). An XSQL "page" is a combination of XSQL tags and SQL statements.

The fragment below shows an XSQL page with a select statement that returns all rows in a table. Variables are used for the connection (@con), the geometry-column (geom), the table (@table), and id (@idcol) are used to make the statement parameterized. A where clause with selection filter for an attribute query or a spatial query to specify a subset of the data set or any other standard SQL statement is possible as well.

```
<?xml version="1.0"?>
<?xml-stylesheet type="text/xsl" href="@xsldoc.xsl"?>
<mymap connection="@con" xmlns:xsql="urn:oracle-xsql">

  <xsql:query rowset-element="" max-rows="5000" >
      SELECT t.@geom.sdo_gtype AS gtype,
                 t.@geom.sdo_elem_info  AS info,
                 t.@geom.sdo_ordinates AS geom,
                 @idcol AS id
      FROM @table t
  </xsql:query>
```

An XSQL page has an .xsql extension. When the Web server receives a request for an XSQL document, the page is passed to the XSQL servlet. The page is processed by the servlet: a connection to the database is made and the select statement is sent to the DBMS. The result set that returns from the database is already in XML format. The second step is then to transform the "raw" XML stream into a X3D or VRML output stream. Because of the XML syntax of X3D, the transformation from Oracle to X3D can easily be handled by XSLT stylesheets (see Appendix B). Therefore, in this prototype we used X3D. With the media type set to "model/x3d+xml" the transformed output stream is recognized by the browser (at client side) as X3D, so that the correct viewer plug-in (we used BS Contact, see link in Reference 124) can be activated for visualization.

The way the XML output stream is transformed into X3D depends on the way the 3D geo-information is stored: when the data are topologically structured, a different XSLT stylesheet must be used than in the case of a non-topological model, when the geometry of each 3D object is stored in the object record itself. In both cases, however, the same basic XSQL technology can be used. For this prototype the 3D spatial information is geometrically structured in the DBMS (Oracle Spatial 9i) using 3D multipolygons.

Getting the attribute information follows the same principle as in the first prototype: with the anchor construction of VRML/X3D, a URL can be called if one clicks on an object, in this case the URL of another XSQL page. The value and attribute name of the link-id are passed to the XSQL page, together with the name of the Oracle table that has to be queried:

```
<Anchor parameter="target=left"
url="http://domain-name/fieldinfo.xsql?table=dh_3d&amp;
idcol=bid&amp;id=1&amp;con=st">
   <Shape>
      <IndexedFaceSet convex="false" solid="false">
         ...
      </IndexedFaceSet>
   </Shape>
</Anchor>
```

The non-spatial information is queried from the DBMS and translated into dynamic HTML. The XSQL servlet on the Web server takes care of most of the steps in the retrieval process: establishing a JDBC connection to the database, sending the SQL queries to the database, and—most important for our purpose—reformatting the SQL response (the result set of the query) into an XML output stream. This output stream can then be presented as (dynamic) HTML (for attribute data), or as SVG (Scalable Vector Graphics for 2D spatial data), or—as in this prototype—as X3D, for 3D visualization. The configuration used in this second prototype is Apache as Web server software with Tomcat 4 as servlet container, XSQL for the server side processing, HTML and JavaScript for the user interface, and BS Contact as X3D viewer plug-in. Tomcat version 4 (and later versions) also contains Web server functionalities; therefore, this prototype could also have been implemented without Apache. The second prototype (Figure 7.7) differs from the first in a number of aspects:

- X3D is used as 3D graphic format instead of VRML.
- The transformation into the 3D format is performed on-the-fly, with real-time access to Oracle (with the possibility of specifying a subset, needed for large data sets).
- Spatial and non-spatial information are not in separate databases or database systems. Both 3D spatial and non-spatial information are stored in Oracle (in this prototype in the same Oracle table, but this is not necessary).

The second prototype is a more platform-independent solution than the first prototype. Because of the Java servlet technology it can be implemented on both Microsoft and Unix Web servers. And although the XSQL Servlet is part of the Oracle XML toolkit, it can also be used to connect to databases other than Oracle (MySQL, PostgreSQL, etc.), provided that these databases can be accessed via JDBC connections.

OGC and Our Prototypes

As seen in Section 5.7.1, OGC has defined Implementation Specifications for several types of Web services. The question is: can these services be used to make our prototypes OGC compliant?

There are several functionalities that are required in our prototypes that are built to visualize and query 3D geo-objects via an Internet client:

- To get sufficient insight in a 3D situation, it should be possible to *navigate* through the 3D representation of reality.
- It should be possible to *identify* objects (i.e., click on objects whereupon information on the objects appears).
- A user should be able to perform a *query*, e.g., give all 3D geo-objects of which the owner is Mr. X.

Navigation using the Web Terrain Service (which returns an image of a 3D view) should in principle be possible. However, for navigation in a 3D model, a sequence of 3D view needs to be generated. This may result in low performance as a result of a lot of unnecessary communication between the Web server and the client.

Identifying and querying of 3D geo-objects can be supported only by Web services that support 3D geo-data and that are able to return the spatial data in vector-format.

At this moment only the Web Feature Service (WFS) returns vector data (in GML), which means that only the WFS could be used to make our prototypes OGC compliant. Because GML also supports 3D features, 3D data can be returned to the client. The WFS not only contains functionalities to visualize and query geo-data, but also supports transactions for the geometry and the alphanumerical attributes, by which it is possible to perform insertions, deletions, and updates.

To be able to visualize, query, and update the geo-data at the client side using an OGC-compliant environment, three architectures are possible (see also Reference 216):

1. Visualizing and querying the 3D geo-data in a WFS enabled client (i.e., which directly supports GML format)
2. The GML data are converted by middleware software or at the client side, to, e.g., X3D or VRML, which may result in low performance compared to our prototype in which the conversion from XML to X3D was performed at the server side
3. Extend GML with a 3D geometry type that is similar to the X3D data type "IndexedFaceSet" (both X3D and GML are based on XML)

7.5 CONCLUSIONS

This chapter started with an overview concerning basic aspects of 3D GIS other than DBMS aspects: organization of 3D data, 3D data collection and object reconstruction, visualization and navigation in 3D environments, 3D analyzing, and 3D editing. Based on this overview it can be concluded that 3D GIS still has to mature. 3D GIS developments are mainly in the area of visualization and animation. Bottlenecks for commercial implementation of 3D GIS are as follows:

- 3D editing in GIS is not yet possible and is traditionally a functionality that is well supported in CAD software but not in GIS.
- There is poor linkage between CAD, traditionally designers of 3D models, and GIS.
- Methods to automatically reconstruct 3D objects are lacking.
- Visualization of 3D information requires special techniques. Characteristics such as physical properties of objects (texture, material, color), behavior (e.g., on-click-open), and different levels of detail representations should also be maintained and organized in DBMSs.
- Virtual Reality and Augmented Reality techniques should be incorporated in GIS software to improve visualization of and navigation in 3D environments.

In this chapter one of the functionalities of 3D GIS was addressed in more detail: accessing 3D geo-information organized in a DBMS (Oracle) by front ends. From the experiments described in this chapter, can be concluded that accessing geo-DBMS by front ends is not yet always straightforward. With the CAD-oriented solution (Bentley's MicroStation GeoGraphics), it is possible to visualize 3D objects rather easily, but when one wants to query or edit objects one has to create features and attach features to the objects (the features and objects refer to the same "thing"), which takes more time and effort and which adjusts the database tables. The Spatial Viewer, which is a Java applet that enables one to query, edit, and post (= update) spatial data in Oracle Spatial without using an MS GG project, is delivered together with MS GG and requires less customization and is therefore easier to use. In addition, where using ArcGIS (a GIS solution), the user must perform actions outside ArcMap/ArcScene before the user is able to access spatial information in ArcMap and ArcScene. Another problem is that vertical polygons stored in Oracle Spatic cannot yet be visualized in ArcScene. An advantage of the ArcGIS solution is that the database tables are basically not changed.

There are some basic differences between the solution of MS GG (also using the Spatial Viewer) and the ArcGIS solution. First, 3D editing using an interface is possible only in CAD-oriented software and not yet in GIS software (z-coordinates can be adjusted only in a special dialog). Second, when accessing the DBMS with MS GG the spatial data is retrieved and copied in the MicroStation design file. When the data are updated, the data must be posted back into the DBMS, although in MS GG it is possible to set an option to immediately post modified elements or to force updates to Oracle Spatial when one closes the dgn (MS GG design) file. When accessing the DBMS with ArcGIS one actually works on the DBMS (without copying the data).

We also described Web-based access to the geo-DBMS, which illustrates how 3D visualization techniques and techniques to query DBMSs via a Web server can be combined. Accessing 3D geo-information via the Internet is appropriate for the 3D cadastre in which easy and open access to the 3D situation, including 3D visualization, is one of the main goals. Web-based access to the geo-DBMS is based on open source software and open standards, which makes it independent of the underlying DBMS. In addition, many users can get access to the data without having to install comprehensive commercial software.

We have built two prototypes to show the possibilities of using Web-based techniques. The prototypes showed how 3D geometry stored in an Oracle database can be converted into dynamic VRML or X3D, and how the 3D objects can be presented in a "simple" Internet browser together with their non-spatial information, which is presented in dynamic HTML. For accessing the spatial and non-spatial information the user does not need to carry out additional actions, in contrast to the Bentley and ESRI solution. The next step is to see whether it is possible to use the OGC Web services to make the prototypes OGC compliant. As seen in this chapter, only the WFS would be suitable for visualizing, querying, and editing 3D geo-objects via the Internet.

The prototypes described showed good potential for accessing 3D geo-objects organized in a DBMS, which is an essential functionality for a 3D cadastre.

8 Integrating 2D Parcels and 3D Objects in One Environment*

The insertion of 3D geo-objects in the cadastral database containing 2D parcels touches on the fundamental issue of combining 2D and 3D geo-objects (geographical features) in one environment: What is the vertical relation between 2D and 3D geo-objects and how can these two sources be integrated in one environment? These issues are addressed in this chapter.

First, we discuss whether absolute coordinates or relative coordinates should be used to define 3D objects (Section 8.1). A case study has been carried out to combine a 3D object (pipeline) with parcels. The aim of the case study is to show possibilities, problems, and conditions of the integration of 3D objects and parcels in one DBMS environment. The case study is presented in Section 8.2.

One of the main findings of the case study is that a height surface per parcel is needed to combine 2D parcels and 3D objects in one environment. Therefore, four TINs (Triangular Irregular Networks) were generated, all representing surface height models based on point heights obtained from laser altimetry, and the last three also including 2D parcels: unconstrained Delaunay TIN, constrained Delaunay TIN, conforming Delaunay TIN, and refined constrained TIN. In Section 8.3 the creation of these TINs is described, together with their data structures and their results. The TINs were stored in the Oracle DBMS, and from this information, some spatial analyses, queries, and vizualizations were performed in the context of the DBMS (Section 8.4).

One of the disadvantages of using a dense laser altimetry data set is the resulting data volume and with that the poor performance of queries. However, due to the "sampling" nature of data obtained with laser altimetry not all points are needed to generate an accurate elevation model (within epsilon tolerance in the same order of magnitude as the original height model and cadastral data). Therefore, we examined how the number of TIN nodes can be reduced by removing nodes that are not significant for the TIN, taking the constraints of the parcel boundaries into account. Section 8.5 describes a method to generate an effective TIN that includes a data structure of 2D objects, in which only the relevant points are used. The first step of this generalization method has been implemented: the filtering of nonsignificant elevation points. The results of this prototype implementation are presented in Section 8.6. The chapter ends with conclusions.

8.1 ABSOLUTE OR RELATIVE COORDINATES

Two possible representations of z-coordinates of 3D geo-objects can be distinguished:

---

*This chapter is based on References 191 and 193.

1. AN ABSOLUTE z-COORDINATE, DEFINED WITHIN THE NATIONAL REFERENCE SYSTEM

When z-coordinates of 3D geo-objects are stored within a national reference system, absolute height has to be assigned to 2D surface parcels to be able to define geometrical and topological relationships between 3D objects and 2D surface parcels, such as above, below, or intersecting. Because 2D parcels need to be defined in 3D space, the complexity of the 2D data increases. Locating 2D parcels in 3D space cannot be done by simply adding one z-coordinate per parcel, as some parcels may contain too much spatial variance for this approach (even in a flat country like the Netherlands).

2. A RELATIVE z-COORDINATE, DEFINED WITH RESPECT TO THE SURFACE

When z-coordinates of the 3D geo-objects are stored with respect to the surface, the current database does not need to be extended with additional z-information on 2D parcels, saving time and data complexity. The z-coordinates of 3D geo-objects known within the national reference system have to be converted into relative coordinates. In this case only the 3D situation in the surrounding of the 3D geo-object needs to be explored (height data are needed only at the location of a 3D object), instead of locating all 2D parcels in 3D space. Maintaining data consistency in case of updates might be difficult, for example, when the surface level changes.

To assign height information to parcels, laserscan data can be used. Laserscan data of the surface are complicated to collect in urban areas (although filtering techniques exist to obtain height at surface level in urban areas based on laserscan data). Therefore, in urban areas with no height variances in surface level, defining z-coordinates of 3D objects with respect to the surface, i.e., using relative z-coordinates (see, for example, Figures 11.2, 11.3, and 11.5) might yield a good representation of the situation. In this case the surface level is the level where $z = 0$. However, in most cases the most sustainable solution is to define 3D objects with absolute z-coordinates. The reasons are, first, because absolute z-coordinates are not influenced by surface changes. Second, the definition of the surface level (the reference level used for values with respect to the surface) is sometimes not clear, especially in dense urban areas with a number of modern constructions with their main entrance at different levels. Finally, when using z-coordinates with respect to the surface, definition of the actual geometry of 3D objects is complicated. In non-flat areas it is therefore not realistic to define 3D geo-objects with respect to the surface.

Once it is decided that it is appropriate to have z-coordinates of 3D objects defined in absolute values, the next issue is how to combine the 3D objects with parcels defined in 2D.

8.2 INTRODUCTION OF A CASE STUDY

A case study was carried out to study the possibilities and constraints of combining a 3D object defined in absolute z-coordinates with parcels, by the integration of point heights and parcel boundaries (Figure 8.1). For this case study, a gas pipeline was used. A Dutch company owning an important network of utility pipelines (hereafter: the

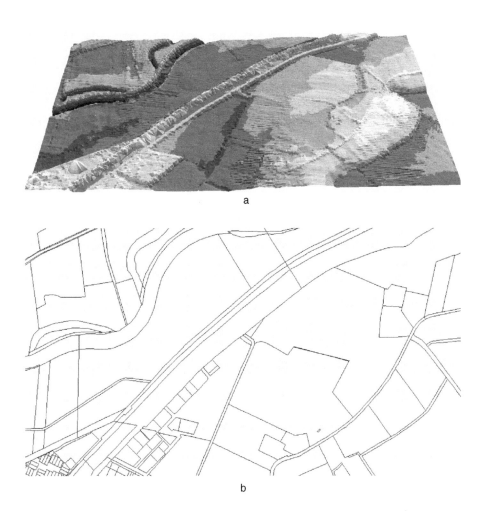

a

b

FIGURE 8.1 (Color figure follows page 176.) Data sets used in this case study. (a) AHN; (b) Cadastral parcels.

"Company") provided us with 3D information on a pipeline in rural area. The 3D information was defined using absolute z-coordinates in the Netherlands National Ordnance Datum: *NAP*).

8.2.1 DESCRIPTION OF DATA SETS

For the case study, two data sets were used: a registration of heights maintained by Rijkswaterstaat and the cadastral registration.

Terrain height points For the terrain elevation model we use a data set representing the DEM (Digital Elevation Model) of the Netherlands, i.e., AHN (*Actueel Hoogtebestand Nederland*).[68] The AHN is a data set of point heights obtained with laser altimetry with a density of at least one point per 16 square meters and in forests a density of at least one point per 36 square meters. The point heights are resampled in a regular tessellation at a resolution of 5 meters. Because only the resampled data set was available, we used this regularly distributed data set. However, the TIN experiments described in this chapter are more appropriate for raw, irregularly distributed laser-scan data. The AHN contains only earth surface points: information such as houses, cars, and vegetation has been filtered out of the AHN. The heights in the AHN differ on average 5 centimeters with the heights in reality.

Parcel boundaries The parcels used are from the cadastral database of the Netherlands. In the cadastral DBMS parcel boundaries are organized in the geometrical structure (polylines) and parcels are topologically stored (see Section 6.2.2). The typical geometrical accuracy is about 10 centimeters. The realized geometry of parcels (i.e., polygons) can be obtained by a function that has been implemented in the Oracle database (see Section 6.2.2).

A test data set was selected from these registrations at the location of the pipeline (represented as polyline).

8.2.2 COMBINING POINT HEIGHTS AND 3D OBJECTS

Without the AHN, the 3D definition of the pipeline in absolute coordinates does not reveal where the 3D object is located with respect to the surface and with respect to the parcels on the surface. Is the pipeline situated above or under the ground, what is the depth of the pipeline? With point heights at the surface level at sufficient density, it is possible to compute the position of the pipeline with respect to the surface. A DEM represented by an unconstrained TIN of the laser altimetry points (see Section 8.3) was used for the extraction of z-coordinates at surface level at the location of the pipeline. The values for "with respect to the surface" for every coordinate of the pipeline could be computed by subtracting the "z" value of the pipeline from the "surface" value at the specific location. The results of these calculations are shown in Table 8.1.

From this table we can see the depth of the pipeline, which shows that the beginning and the end of this pipeline are located above the surface, which is true in reality (the units are in meters while the diameter of the pipeline is 45 centimeters).

TABLE 8.1
Results of Integrating 3D Pipeline with DEM

x	y	z_pipeline	surface_level	with_respect_to_surface
242850.36	512938.67	10.44	9.18	1.26
242849.52	512939.37	10.35	9.18	1.16
242847.21	512941.25	10.38	9.17	1.21
242844.80	512943.11	10.23	9.16	1.07
242843.01	512944.55	8.89	9.17	−0.28
242840.47	512946.54	7.33	9.17	−1.84
242820.76	512962.44	7.16	9.12	−1.96
242811.67	512969.86	6.93	8.96	−2.03
....	....	....	....	....
243433.04	516518.54	7.86	9.11	−1.23
243437.08	516499.08	7.89	9.04	−1.15
243437.59	516498.49	8.10	9.00	−0.90
243438.11	516498.19	8.37	8.99	−0.62
243438.58	516498.10	8.65	8.99	−0.34
243439.39	516498.22	9.16	9.01	0.15
243440.10	516498.36	9.58	9.02	0.56
243441.29	516498.62	9.99	9.04	0.95
243441.40	516498.64	10.10	9.00	1.05
243442.28	516498.83	10.27	9.04	1.23
243445.85	516499.61	10.37	9.07	1.30
243448.12	516500.04	10.46	9.10	1.36

8.2.3 ASSIGNING HEIGHT TO PARCELS

Figure 8.2 is a combination of 2D parcel boundaries (at the $z = 0$ plane) and the pipeline defined in 3D with absolute z-coordinates. The dashed lined shows the projection of the 3D pipeline on the plane where the z-coordinates equal zero (the plane where the 2D parcel boundaries are positioned). The 3D pipeline (which has absolute z-coordinates between +5 and +10 meters) is drawn above the parcel boundaries ($z = \phi$). However, this is not correct because, apart from the entrances, the pipeline is located below the surface.

The alternative is either to locate the parcels in 3D space or to use relative heights (last column in Table 8.1). Because we already concluded that using absolute z-coordinates for defining 3D objects is more appropriate for a 3D cadastre, we need to locate the parcels in 3D space. Using one z-coordinate for each parcel is not sufficient. Therefore, z-coordinates were assigned to the vertices describing the parcel boundaries in order to locate the parcel boundaries in 3D. The DEM represented by an unconstrained TIN was used to extract z-coordinates for the vertices describing parcels (see Section 8.3). Figure 8.3 shows different visualizations of the 3D pipeline and the 3D parcel boundaries. In Figure 8.3a, the 2D parcel boundaries are drawn with dashed lines (on plane z=0) in combination with 3D parcel boundaries. In Figure 8.3b the 3D parcel boundaries are drawn with the 3D pipeline (in absolute coordinates),

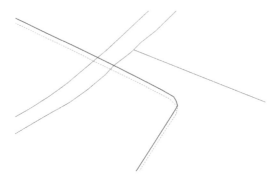

FIGURE 8.2 Combining 2D parcels boundaries with a pipeline defined with absolute 3D coordinates (in bold) and the projection of the pipeline on $z = \phi$ dashed.

which correctly reflects the real situation. Figure 8.3c is the same as Figure 8.3a, but now the pipeline in 3D is inserted with the projection of the pipeline on the $z = 0$ plane (dashed line), while sticks indicate the distance between the pipeline and the $z = 0$ plane.

On the locations of the parcel boundaries it is now possible to determine the depth (or height) of the pipeline. Interaction with the views (rotating, zooming, etc.) helps us better understand the situation. However, within one parcel it is still not clear where the pipeline is located. Therefore, the parcel surface needs to be obtained. Having a height surface of parcels, it is possible to position parcels in 3D in order to integrate

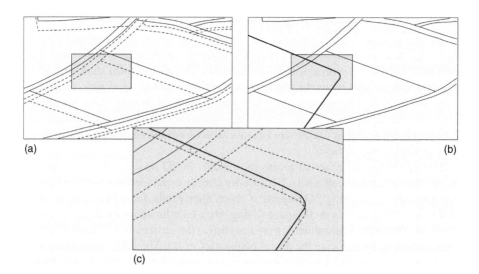

FIGURE 8.3 Parcel boundaries defined in 3D give insight to where the pipeline is positioned with respect to the surface.

3D geo-objects (defined in absolute height values) with the cadastral map and to extend the cadastral map in 3D using the 2.5D representation of parcels.

The height surface of parcels can be obtained by the integration of point heights and parcel boundaries.

8.3 INTEGRATED TINS OF POINT HEIGHTS AND PARCELS

First, it should be noted that there is a close relationship between DEMs (2.5D representations), based on, for example, raw laser altimetry point data, and the topographic objects or features embedded in the terrain. Feature extraction techniques seek to obtain the 2D geometry and heights for certain types of topographic objects such as buildings out of the DEMs. There are methods for object recognition in TINs in which the selection of an object (e.g., building roofs, flat terrain between buildings) corresponds to planar surfaces.[57] This technique can be used for 3D building reconstruction from laser altimetry, which is not of particular interest for the research presented in this book.

On the other hand, 2D objects coming from another source, such as a cadastral or topographic map, can explicitly be incorporated as part of the TIN structure that represents a height surface in order to integrate height information and 2D geo-data.[99] In this case the TIN structure is based on both 2D objects and point heights. The data structure of a planar partition of 2D objects is incorporated in the TIN structure. Within this data structure, the 2D objects are identifiable in the TIN and obtainable from the TIN, as a selection of triangles that yield 2.5D surfaces of individual 2D objects. To explore the possibilities of including a data set defined in a 2D planar partition in a TIN structure, four TINs, all representing height models and the last three also including 2D objects, were generated:[182]

1. Unconstrained Delaunay TIN, based on AHN point heights only
2. Constrained Delaunay TIN, based on AHN point heights and constraints, which are the original edges from the 2D objects (parcel boundaries), without changing the input edges
3. Conforming Delaunay TIN, based on AHN point heights and constraints (again edges of the 2D objects), now also Steiner points are added on long edges during the triangulation process to improve the triangle structure
4. Refined constrained Delaunay TIN, based on AHN point heights and constraints, which are the original edges from the 2D objects that are subdivided before the triangulation process

8.3.1 UNCONSTRAINED TIN

First, a TIN was generated using only the point data. The triangulation was performed outside the DBMS because TINs (and triangulation) are not yet supported within DBMSs. The ideal case would be just to store the point heights and the parcel boundaries in the DBMS and to generate the TIN of the area of interest on the user's request in the DBMS, without explicitly storing the TIN structure in the DBMS.

The representation of the implicit TIN could then be obtained via a view. This would be more efficient and less prone to decrease in quality because no data transfer and conversion would be needed from DBMS to TIN software and back. In the future, a distributed DBMS structure may be possible within the Geo-Information Infrastructure (GII). An integrated view, based on two different databases (as the point heights and cadastral data are maintained by different organizations in different databases) may be feasible from a technical perspective. However, in this integrated approach, on-the-fly triangulation should not take too long; otherwise, generating TIN on request will not work from the usability point of view. In our research we stored copies of all data sets in one single DBMS.

The TIN was generated by means of triangulation software called Triangle.[181] Triangle software was used, as it offers many types of triangulations and control parameters and in addition it is freeware software, written in C. There is a program by which it is possible to use it directly on the command line. The input as well as the output files are easily accessible as they are ascii files. Triangulation software implemented as part of GIS or CAD packages such as Geopak (Bentley[11]) or the 3D Analyst extension of ArcGIS (ESRI[46]) have their own internal data structure, especially for the produced TINs, which makes this triangulation software less flexible. In addition, Triangle supports more types of TINs; e.g., 3D Analyst does not offer support for constrained TINs. This is why these applications were not used in the first part of this research. Later on in this research (see Section 8.6), 3D Analyst was used as it has an easy graphical user interface and it supports the conforming TINs, which is suitable for the purpose of our research at this stage.

CGAL,[22] a freeware C++ library with computational geometry functions, does not have the option of building conforming TINs (only unconstrained and constrained TINs are supported) and in addition one still has to create a program (based on the library), which is the reason CGAL was not used.

In our test case, first an unconstrained TIN was generated with Delaunay triangulation (see for an explanation Reference 224). The Delaunay triangulation results in triangles that fulfill the "empty circle criterion," which means that the circumcircle around every triangle contains no vertices of the triangulation other than the three vertices that define the triangle. In general, this results in good and numerically stable polygons.

It should be noted that Delaunay TINs are not unique when more than three points are located on a circle. Further, it should be noted that the Delaunay TINs are computed in 2D and may therefore be suboptimal for true elevation data. The z-value of points is not taken into account in the triangulation process, but added afterwards. This is not straightforward if one realizes that the TIN is computed for an elevation model in which the z-value is very important.[210]

The resulting TIN (containing x-,y-,z-coordinates on point data and the TIN triangles defined with references to those points) are stored in the DBMS (Oracle Spatial 9i[152]) in a topological structure. The UML model of the TIN is shown in Figure 8.4.

In the topological structure two tables are stored: a table with faces (TIN triangles) = "tin", and a table with nodes = "tin_vertex." Note that edges are not stored explicitly. The TIN triangle table contains references to the ids in the node table (three references for every triangle):

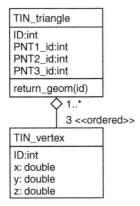

FIGURE 8.4 UML class diagram of TIN data structure.

```
SQL> DESCRIBE tin

Name              Type
----              -----
ID                NUMBER(8)
PNT1_ID           NUMBER(8)
PNT2_ID           NUMBER(8)
PNT3_ID           NUMBER(8)
```

In the integrated terrain elevation and object model defined with a constrained TIN, a conforming TIN, or a refined conforming TIN, this table could also contain the object_id (e.g., parcel number) indicating to which object the triangle belongs. In the second table the coordinates of the points are stored together with their ids:

```
SQL> DESCRIBE tin_vertex

Name              Type
----              -----
ID                NUMBER(8)
LOCATION          MDSYS.SDO_GEOMETRY
Z                 NUMBER(10)
```

In this way, every point is stored only once (sdo_geometry is the Oracle spatial data type, which could also represent 3D points). A function has been written to generate ("realize") the geometry of triangles (3D polygons) based on the topological tables. The function returns a 3D polygon of type sdo_geometry. The geometry is represented as a view on the topological structure by means of the following SQL statement:

```
CREATE VIEW tin_geom AS SELECT id, return_geom(id) shape FROM tin;
```

The selection of triangles from the unconstrained TIN (partly) overlapping one parcel surface represents an area larger than the parcel itself since triangles cross parcel boundaries (Figure 8.5). Therefore, to improve the selection of a parcel surface, a constrained TIN was generated.

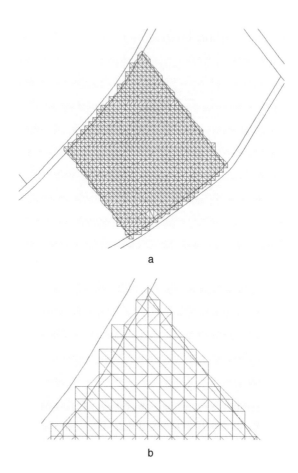

FIGURE 8.5 A parcel surface extracted from the DBMS based on an unconstrained TIN. Note that parcel boundaries are shown as well. (a) Whole parcel surface; (b) zoom in on upper-right corner.

8.3.2 CONSTRAINED TIN

To obtain a more precise parcel surface, a constrained TIN was generated, using the parcel boundaries as constraints. Again, the Triangle software was used (outside the DBMS). At first, we did not divide the parcel boundaries (by adding points to the interior of constrained edges in order to fulfill the Delaunay criterion; see Section 8.3.3). Therefore, the original boundaries in the TIN were preserved. We assigned z-coordinates to the nodes of parcel boundaries by projecting them in the unconstrained TIN. The constrained TIN is again stored in a topological structure, with a geometrical view on top of it. In the constrained TIN (Figure 8.6) each triangle belongs to one parcel only and therefore the selection of triangles exactly equals the area of a parcel. To select a parcel object from the TIN (as a set of triangles), the constrained TIN is more appropriate than the unconstrained TIN, because the data structure of the planar partition of parcels is incorporated in the constrained

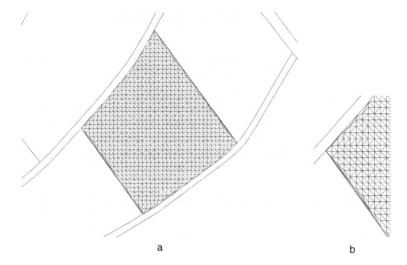

a   b

FIGURE 8.6 A parcel surface based on a constrained TIN. (a) Whole parcel surface; (b) zoom in on upper-left corner.

TIN. However, as can be seen in Figure 8.6, keeping the edges undivided leads to elongated triangles near the location of parcel boundaries. This has two important drawbacks. First, the very flat elongated triangles may be numerically unstable (not robust, as small changes in the coordinates may cause errors) and the visualization is unpleasant. Second, and maybe even more important, a long original parcel boundary will remain a straight line in 3D even when the terrain is hilly. There are no intermediate points on the parcel boundaries, so it is not possible to represent height variance across the parcel boundaries.

8.3.3 CONFORMING TIN

Keeping the original edges in the constrained TIN undivided in the triangulation process leads to elongated triangles if parcel boundaries are much longer than the average distance between DEM points (5 meters), which is the case in using parcel boundaries with the AHN data set. An alternative to the constrained TIN may be the conforming TIN. The computation starts with a constrained TIN, but every constrained edge that has a triangle to the left or right not satisfying the empty circle condition is recursively subdivided by adding so-called Steiner points (and locally recomputing the TIN with the two new constrained edges). The recursion stops when all triangles, including the ones with parts of the constrained edges, satisfy the empty circumcircle criterion (the Delaunay property). The conforming TIN has both the Delaunay property and the advantage that all constrained edges are present, possibly subdivided in parts, in the resulting TIN.

The software Triangle uses the Ruppert's Delaunay refinement algorithm to produce conforming TINs.[176] To further improve and control the shape of triangles, and the overall mesh, two additional parameters can be set (and it is possible to specify that the result is a conforming TIN):

1. By specifying a minimum angle for the triangles (in 2D)
2. By specifying a maximum area for the triangles (in 2D)

We did experiments with the data set to evaluate the two options. First, we generated a conforming TIN by setting the minimum angle for triangles and we chose a threshold of 10°. Using the second option we generated a conforming TIN imposing a maximum triangle area. No triangle is generated larger than the maximum triangle area. The density of the height points is one point per 25 square meters. In case of grid-organized data (as in this case) the number of triangles in a TIN is usually approximately twice as large as the number of points; i.e., triangles have an area of 12.5 square meters on average. We therefore decided to set the maximum area on 25 square meters (see Figure 8.7).

The disadvantage of both the minimum angle and the maximum area method is that data points are not only inserted on the edges of the parcel boundaries, but also in the mesh itself. Note that this problem becomes bigger if there are significant "gaps" in the laserscan data set. Height points are added but these height points do not contain additional information, since the heights of these added points are calculated during the triangulation process. Therefore, it was concluded that for our purpose the additional minimum angle and maximum area methods are not beneficial and we used the normal conforming TIN. Figure 8.8 shows a conforming TIN covering several parcels (different shades of gray). To improve visualization the height has been exaggerated (ten times).

8.3.4 Refined Constrained TIN

However, a normal conforming TIN also has its drawbacks compared to a constrained TIN. In case of two very close parallel constrained edges, a large number of very small triangles are generated while these constrained edges are split in many very

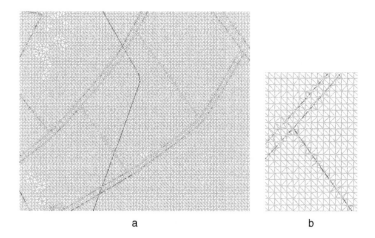

a   b

FIGURE 8.7 A height model based on a conformal TIN. (a) Whole parcel surface; (b) zoom in on upper-right corner of query parcel.

FIGURE 8.8 Conforming TIN in which point heights and 2D planar partition of parcels are integrated.

small edges (Figure 8.9). This can also happen when AHN points are very close to the constrained edges. These small triangles have no use, as they do not reflect any height difference (at least the height differences cannot be derived from the AHN points) and they also do not reflect additional object information.

A solution for this is splitting the constrained edges, before inserting them, into parts not larger than two or three times the average distance between neighbor AHN points (e.g., 10 meters) and then computing the normal constrained TIN. In this way, on the one hand the too-flat triangles of the constrained TIN are avoided (problem of very long constrained edges), and on the other hand the too-small triangles of the conforming TIN are avoided.

Figure 8.10 shows the refined constrained TIN for one parcel. The edges of the parcel boundaries were split into parts of at most 10 meters. These edges were then used as constraints in the triangulation, which resulted in a refined constrained TIN. This improves the shape of triangles considerably (too-flat and too-small triangles are avoided). Moreover, since points are added on the parcel boundaries for which the

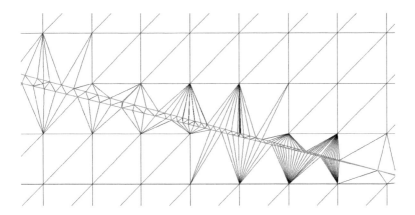

FIGURE 8.9 In some cases a conforming TIN results in very small triangles.

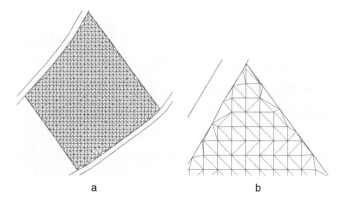

a   b

FIGURE 8.10 A parcel surface based on a refined constrained TIN. (a) Whole parcel surface;
(b) zoom in on upper-right corner.

height has been deduced based on the unconstrained TIN, it is possible to represent
more variation in height across a parcel boundary.

8.4 ANALYZING AND QUERYING PARCEL SURFACES

The actual extraction of a parcel surface is performed within the Oracle DBMS. In
this process all triangles that are covered by one parcel are selected by means of a
spatial query. To select these triangles, first realization of the geometries of triangles
needs to be performed. To illustrate the query to extract a parcel surface from the
DBMS, the refined constrained TIN has been used. In these queries we used the
realized geometries of parcels.

To speed the query, first a function-based index was built on the TIN table (R-tree
index):

```
INSERT INTO user_sdo_geom_metadata VALUES(`TIN_R', `return_geom(id)',
mdsys.sdo_dim_array(
mdsys.sdo_dim_element(`X', 0, 254330, .001),
mdsys.sdo_dim_element(`Y', 0, 503929, .001)), NULL
);
CREATE INDEX tin_idx ON tin_r(RETURN_GEOM(ID)) INDEXTYPE IS mdsys.spatial_index;
```

The spatial query to find all points or triangles that are located within one parcel
can be performed in two ways (in Oracle Spatial terms): (1) with the spatial operator
(sdo_relate) that uses a spatial index and (2) with the spatial function (sdo_geom.relate)
that does not use a spatial index (see Section 6.3). The query to select triangles that
are within a specific query parcel (number 4589, municipality GBG00, section D)
using the spatial function is:

```
SELECT tin_r.id, return_geom(tin_r.id) shape
FROM tin_r, parcels par
WHERE par.parcel=` 4589' AND par.municip=`GBG00' AND par.section=` D' AND
sdo_geom.relate(par.geom, `COVEREDBY+INSIDE',return_geom(tin_r.id),1)='TRUE';
```

For the unconstrained TIN we used the option ANYINTERACT instead of COVEREDBY+INSIDE, because otherwise we miss the triangles that cross parcel boundaries. The parameter ANYINTERACT returns TRUE if the two geometries are not disjoint. Two objects are DISJOINT when the objects have no common boundary or interior points.

3D Area of Parcel Surface

The cadastral map is a 2D map containing projection of parcels. Consequently, the cadastral map does not contain the true area of surface parcels. In mountainous countries the true area of parcels may be needed, as tax rates are based on the area of parcels. The integrated TIN based on point heights and parcel boundaries provides the possibility of obtaining the true area of a parcel.

The area of a parcel in 3D space can be computed by summing the true area of all triangles covering one parcel in 3D space. Chapter 6 showed that DBMSs do not support 3D data types and consequently they also do not contain functions to calculate the area in 3D. To be able to compute the area of triangles in 3D in Oracle, the function "area3D" that was implemented as part of the 3D geometrical primitive (Section 6.4) was used. The 3D area calculation could therefore be performed inside the DBMS.

First we calculated the 2D area of the original parcel polygon. The query parcel is the parcel with a small "hill" on it (Figure 8.8):

```
SELECT sdo_geom.sdo_area(geom, 0.1) FROM parcels
WHERE parcel =` 4589' AND municip=`GBG00' AND section=` D';
```

The area in 2D is 6737 square meters. The 3D area of the same parcel, which resulted in 6781 square meters, is performed with the following query:

```
SELECT sum(area3d(return_geom(id)) FROM tin_r tin, parcels par
WHERE parcel=` 4589' AND municip=`GBG00' AND section=` D'
AND
sdo_geom.relate(par.geom, `COVEREDBY+INSIDE',return_geom(tin.id),0.1)=`TRUE';
```

As can be seen from these results, the difference between the projected area and the real area in 3D of this parcel is 44 square meters.

Other queries can be performed as well, e.g., find steepest triangle, find all triangles pointing to the south, or find the highest (lowest) point in this parcel:

```
SELECT MAX(z), MIN(z) FROM tin_vertex, parcels par
WHERE par.parcel=` 4589' AND par.municip=`GBG00' AND par.section=` D'
AND
sdo_geom.relate(par.geom, `COVEREDBY+INSIDE',location,0.1)=`TRUE';
```

```
   MAX(Z)    MIN(Z)
   ------   ---------
   14.24     10.027
```

8.5   GENERALIZATION OF THE INTEGRATED TIN

Both the conforming TIN and the refined constrained TIN (with constraints based on subdivided parcel boundaries in order to avoid long straight lines) look promising: the triangles are well shaped (not too flat and, in case of the conforming TIN, the Delaunay criterion is fulfilled) and points are added on parcel boundaries in order to represent more height variance on them. However, after some analyses we suspected that in both cases far too many points are used to represent the surface TIN with the same horizontal and vertical accuracy as the input data sets (AHN points and cadastral map). A problem with huge data sets is the resulting data volume and, with that, poor performance of queries and analyses. Therefore, filtering of the data set aiming at data reduction (generalization) is needed.

The filtering aiming at data reduction, i.e., generalization, is based on filtering the TIN structure and not the point heights themselves. The filtering can use the characteristics of the height surface. On location with little variance in height, points can be removed while on the location with higher variances points are maintained to define the variance in height accurately. Important advantages of data reduction in a TIN structure are that it can be used on irregularly distributed points and that locations with high height variance will remain as such in the new data set. Unfortunately, we were not able to start with an irregularly distributed data set, by which we were not able to use all advantages of the filtering performed on the TIN structure. However, the result data set is an irregularly distributed data set.

This section describes two methods to improve the initial integrated height and object model: a detailed-to-coarse approach (Section 8.5.1) and a coarse-to-detailed approach (Section 8.5.2). In Section 8.5.3 a more advanced generalization method of the integrated model is discussed (that is, more than based on height only).

8.5.1   Detailed-to-Coarse Approach

The first method starts with the complete integrated model. From this model a number of non-relevant point heights are removed while maintaining the significant points, e.g., removing the points where the normal vectors of the incident triangles have a small maximum angle. After removing such a point, the triangulation is locally corrected and it is explicitly checked if the height difference at the location of the removed point in the new TIN is within this tolerance. If so, the point is indeed not significant for the TIN and can be removed. In this process the parcel boundaries are still needed as constraints, because the aim is to be able to select a parcel surface from the TIN.

The prototype implementation is based on this method (Section 8.6). The result of the generalization using the prototype is shown in Figure 8.11.

8.5.2   Coarse-to-Detailed Approach

The procedure described above starts with all available details and then tries to remove some of the less relevant details, which is not always easy. An alternative method would be starting with a very low detail model and then adding points where the errors are the largest. The initial model could be just the constraints (with estimated

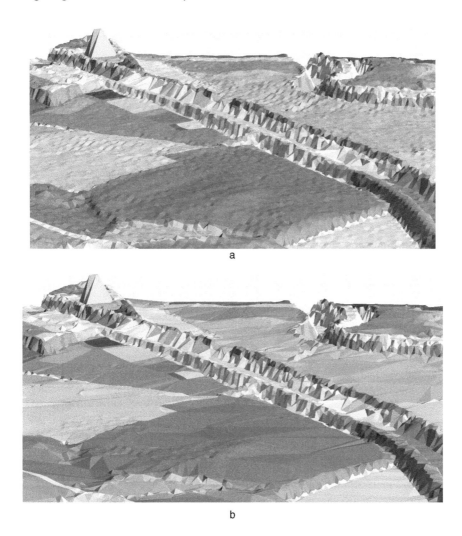

a

b

FIGURE 8.11 (**Color figure follows page 176.**) Conforming TIN in which point heights and 2D planar partition of parcels are integrated, (a) before and (b) after filtering.

z-values at every vertex of the parcel boundary) inserted in a conforming or refined constrained TIN. In the next step the AHN height point with the largest distance to this surface is located. If this point is within eps_vert distance from the surface (maximum tolerance in the vertical direction defined in epsilon tolerance), then the model already satisfies the accuracy requirements. If this point is not within the tolerance, then it is added to the TIN (and the TIN is re-triangulated under the TIN conditions). This procedure is repeated until all AHN point heights are within the tolerance distance. This procedure is a kind of 2.5D counterpart of the well-known Douglas-Peucker[33] line generalization.

8.5.3 INTEGRATED HEIGHT AND OBJECT GENERALIZATION

Thus far, only the height was taken into consideration during the generalization process, both in the detailed-to-coarse and coarse-to-detailed approach. However, as the model is supposed to be an integrated model of height and objects, the objects should also participate in the generalization. Therefore, the integrated height and object model could be further generalized by taking into account both the elevation aspect and the 2D objects at the same time. It is already possible to separate generalization of the terrain model[15,19,84,157] and 2D objects.[33,128,134,159,172,186,225] However, the integrated generalization of the height and object model makes this model also well suited for other resolutions (scales) or even in a multiresolution context.

Starting with the detailed-to-coarse approach, one could identify the following steps:

Step 0: Integrate raw elevation model (AHN) and objects (parcel boundaries) in a (conforming or refined constrained) TIN (see Section 8.3).

Step 1: Improve the efficiency of the TIN created in step 0 by removing AHN points from the TIN until this is not longer possible given the maximum tolerance value in the vertical direction: eps_vert_1 (as described in Section 8.5.1). Note that this tolerance could be adjusted for different circumstances, but the initial value should be the same size as the accuracy of the input data.

Step 2: Now also start generalization of the object boundaries, for example, with the Douglas-Peucker line generalization algorithm, by removing those boundary points that do not contribute significantly to the shape of the boundary. This can be done in 2D (standard Douglas-Peucker), but it is better to apply this algorithm in 3D. Keep on removing points until this is impossible within the given tolerance in the horizontal direction: eps_hor_1. After this line generalization of the constraints, re-triangulate the TIN according to the rules as in step 0 (of a conforming or refined constrained TIN).

Step 3: Finally, for multiresolution purposes, also start aggregating the objects, for example, in our case: parcels to sections (and the next aggregation level would be sections to municipalities, followed by municipalities to provinces, etc.). In fact this removes some of the constrained edges (original parcel boundaries) from the input of the integrated model. Repeat steps 1 and 2 with other values for the epsilon tolerances at every aggregation level with their own tolerances in the vertical and horizontal direction: eps_vert_2, eps_hor_2 (at the section level), eps_vert_3, eps_hor_3 (at the municipality level).

8.6 GENERALIZATION PROTOTYPE

The first steps (step 0 and step 1) of the generalization method have been implemented in a prototype. The fundamental idea of the implemented filtering method is to detect characteristic points (detailed-to-coarse approach). Points that are not characteristic, i.e., they do not contribute significantly to the height surface, will be removed. The question of whether points are characteristic in the prototype is dependent on the following conditions:

- A point is characteristic if the slope of two neighboring triangles of the point are significantly different.[19] To detect this, the normal vectors for the neighboring triangles are determined and compared. If the difference is larger than a given threshold angle, the point will be defined as characteristic and not be removed from the TIN.
- Local minima and maxima are also characteristic points of a TIN. If two neighboring triangles are in the same direction in the first condition, the change in angle is less important than where the change in angle demarcates a top or a valley. Therefore, a smaller threshold angle is used when the specific point is a local minimum or maximum. A minimum or maximum is the case when the azimuths of the slope of two neighboring triangles are opposite of each other, which can be determined by calculating the differences in the azimuths. If the difference is larger than a given threshold value, a smaller threshold angle is used in the first condition.

Based on these conditions a point is maintained or removed. If two neighboring triangles of one point already fulfill one of the criteria, the point will be maintained. The next step is to look if the removal of a point can be justified. This test is done by calculating the height difference at the location of the removed point between the original TIN and the new TIN (which is generated based on the reduced data set). If this difference is larger than a threshold value, the removed point is re-added. After this step the data reduction is performed again; i.e., the data reduction is an iterative process until the process is stopped on user request, preferably when a more or less stable data set is obtained (e.g., when all points have been found to be characteristic).

The prototype has been implemented in the 3D Analyst extension of ArcView (ESRI) using the macro language of ArcView (avenue). In ArcView the TIN is recognized as an object and therefore the TIN data structure can be used directly in the reduction algorithm, and in addition the results can easily be visualized. This prototype already shows the possibility of data reduction on a TIN, but should be implemented as part of the database in the future, once a TIN data structure is supported as data type in a geo-DBMS.

For our initial test, the data reduction is performed on the unconstrained TIN. This means that the 2D objects and the point heights are kept separately during the data reduction process in order to get a first impression of the achievable results. Incorporating the constraints at this stage would have made the data reduction process more complex. Figure 8.11 shows the conforming TIN of our test data set, after filtering. We did experiments with different parameters. The parameters that showed best results were as follows:

- Minimum angle between two neighboring triangles to be a characteristic point if a point is not a top or a valley: 4.5°
- Minimum angle between two neighboring triangles to be a characteristic point in case of a top or valley: 3°
- Difference in azimuth between two neighboring triangles to determine if two triangles are opposite of each other: 120°
- Maximum allowed difference in height to determine if a removed point should be re-added: 0.25 meters

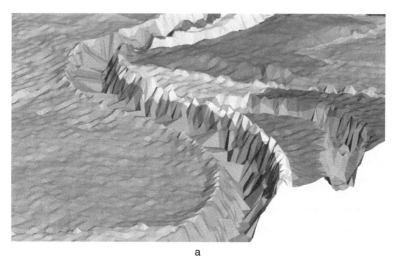

a

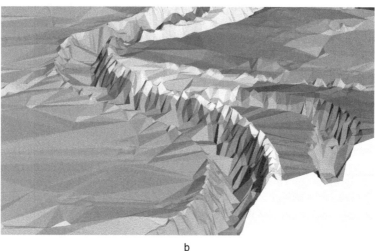

b

FIGURE 8.12 Detail of filtering results: before (a) and after (b) data reduction.

Apart from the minimum angle, the chosen parameters are based on previous research.[160]

The data set used in this example covers an area of 1450 by 800 meters and contains 44,279 AHN points (maximum z-value 14.2 meters, minimum z-value 6.7 meters, mean 9.5 meters). Three iterations steps were used to filter the data set. Figure 8.12 and Figure 8.13 clearly illustrate how the filtering maintained all terrain shapes but reduced the number of points substantially (height is exaggerated ten times).

In the first iterative step, 34,457 AHN points were removed (9822 were considered to be characteristic). The average height difference between the original and the new TIN was 0.09 meter. A total of 3243 points were re-added since they exceeded

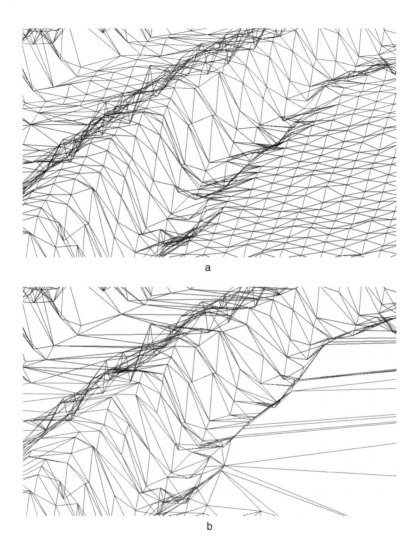

FIGURE 8.13 Detail of filtering results: before (a) and after (b) data reduction.

the height difference of 0.25 meter, which resulted in 13,065 points after the first step. After the second iteration step 8697 points were determined as characteristic, the average height difference between the new and the original point was again 0.09 meter; 3469 points were re-added and this all resulted in 12,166 points. After the third iteration step, 8455 points were considered to be characteristic (average height difference 0.09 meter) and 3529 points were re-added. After this step the data reduction process was stopped. The results of the data reduction process are listed in the Table 8.2.

After the total data reduction process 11,984 points from the original 44,279 points were maintained. This is a reduction of 73%. As can be seen from Figure 8.14, points

TABLE 8.2
Results of the Data Reduction Process After Three Iterations

it_step	#points	#rem points	#char points	#re-added points	red_rate
1	44,279	34,457	9,822	3,243	70%
2	13,065	4,368	8,697	3,469	73%
3	12,166	3.711	8,455	3,529	73%

were removed from areas with little height variance, while density of point heights in areas with high height variance (e.g., on the dikes) is still high.

The experiments with the prototype on the selected data set yielded a number of conclusions. The minimum angle needs to be adjusted to characteristics of the terrain: in case of many points in relative flat area, the minimum angles need to be small in order to avoid clusters of removed points being re-added because they exceed the maximum allowed height difference. A small minimum angle avoids removing at least one of the points in the cluster. In this case the new TIN is closer to the original TIN; therefore removed points do not exceed the maximum height difference condition and they are not re-added. In terrain with more height variance, larger minimum angles are needed in order to remove more points. In mixed terrain (with both areas with low height variance and areas with high height variance, which will often be the case), a balance should be found. In the future one could think of an implementation that can differ the parameters during one data reduction process based on the local height variance.

8.7 CONCLUSIONS

A basic aspect of a 3D cadastre is to combine 3D objects with parcels in the DBMS. This combination makes it possible to indicate where a 3D object is located with respect to the surface level (what is the depth/height of a 3D object at this location?) and with respect to parcels on the surface.

FIGURE 8.14 Result of data set after data reduction (points not removed are black).

It was concluded that defining 3D objects with absolute z-coordinates (instead of using z-coordinates with respect to the surface) is the most sustainable way of defining 3D objects. By using absolute values for 3D objects, height surfaces of parcels are needed to be able to combine the parcels and the 3D objects.

Incorporating the planar partition of 2D objects, e.g., the cadastral map, into a height surface makes it possible to extract the 2.5D surfaces of 2D objects and to visualize 2D maps in a 3D environment by using 2.5D representations.

As described and discussed in Section 8.3 it is not easy and straightforward to create a good integrated elevation and object model. Several alternatives were investigated: unconstrained Delaunay TIN, constrained TIN, conforming TIN, and finally refined constrained TIN. After some analyses, the most promising solution, the refined constrained TIN, was selected and applied with success to our test case with real-world data: AHN height points and parcel boundaries.

The integrated model, however, contains too many AHN points, which do not contribute much to the actual terrain description. Therefore, we proposed a method to generalize the integrated model. This method takes both the elevation aspect and the 2D objects into account at the same time. We implemented the first steps of this method into a prototype. In this prototype noncharacteristic points are removed from the (unconstrained) TIN in an iterative generalization process. As can be concluded from experiments with the prototype, it is possible to determine important terrain characteristics by using a simple criterion (difference in angle of neighboring triangles). With this method it is possible to reduce the data set considerably. The test data contained about four times fewer points after filtering, but still within the epsilon tolerance of the same size as the quality of the original input data sets. On the other hand, significant information on the height surface is still available in the TIN. The initial filtering yielded therefore a much-improved integrated model. Improvements can be expected when removed points are not re-added collectively but one-by-one or when the used parameters can differ during one data reduction process, based on local terrain characteristics.

The integrated model is a good basis to obtain a 2.5D representation of 2D parcels (2D parcels draped over a height surface). This is required when combining 3D objects and 2D parcels in one environment.

Part III

Models for a 3D Cadastre

9 Conceptual Model for a 3D Cadastre*

In the previous chapters the need for a 3D cadastre and the conceptual and technical framework for modeling 2D and 3D situations were studied. These chapters sketched the legal, cadastral, and technical frameworks where a 3D cadastre, to some extent, should fit in.

Based on the theory and findings in the previous chapters, this chapter details a design of a conceptual schema for a 3D cadastre. Three possible concepts (with several alternatives) have been distinguished to register 3D situations. These three concepts are introduced in Section 9.1. The three conceptual models for 3D cadastral registration with their alternatives are further discussed and completed in Sections 9.2 to 9.4.

The solutions proposed in this chapter are considered both using cadastral criteria and technological criteria in Section 9.5. Based on these considerations the best concepts for 3D registration are selected. The chapter ends with conclusions.

9.1 INTRODUCTION OF POSSIBLE SOLUTIONS

The term "3D cadastre" can be interpreted in many ways ranging from a full 3D cadastre supporting volume parcels, to the current cadastre in which limited information is maintained on 3D situations. Here three fundamental concepts are distinguished (with several alternatives): the most advanced solution, the simplest solution, and one in between in which 3D situations are still registered within current cadastral and technical frameworks:

- Full 3D cadastre:
  - Alternative 1: Combination of infinite parcel columns and volume parcels, (i.e., combined 2D/3D alternative)
  - Alternative 2: Only parcels are supported that are bounded in three dimensions (volume parcels)
- Hybrid cadastre:
  - Alternative 1: Registration of 2D parcels in all cases of real property registration and additional registration of 3D legal space in the case of 3D property units
  - Alternative 2: Registration of 2D parcels in all cases of real property registration and additional registration of physical objects
- 3D tags linked to parcels in current 2D cadastral registrations

---

*Part of this chapter is based on Reference 194.

9.1.1   A Full 3D Cadastre

This means introduction of the concept of (property) rights in 3D space. The 3D space (universe) is subdivided into volume parcels partitioning the 3D space. The legal basis, real estate transaction protocols, and the cadastral registration should support the establishment and conveyance of 3D rights. The 2D cadastral map does not lay down any restrictions on 3D rights; i.e., rights that entitle persons to volumes are not related to the surface configuration. Rights and restrictions are explicitly related to volumes. Apartment units will be real estate objects defined in 3D, on which a subject can have a right in rem. The full 3D cadastre requires a change in the legal way of thinking as well as in the cadastral and technical framework. For a full 3D cadastre, the same UML model as described in Section 2.1 applies. However, the real estate object may now also be defined in 3D. Two alternatives are distinguished for the full 3D cadastre. In the first alternative volume parcels (bounded parcels) are established only in 3D situations, and therefore it is still possible to establish parcels that are defined with boundaries on the surface. The first alternative starts with the conversion of the conventional representation of parcels into the third dimension: a parcel defined by the boundary on the surface is converted into an infinite (or actually indefinite) parcel column that intersects with the surface at the location of the parcel boundary. In the first alternative, two types of real estate objects are distinguished: infinite parcel columns (which still apply in "classic" 2D situations) and volume parcels. In a complete implementation of a full 3D cadastre (second alternative), the only real estate objects that are recognized by the cadastre are volume parcels (bounded in all dimensions) and the volume parcels form a complete partition of space. In the second alternative of the full 3D cadastre, it is no longer possible to entitle persons to infinite parcel columns defined by boundaries on the surface, but only to well-defined, totally bounded and surveyed volumes.

9.1.2   A Hybrid Solution

This means preservation of the 2D cadastre and the integration of the registration of the situation in 3D by registering 3D situations integrated and being part of the 2D cadastral geographical data set. This results in a hybrid solution of the legal registration (2D parcels) and a registration of the 3D situation. The separate registration of the legal and the 3D situation are combined and integrated. The cadastral registration of the 3D situation gives insight, but is not legally binding: the exact legal situation has still to be derived from authentic documents (deeds, survey sheets) recorded in the land registration. In those deeds, both the buyers and sellers have to agree on the description of the volume to which the new owner is entitled. This description can then be used in the 3D registration. The 3D representation (see Figure 9.1) can be either the volume to which a person is entitled (first alternative) or a physical object itself (second alternative). The first alternative implies the 3D registration of rights that are already registered and that concern 3D situations using right-volumes. This alternative is seen as a tool to gain insight into the 3D aspect of rights, i.e., visualization of rights in 3D as part of the cadastral geographical data set which can consequently be queried. The second alternative is the registration of physical objects themselves by which constructions are integrated in the cadastral geographical data set in the

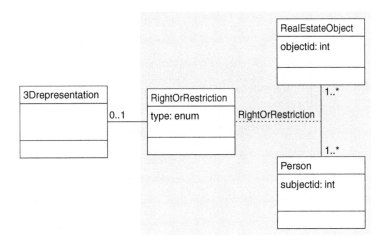

FIGURE 9.1 UML class diagram of the hybrid cadastre. The 3D representation refers to either a volume to which a person is entitled or a physical object.

same way as buildings in the current cadastral registration: in addition to parcels to clarify the real situation. In the case of right volumes (first alternative), the parcel is the starting point of registration (which limited rights are established on this parcel?), while in the case of 3D physical objects (second alternative) a physical object is the starting point of registration. In both alternatives the legal and cadastral concept of ownership and property of land is not changed as in the full 3D cadastre: rights are always established and registered on 2D parcels, while an owner of a parcel can be restricted in using the whole infinite parcel column by limited rights and legal notifications. Consequently, rights for 3D property situations are established in the same way as in current surface-oriented cadastral registrations. The difference is the way these rights are registered (and visible) in the cadastral registration.

9.1.3 3D Tags in Current 2D Cadastral Registrations

This means preservation of the 2D cadastre with external references to (digital or analog) representations of 3D situations (Figure 9.2). Complex 3D situations are registered using *ad hoc* solutions within current 2D registration possibilities, while every right that is registered can be attributed with a reference to a 3D representation. The difference with the hybrid cadastre is that the 3D representations are maintained separately, not integrated with the cadastral geographical data set.

9.1.4 Retirement at Conceptual Match

In the following sections, we further concretize these conceptual models for a 3D cadastre. We will start with the solution that requires the fewest fundamental changes of the current cadastral concept: 3D tags in the current registration (Section 9.2), followed by an elaboration on the hybrid approach (Section 9.3). The conceptual model for a full 3D cadastre is further completed in Section 9.4.

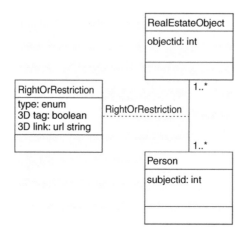

FIGURE 9.2 UML class diagram of 2D cadastre with tags to 3D situations.

9.2 A 2D CADASTRE WITH 3D TAGS

In the "3D tag" solution, real rights to real estate are always established and registered on 2D parcels. However, the notification of the existence of a 3D situation can be added to the registration by registering a 3D tag on the parcel. This means that every parcel that has more than one person entitled to it can be indicated as a 3D situation. In addition to the tag, a reference can be added to a legal document or to a drawing that illustrates the situation. The reference can be implemented in various ways. The simplest solution is just to tag 3D situations in the cadastral registration whereupon the user has to consult the documents in the land registration to find detailed information. A more advanced option is to add a reference to a 3D (digital) description maintained in the cadastral registration. The description is maintained in the cadastral registration in analog or digital form (e.g., a CAD drawing). In the latter case the information can be included as a file in the cadastral database. The projected outlines of the 3D physical object can be inserted into the cadastral geographical data set. The main difference with the hybrid solution is that drawings of 3D situations can be examined only per parcel: no integrated view on the whole situation is possible. Furthermore, the 3D situations can only be visualized and not queried since the property units indicated in the drawings are not linked to the administrative database. This registration is more or less similar to current practice of subsurface constructions in the Netherlands where subsurface constructions can be indicated using an OB (underground construction) code. This OB notification does not clarify the legal situation; it is just an indication of the factual situation.

9.3 HYBRID APPROACH

The hybrid approach consists of a registration of 3D situations in addition to, and integrated with, the existing 2D parcel registration. To effectuate this approach two alternatives have been designed. The first alternative focuses on improving insight in the 3D extent of rights (Section 9.3.1) and the second alternative focuses on the registration of physical constructions (physical objects) (Section 9.3.2).

9.3.1 REGISTRATION OF RIGHT-VOLUMES

A right-volume is a 3D representation of the legal space related to a limited right or apartment right that is established on a parcel and concerns a 3D situation, for example, a right of superficies established for a tunnel. The right of superficies, established for a tunnel, refers to a volume below the surface. The landowner is restricted in using the whole parcel column and the volume that is "subtracted" from this parcel column is visualized in 3D as a right-volume as part of the cadastral map in a 3D environment. The cadastral map should then be converted into 2.5D (see Chapter 8). Right-volumes refer only to "positive" right-volumes. If a person obtains a right for a bounded volume on a parcel (positive right-volume), this volume is subtracted from the parcel column owned by the landowner (negative right-volume). The right-volume is registered only for the person who is entitled to the bounded volume, while the spatial extent of the property of the bare owner can be derived from the registered information.

One should note that a right-volume is a different entity than 3D right as used in the full 3D cadastre, since the traditional legal framework is not changed. Rights are still always established (and registered) on surface parcels, while in the full 3D cadastre, in case of a 3D right, a person is explicitly entitled to a well-defined volume (real estate object defined in 3D), which is no longer related to surface parcels.

The boundary of the 3D representation of a right-volume starts with the parcel boundary since a real right in most countries is always established on a complete parcel. If more detail is required, e.g., when a parcel intersects with two tunnels in opposite corners of the parcel, the parcel needs to be subdivided, which is also practiced in many current cadastral registrations. A right-volume is extended into 3D (extruded) by means of defining the upper and lower limits of the right. The upper and lower limits of right-volumes are initially defined with horizontal planes. This type of registration is sufficient to warn the user that the landowner is restricted in using the whole parcel column. It also gives an indication on the space to which the limited right applies. More precise information (with legal status) can be obtained from deeds and survey plans archived in the land registration. The question can be posed if more detail on upper and lower limits of right-volumes (variance in z-level) is needed. If it is possible to register a real right on only part of a parcel, a right-volume can be defined as a polyhedron located anywhere within a parcel. The first aims of right-volumes are to warn the users that something is located above or below the surface and to indicate approximately the space where this "something" is located. The prototype implementations will show if the simple definition of right-volumes already satisfies these aims and to what extent.

In deed registrations legal conclusions can be drawn only from deeds and not from the cadastral registration. However, most frequent use will be based on querying the cadastral registration without examining the source document (deed or survey document). Therefore, the quality of the 3D representations should be exact enough for practical use.

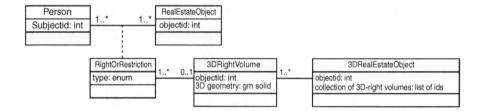

FIGURE 9.3 UML class diagram of right-volumes.

All the parties involved should agree on the upper and lower levels of the right-volumes. The levels should be laid down precisely in the concerning deeds and survey plans. Based on this information the right-volumes can be generated and inserted in the cadastral registration.

The right-volumes that are maintained are associated with a registered right. The collection of right-volumes that make up a whole 3D real estate object (e.g., one tunnel) is also maintained, and the right-volumes contain references to this whole real estate object. This is done because then all right-volumes belonging to one real estate object can be derived (with an administrative and not a spatial query). One cannot perform this query in the current cadastral registration, as there are no references to the whole real estate object maintained.

The UML class diagram of right-volumes is shown in Figure 9.3. For every right that is established on a parcel and that concerns a 3D property situation (more users on a parcel) a right-volume is maintained. The right-volume is only referenced as positive right-volume (for the holder of the right) and not for the holder of the ownership right that is restricted by the right-volume (bare owner). The right-volume is a 3D representation of the right, of which the geometry is maintained in the DBMS as type gm_solid, which is a geometry type defined by OGC and ISO;[144] see Figure 9.4. As one sees, this data model needs only little adjustment compared to the current cadastral data model (see Figure 2.1).

The most basic improvement of the registration of right-volumes compared to current 2D cadastral registrations is that the 3D extent of rights can be visualized in one integrated view in the cadastral map and not only per parcel in isolated visualizations. Furthermore, the 3D situations can be queried since the right-volumes are linked to non-spatial information in the cadastral database in contrast to the (scanned) drawings available in a cadastre containing only tags to 3D situations. However, the solution has also two drawbacks, especially when physical objects cross parcel boundaries.

A right-volume can only be registered when a right is established that will be registered in the cadastral registration, e.g., in the case of a limited real right or in the case of a real estate division in apartment rights. In other cases, a right-volume will not be registered. For that reason it is possible that the 3D location of the whole real estate object is not (and does not have to be) completely known in the cadastral registration. This can be illustrated by the example of a railway tunnel. This tunnel is built in the underground of six parcels. The owner of the tunnel (the company T) is also full owner of two of these parcels. The other four parcels are owned by respectively A, B, C, and D. For each of these parcels a right of superficies is established. In this

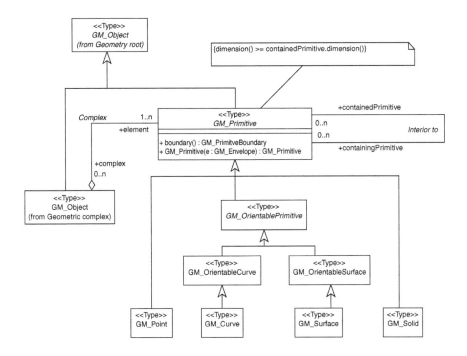

FIGURE 9.4 UML class diagram of gm_primitive as defined by OGC and ISO. (From OpenGIS Consortium, Project Document Number 01–101, Wayland, Massachusetts, 2001 copyright OGC.)

case a right-volume because of the tunnel is registered for four surface parcels. Not for the two parcels owned by T. Consequently the tunnel, which is registered by means of right-volumes, will not be locatable on the cadastral map in detail at all locations (see also Figure 4.1c). Another drawback of right-volumes is that it cannot correctly reflect all 3D situations because of the simple representation of horizontal boundaries, e.g., in the case there are two tunnels above each other (road and metro) at the same location, both with varying z. The simple representations of right-volumes (with horizontal upper and lower boundaries) would intersect each other, while the legal spaces do not intersect (see also Figure 10.1).

To meet these complications, a second alternative of the hybrid cadastre was designed, which focuses on the registration of physical objects.

9.3.2 REGISTRATION OF 3D PHYSICAL OBJECTS

Insight into 3D situations, especially in the case of construction crossing several parcels, were improved considerably, if the actual location of physical objects were available in the registration. With this information, "gaps" as in the case of right-volumes could be avoided. A possible solution to have the spatial extent of a whole physical object in the cadastral registration could be to register one volume parcel enclosing the legal space of the physical object, which is actually the combined 2D/3D alternative of a full 3D cadastre (see next section). However, a solution that fits within traditional legal (2D) frameworks is the registration of the complete construction (tunnel, pipeline) itself with a spatial description of the object. The registration of physical objects is independent of the question whether there have been rights established and registered on the intersecting parcels. The physical objects are added for the same purpose in the cadastral geographical data set as buildings: to link cadastral registration with representations of reality (i.e., topography) for orientation and reference purposes. A physical object is a construction above or below the surface that may cross parcel boundaries. In the case of physical objects, the objects themselves are registered and not the 3D legal space (as in the first alternative of the hybrid solution). The legal space is the space to which the holder of a physical object wants to have a right to ensure the property of the object, which is usually larger than the physical extent of the object itself (for example, including a safety zone). In general, the holder of a 3D physical object is the person or organization responsible for the 3D physical object. The holder has an economic ownership of the construction (right of exploitation) and benefits from the construction but also pays the costs for maintenance and replacements. The main objective of the registration of physical objects is to reflect the construction itself. This information can be then used to examine the legal status of the situation.

The registration of physical objects can be compared with the registration of telecom networks in the Netherlands (since June 2003). A centerline of the network (possibly with information on a zone indicating the accuracy or the width of the network) is offered for registration at the Kadaster (in 2D and thus it is not clear if the network is located above or below the surface). The network is registered in the land registration using a drawing of the situation and in the cadastral registration using one or several anchor parcels while a legal notification for all intersecting parcels can be registered voluntarily. At this moment the spatial information on the network is not added in the cadastral geographical data set. Therefore, the user still has to consult the legal document and the drawing archived in the land registration. If the spatial information of networks were available in the cadastral registration, the networks could be used for orientation. Registration of legal notifications on all intersecting parcels would no longer be necessary.

A registration of 3D physical objects needs to be organized and maintained and this registration will become a cadastral task. For the implementation of this registration, either a finite list of objects that need to be registered has to be made or the registration could be voluntary, as is currently the case for telecom networks based on the idea that such a registration offers benefits to the holders of 3D physical objects. In the cadastral registration spatial as well as non-spatial information on the whole 3D physical object is maintained. This information could be maintained directly, but preferably via the GII by which the information on physical objects can remain at their source (see Section 4.2). A 3D physical object can be queried as a whole. For example, which parcels are intersecting with the projection of a 3D physical object (this is a spatial query)? Which rights are established on these parcels? Who are the associated persons?

The solution of registering 3D physical objects (including geometry in 3D) meets the need of a 3D cadastre to register constructions themselves, or at least to have the

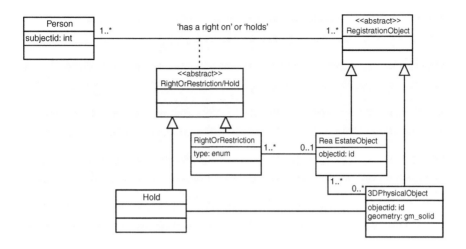

FIGURE 9.5 UML class diagram of 3D physical objects.

location of physical objects available in the cadastral registration (and included in the cadastral map). A 3D description of physical objects can be used if the cadastral map is available in 2.5D. A limited right still needs to be established on the intersecting parcels referring to the physical object to explicitly secure the legal status since the 2D parcel is still the basic entrance for establishing real rights and for the cadastral registration. However, the parcels do not need to be divided into smaller parcels, since the exact location is known in the registration. In addition, the information on the physical object needs to be maintained only once, instead of with every intersecting parcel. Since the physical objects are integrated in the cadastral geographical data set, the real situation is much better reflected than in the current cadastral registration. For the registration of 3D physical objects the UML class diagram in Figure 9.5 applies.

Apart from parcels (real estate objects), 3D physical objects are also registration objects. Rights and limited rights are still registered on real estate objects (2D parcels in this case). The only right that a person can get on a 3D physical object is that he or she can become the holder of this object. A 3D physical object is not a specialization of real estate objects: 3D physical objects are maintained in addition to parcels and parcels are still the basic entity of registration.

A basic complication that is not met by either solution is that a 2D parcel is still the base for registration, implying that the legal status of constructions cannot directly be established and not directly be registered on the construction or volume itself. Surface parcels (defined in 2D or 2.5D) are still always needed to ensure the legal status in 3D.

9.4   FULL 3D CADASTRE

To meet the cadastral needs at a more fundamental level, the concept of 2D parcel should be reconsidered, as well as the changing role of cadastral registration. As was seen in Chapter 1, nowadays, cadastral registration not only focuses on the registration of ownership of real estate, but also serves other tasks (used by both private and public sectors in land development, urban and rural planning, land management, and environmental monitoring). In the full 3D cadastre the concept of 2D parcels as the only basis for registration is abandoned. The registration object in the full 3D cadastre has a wider meaning. It may include areas or volumes, not necessarily coinciding with 3D ownership boundaries of land, e.g., a forest protection zone. This is similar to the term "legal land object" as defined in Reference 49 and 'RealEstateObject' in Reference 98.

In the full 3D cadastre, rights are no longer established on parcels, but on well-defined, surveyed volumes. This is the basic difference from the hybrid solution, which still holds to a 2D, but implicit 3D, registration. For the full 3D cadastre, two alternatives are distinguished: (1) a 3D cadastre in which a 3D real estate object is either an infinite parcel column, defined by a surface parcel, or a volume parcel and (2) a real estate object is always defined as a bounded volume parcel.

9.4.1 COMBINED 2D/3D ALTERNATIVE

The combined 2D/3D alternative starts with the currently registered parcels, which are converted into infinite parcel columns. In addition to infinite parcel columns, volume parcels are distinguished.

In this solution, the real estate objects can be:

- "Traditional" parcels, representing either infinite parcel columns, or columns of space of which volume parcels have been subtracted: these parcels are actually defined in 3D (based on the 2.5D surface representation)
- Volume parcels
- Restriction areas (only defined in 2D)
- Restriction volumes (defined in 3D)

The UML class diagram of this solution is shown in Figure 9.6 (see also Reference 98). In a full 3D cadastral registration, implemented according to this model, an instance of a parcel always exists, which is the basis of the cadastral registration. A volume parcel is only established if a bounded space is subtracted from a parcel column defined by the boundaries on the surface. Consequently, in the fictive case in which no stratified properties exist, this full 3D cadastre would contain only infinite parcel columns defined by boundaries on the surface that form a full 2.5D partition.

The collection of the 2.5D surfaces of parcels (parcels draped over a height surface) explicitly covers the whole surface (without overlaps and gaps). This is a very important concept in cadastral registration in order to avoid inconsistencies. A "Parcel" implies the whole 3D column above and below the surface or what is left after volume parcels have been subtracted from the parcel column. The geometry of the volume parcel defines a bounded space in 3D. Consequently, a complete space partition is defined by the infinite parcel columns and the volume parcels. One volume parcel can be established crossing several parcels.

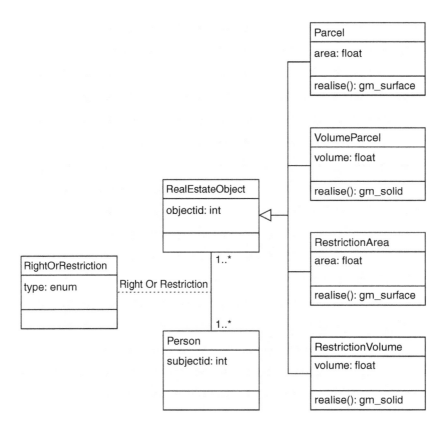

FIGURE 9.6 UML class diagram of a full 3D cadastre that supports both infinite parcel columns and volume parcels. The parcel objects are part of a 2.5D partition.

Important constraints for the full 3D cadastre are as follows:

- Projection of parcels should form a full partition of the 2.5D earth surface.
- Volume parcels may not intersect other volume parcels in 3D.

Because of the different meaning of restriction areas and restriction volumes, restriction areas may intersect other restriction areas (e.g., a forest protection zone may intersect a groundwater protection zone), and restriction volumes may intersect other restriction volumes. For example, a 3D volume that indicates severe soil pollution may intersect with a volume that indicates the presence of a monument imposed by the Law on Monuments.

To be able to register the parcels, volume parcels, restriction areas, and restriction volumes in the cadastral registration, all real estate objects must have a survey document, which should make clear what space the real estate object refers to. The 3D information in these survey documents can then be integrated in the cadastral geographical data set, which will be a mix of 2.5D objects (surface parcels and restriction areas) and 3D objects.

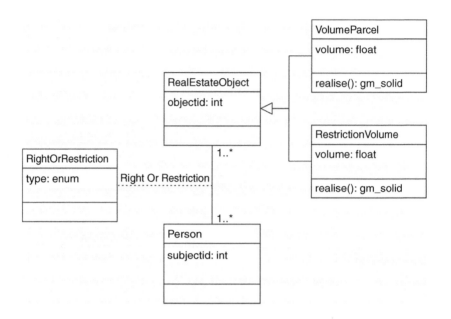

FIGURE 9.7 UML class diagram of a full 3D cadastre that supports only volume parcels. The volume parcels are part of a space partition.

9.4.2 PURE 3D CADASTRE

In the pure 3D cadastre that only supports volume parcels, the concept of 2D parcels (or infinite parcels that are defined by parcel boundaries on the surface) is totally abandoned (Figure 9.7). It is no longer possible to establish an ownership right on an infinite parcel column defined by boundaries on the surface.

Property rights to real estate objects can only be related to volume parcels that are fully defined and bounded in 3D. Consequently, open (unbounded) parcels do not exist. The volume parcels, which are the basis for registration, form a full partition of 3D space without gaps or overlaps. This requires a change in the legal framework since ownership no longer reaches as high or as low as a user has possible interest, but should always be explicitly limited in height and depth. When starting such a cadastre, one could think of limiting the already registered ownership of parcels in height and depth, e.g., reaching from 100 meters below the surface to 500 meters above the surface.

In addition to volume parcels, restriction volumes are registered that may intersect volume parcels as in the combined 2D/3D alternative. Every limited right and restriction that are established and registered in the cadastral registration should be accompanied with a 3D spatial description defined in a 3D survey document. The cadastral map is fully 3D since it only contains volumetric objects (volume parcels bounded with 3D boundaries).

In this solution cadastral registration of the whole country is converted into 3D.

9.5 EVALUATING THE CONCEPTUAL MODELS

In this section the proposed solution is considered, both from a cadastral point of view (Section 9.5.1) and a technical point of view (Section 9.5.2). Section 9.5.3 concludes with the optimal solution for a 3D cadastre.

9.5.1 SOLUTIONS SEEN FROM A CADASTRAL POINT OF VIEW

Cadastral Objectives

As concluded in Part I of this book, the main objective of cadastral registration is to warrant legal security in real estate transactions. This means that stratified property has to be registered in a correct way and that the registration should provide insight into the actual legal situation in a simple, straightforward, and sustainable manner (i.e., the cadastre should support optimal accessibility and maintainability). At this moment the accessibility of the registration in 3D situations is poor. At first sight even the professional user (notary, real estate agent, or cadastral employee) may not be aware of or completely understand the 3D situation, not to mention the public at large and non-cadastral specialists (e.g., planners and contractors). The better the accessibility of the registration in complex 3D situations, the better the legal security of the real estate is warranted. The main objective of a 3D cadastre focuses therefore basically on improving the information available in the cadastral registration in 3D situations (see also Chapter 4).

Cadastral Considerations on Proposed Solutions

A full 3D approach would solve a lot of problems: the basic entity of a cadastre is no longer a 2D parcel, by which all 3D situations have to be projected on a 2D cadastral map, but a volume. This offers better possibilities to reflect the real situation, as rights to real estate always have been related to a volume and not to just an area. However, the full 3D approach results in a renewal of the cadastral registration in which the concepts of rights in 3D and of the right of ownership need to be reconsidered. Within the current legal framework it is already possible to establish stratified property (3D property units); however, within this framework the ownership of real estate is mostly still land (surface) oriented.

In the first alternative of the full 3D cadastre (combination of infinite parcel columns and volume parcels) the new concept of the right of ownership of a parcel could include all space above and under the surface with possibly volumes subtracted to which other persons are entitled with a right of ownership. The question whether the legal framework can simply adopt this concept, without any complication, is dependent on the background of the specific cadastre. As seen in Chapter 3, some countries have already introduced the concept of multilevel ownership. However, in the Netherlands, the introduction of legal space that is no longer related to surface parcels may cause more complications.

The approach where 3D situations are stored in the 2D cadastre (hybrid solution) is advantageous from the point of view of accessibility, compared to the current situation. Both the 2D and 3D information is directly available and can be integrated, while this solution requires only minor changes in cadastral registration (and only in 3D situations) and no changes in the legal framework. The legal status of real estate is still strongly related to 2D land parcels, and not to 3D volumes, as in the full 3D cadastre approach.

The approach with external references to 3D situations is followed at the moment, apart from the fact that 3D situations are not stored in the cadastral database as so-called local files, but separately on paper (and recently on scanned) drawings. This registration has proved to be practical with apartment rights and could be improved by the inclusion of digital 3D drawings in the cadastral database. In addition, making the digital scanned deeds, including drawings, accessible through the cadastral database, will improve accessibility of information in 3D situations. Given the current cadastral data model, this option is a good starting point, but not a sustainable option for the future. The basic disadvantage is that the spatial and non-spatial information of the 3D property situation cannot be integrated with the cadastral registration.

9.5.2 Solutions Seen from a Technical Point of View

When looking at the solutions from a technical point of view the basic questions are how to support 3D spatial features in the current cadastral geo-DBMS, how to access this spatial information by all kinds of front ends, and how to represent parcel boundaries in 2.5D. The answers to these questions depend on technological possibilities and developments.

This subsection starts with a description of the optimal technical environment of a 3D cadastre followed by a description of the state of the art summarized from Chapter 6, Chapter 7, and Chapter 8. Based on these two aspects the technical perspective on the proposed solutions is given.

Technical Implementation of a 3D Cadastre: The Optimal Solution

The integrated architecture in which geometrical and topological information as well as administrative information on objects are stored and maintained in one integrated geo-DBMS should be the starting point for a 3D cadastre, as this offers best maintenance (consistency, integrity) possibilities.

An ideal case would be to have spatial information on all objects relevant to the cadastre (physical objects, objects representing legal space and height surfaces of parcels) in 3D space available in the database. The support of 2D, 2.5D, and 3D data types in the DBMS will offer the integrated storage of spatial data within the DBMS and spatial functions in 2D and 3D at SQL level in order to keep a consistent data set. The support of spatial data in a geo-DBMS includes spatial data types (geometry and topology), spatial operators (or functions), spatial indexing and clustering, and topological structure management of both planar and volumetric partition. All the spatial information that is maintained in DBMSs should be accessible by all kinds of front ends (GIS, CAD, Web-based front ends).

Technical Implementation of a 3D Cadastre: The State of the Art

As can be concluded from Chapter 6, mainstream geo-DBMSs have implemented spatial data types and spatial functions more or less similar to the OpenGIS Simple Features Specification for SQL. However, these implementations are basically 2D, with the possibility of storing 3D coordinates, and mainly focus on the geometrical primitive. OGC still works on extending the Simple Feature Specification for SQL to support topological structure and 3D geo-objects.

In the area of topology many concepts have been developed (both for 2D and 3D). However, extensive 2D topology structure management (partitions and linear networks) have only recently become available within some DBMSs. Therefore, it is still difficult to update geometry in DBMSs, because of the risk of inconsistencies. Standard support for 3D topology will still take years. In the mean time, the topology structure could be supported at application level while storing the results in the DBMS.

Although 3D geometrical primitives and 3D topological structures are not yet available within mainstream DBMSs, Chapter 6 proved the potentials of user-defined solutions. Chapter 7 showed that the user-defined solutions could also be accessed by several types of front ends, although 3D GIS functionalities in general still need to mature.

Initially steps to maintain an effective integrated height and parcel model in the DBMS were taken in this research and already showed potential in Chapter 8, although the integrated model still needs further improvements.

Apart from the modeling aspects, also the collection and insertion of 3D information should be considered as well as the conversion from 2D parcels to 2.5D representations of parcels. Although it is becoming easier to collect data in 3D (by means of video, laser scanning, and GPS), it will take a lot of effort to collect all the data needed for registering 3D situations. Collecting information on physical objects could benefit from automatic object reconstruction techniques, although complete automatic reconstruction implementations do not yet exist (see Section 7.1). On the other hand, the 3D objects of interest to the cadastre do not always relate per se to physical objects, e.g., a volume to which a right applies does not necessarily correspond completely with the 3D extent of the physical object, because a safety zone also may be included (which might require a buffer operation in 3D). It may be difficult to survey such a situation. The use of CAD designs to represent 3D geo-objects also needs further research.

Technical Considerations on Proposed Solutions

A full 3D cadastre is comprehensive from a technical point of view. It requires the integration of 3D surveyed data in a 3D topological structure (initially defined by parcel columns). Implementations for full 3D support in DBMS (geometry as well as topology) have just started and do not yet exist. A full 3D cadastre will therefore be dependent on user-defined implementations, which will not necessarily be a problem, as we have learned from experiences with the Dutch cadastral database in which both history and topology are maintained successfully, although not supported at the DBMS level.[129] The combined 2D/3D alternative of a full 3D cadastre is less complex than the pure 3D cadastre alternative, since a full volume partition is not needed. The collection of 3D data when the parcel *is* bounded will take considerably more effort than in the traditional 2D case, in which only the 2D boundary needs to be surveyed. The content of the current cadastral database containing 2D parcel boundaries is the result of surveying that has been carried out since the beginning of the 19th century. The collection of bounded parcel geometries in 3D has to begin from scratch. Consequently, it may take years until a serious 3D cadastral database is a reality. The integrated view based on 2D parcel boundaries and point heights is needed in the combined 2D/3D alternative of a full 3D cadastre. Chapter 8 showed the potentials for this integrated model, but it also showed that such an integrated view still needs further development.

The hybrid solution, with the current 2D cadastre as starting point (with infinite parcel columns defined with boundaries on the surface) and an extension to register 3D situations, seems a feasible solution for the medium-term future. 2D spatial objects are supported in DBMSs and 2D data are available in large amounts and are often still sufficient. The implementation of an extension to maintain 3D spatial features, having also non-spatial attributes, seems possible as well as the possibility to maintain a 2.5D surface of parcels. In addition, the hybrid solution does not need a full volume partition of space. The implementation of the hybrid cadastre will be based on techniques available to represent 3D spatial features and on new developments based on research.

The 2D classical registration with tags to 3D situations is current practice but seems not to be a sustainable solution for the future. The database contains references to paper or digital drawings (or files), instead of integrating the 3D situation as 3D spatial features in the geo-DBMS. The technical problem of this solution is that the DBMS cannot guarantee consistency (do two 3D situations overlap?), nor can the 3D situation be queried in a combined environment with 2D parcels or other 3D situations.

9.5.3 The Optimal Solution for a 3D Cadastre

As can be concluded from the above, the option "3D tags in the current cadastral registration" is a solution that works, as current practice proves; however, it has some basic limitations. The solution cannot provide one 3D overview of a cadastral map integrated with 3D property situations: 3D situations can only be examined per parcel, i.e., isolated from each other. This solution therefore does not give a base for efficient and sustainable registration in the future.

A full 3D cadastre offers solutions on the basic concept of the cadastre. With this solution ownership boundaries in 3D can be established and division of ownership in all directions can be defined and registered in the cadastral registration. However, the question can be posed whether a full 3D cadastre that supports only volume parcels (pure 3D cadastre alternative) is realistic for cadastral registrations that have a long history and already contain a lot of information that is related to 2D parcels. In addition, the 2D parcels still suffice in many cases. Technologically, it is possible to convert the unbounded parcel columns into bounded volumes; however, this conversion may encounter a number of complications within the legal and cadastral frameworks. The pure 3D cadastre alternative requires a total renewal of the cadastre, also in 2D situations, while the first alternative of a full 3D cadastre still has a strong link with current cadastral registration: traditional 2D situations (parcels with only one person entitled to each) can be kept largely unchanged. From a practical point of view, a 3D cadastre is mostly needed in densely built-up areas. For most of the country, however, a "classical" 2D cadastre based on 2D parcels serves its purpose well. Therefore, the pure 3D cadastre alternative is not seen as a feasible solution. The combined 2D/3D alternative of a full 3D cadastre offers the best opportunities to solve the complications of current 3D registration. Therefore, the combined 2D/3D concept of a full 3D cadastre will be the aim of this research.

Although the combined 2D/3D concept of a full 3D cadastre is the final stage where most problems of 3D registration are solved and although from a legal point of view this solution has already shown potential in some countries and states (Norway, Sweden, Queensland, British Columbia, the United States, Argentina), it might take some time before this concept can be adopted in legal frameworks in other countries, as in the Netherlands. The concept reconsiders the basics of cadastral registration: the concept of 2D parcels is abandoned since in a full 3D cadastre it is possible to bound the ownership of real estate in the third dimension, while the ownership of real estate is no longer (always) related to surface parcels.

Therefore, we also focus on a feasible solution for cadastral registrations in legal frameworks that are still surface oriented, i.e., frameworks in which 3D property units can be established only by imposing rights on the intersecting surface parcels. In those registrations the hybrid solution shows potential for the medium-term future, as it already meets the most important need of 3D cadastral registration: improve insight in 3D situations.

In conclusion, we start with the implementation of a registration of 3D situations in current 2D cadastral registrations, which is similar to the hybrid concept. Subsequently, we also look at the implementation of a full 3D cadastre, which not only meets the complications of cadastral registration in the medium-term future, but also reconsiders the concept of the 2D parcel as the basic cadastral entity.

9.6 CONCLUSIONS

In this Chapter 3 solutions for a 3D cadastre were studied: a full 3D cadastre with two alternatives, a hybrid 3D cadastre with two alternatives, and a cadastral registration that contains tags to 3D situations and links to 3D representations. The UML class diagrams for the proposed models were also given.

For the full 3D cadastre, the following two alternatives were introduced:

1. A 3D cadastre that combines the registration of volume parcels with the registration of infinite parcel defined by parcel boundaries on the 2.5D surface. The basic registration entities are infinitive parcels columns or original infinite parcel columns from which a volume parcel has been subtracted (parcels) and volume parcels.
2. A 3D cadastre that only supports volume parcels. The volume parcels form a full partition of space, without any gaps or overlaps.

The two alternatives to effectuate the hybrid approach are based on the fundamental needs for a 3D cadastre, i.e., to register the spatial extent of rights and to be able to reflect constructions themselves in the cadastral registration. The proposed concepts are the registration of right-volumes, in which the right that entitles a person to a volume is the starting point for registration, and the registration of 3D physical objects, in which the physical object is the starting point for registration. A right-volume is a 3D description of a right (legal space) that has been established for a 3D situation. The 3D description covers the complete 2D parcel with limitations in height and depth by horizontal planes. This 3D description is integrated in the cadastral map. The user can see that something is located above or below the surface including the approximate location. Precise information can be obtained from deeds and survey plans archived in the land registration. Registration of right-volumes requires only little adjustment in the current cadastre, although this registration may lead to gaps in the visualization of the 3D situation. The gaps occur since only right-volumes are registered when the 3D situation has led to a cadastral recording. Other cases are, for example, cases of not-registered personal rights (short lease) or obligations to tolerate constructions for public good that follow from general laws. The gaps are solved by a registration of physical objects. The registration of physical objects maintains physical objects above or below the surface, which mostly cross parcel boundaries. The spatial extent of physical objects is integrated in the cadastral geographical data set, or accessed via the GII, by which it is possible to use this information to support the cadastral tasks.

The proposed solutions were considered both from a cadastral point of view and a technical point of view. Based on these considerations, the full 3D cadastre approach was selected as the most optimal solution. The second alternative (the full 3D space partitioning) seems less realistic, as it requires a total renewal of cadastral registration while 2D parcels still suffice in many cases. Therefore, the first alternative is selected as the conceptual model that meets the registration of 3D situations in the most optimal way, seen from the legal, the cadastral, and the technical point of view. Technology developments have progressed in such a way that it is realistic to study the possibilities of maintaining and accessing 3D geometrical primitives in a DBMS and to maintain a 2.5D representation of parcels.

The full 3D cadastre is a step too far for the near- and medium-term future for some cadastral registrations, because the legal framework needs to be adjusted, which will encounter complications. It requires a reconsideration of the basic concept of cadastral registration: cadastral registration is no longer focused on land but on volumes. Therefore, the hybrid approach is also further considered in this book.

The remainder of this book focuses on the proposed solutions that show best potentials: the first alternative of a full 3D cadastre (parcel columns and volume parcels) and both alternatives of the hybrid solution. The selected alternatives are assessed on the concepts, the implications, and the implementations to come to optimal recommendations for a 3D cadastre.

10 Logical Model for a 3D Cadastre

In Chapter 9, the conceptual models for several alternatives for a 3D cadastre were described and evaluated. The alternative that showed best potentials is the full 3D cadastre in which parcel registration (where parcels imply an infinite parcel column or what is left after subtracting the intersecting volume parcel or parcels) is combined with the registration of volume parcels. However, the implementation of this concept of a full 3D cadastre might be very complicated in some cadastral registrations due to the required changes in the legal concept of ownership of land. Therefore, the hybrid solution is also further considered in this book. The hybrid solution, in which 3D situations are registered in addition to 2D parcels in order to improve insight into the 3D situation, showed best potentials for cadastral registrations that are still land (surface) oriented. This solution was translated into two alternatives: the registration of right-volumes and the registration of 3D physical objects. Both alternatives are further examined in this book.

The next step after the design of the conceptual models is the translation of the conceptual models into a logical model, i.e., database structures. In Chapter 6, the object relational model was selected as the most appropriate database model for the 3D cadastre implementation. In this chapter considerations are described that have to be taken into account when the conceptual models of right-volumes, 3D physical objects and the selected alternative of the full 3D cadastre (parcel columns and volume parcels) are translated into an object relational database structure (Section 10.1, 10.2 and 10.3). To actually effectuate the 3D cadastre the logical model can be populated with instances (data). Section 10.4 describes the 4D aspects of the logical models of the three selected alternatives: how to maintain history in a 3D cadastre. In Chapter 11 the logical model is filled with data in prototypes that are applied to the case studies introduced earlier in this book (Chapter 3). The chapter ends with concluding remarks.

10.1 RIGHT-VOLUMES IN THE DBMS

This section describes considerations for the two main parts of the logical model for right-volumes: the spatial data model (Section 10.1.1) and the administrative data model (Section 10.1.2). The data models need to be populated with data. Considerations for spatial data collection for right-volumes are described in Section 10.1.3.

The implemented logical model makes it possible to perform queries that are required to meet the need for a 3D cadastre. In Section 10.1.4 the queries are described that are possible when right-volumes are maintained.

10.1.1 SPATIAL DATA MODEL

The 3D description of right-volumes initially starts with parcel boundaries in 2D. Parcel boundaries are extended into the vertical dimension using the upper and lower

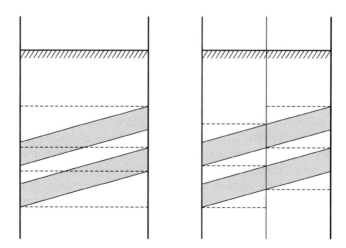

FIGURE 10.1 Subdivision of parcels (right) may solve the problem of two overlapping right-volumes within one parcel (left).

limits of rights. This spatial definition of right-volumes results in 3D volumetric features bounded with flat faces. The upper and lower boundaries of right-volumes are defined by horizontal planes. It should be noted, once again, that the representation of one z-value per upper or lower boundary is restrictive, especially when the terrain itself has relief.

To implement the spatial data model of this concept, a table is introduced (right-volume table) that contains for every parcel the different height levels of properties piled on one parcel (z-list). The z-list contains n z-values corresponding to $n - 1$ consecutive ranges associated with the parcel. The z-values should be preferably defined in absolute values. The z-values are stored as an Abstract Data Type (array).

Because of the simple definition of right-volumes (the footprint always covers the complete parcel), right-volumes within one parcel could overlap, which does not correctly reflect the real situation. This could be solved in some cases by subdividing parcels in such a way that the overlap of right-volumes within one parcel is solved (Figure 10.1). However, this subdivision does not work in all cases, e.g., where the two sloping levels touch. In addition, the question can be posed if this is a feasible solution to this problem.

As seen in Chapter 9, only positive right-volumes will be registered. Therefore, the remainder of parcel columns will not be registered with right-volumes. For example, when a tunnel intersects a parcel that is owned by a private person and a right of superficies has been established on the parcel to hold the tunnel, only one right-volume is registered, referring to the right established for the tunnel and no right-volume referring to the upper and lower "open" space. More generally, this means that a right-volume for the bare owner (person who holds the ownership that is encumbered with limited real rights) will not be registered. The space to which the bare owner is entitled can be found by subtracting all positive right-volumes that have been registered above and below the parcel. It is possible that the space that is left for the bare owner may also include space between two right-volumes, e.g., when a tunnel would be drilled below The Hague Central Station (see Section 3.1.3) and a right of superficies would be established for the tunnel. The holder of the railway platforms (in this case the bare owner of the parcel) would in that case own the space between the right-volume of the bus and tram station and the right-volume of the tunnel. Since the right-volumes are defined with a list of z-values, this space between is also included in the z-list.

Geometry

3D (volumetric) data types are future work for DBMSs. Still the advantages of current techniques can be used. The polyhedron approach as it is currently available in the DBMS is appropriate for defining right-volumes since the geometry of polyhedrons is similar to the way right-volumes are spatially defined (existing of flat faces). A right-volume is built by starting with the list of coordinates of the whole parcel ring. A vertical face is generated between every two coordinates, using the upper and lower limits of the right. The right-volume is closed by two horizontal faces: one on top and one below.

Based on the right-volume table containing the z-lists, and the parcel boundaries a 3D geometrical representation of the right-volumes can be generated in three ways (the first two are available within current techniques, and the last one is the self-implemented solution):

- Define a right-volume as a set of polygons defined in 3D, this is partly a topological solution, since faces can be shared (see Section 6.1.2).
- Define a right-volume as one multipolygon defined in 3D (see Section 6.1.2).
- Using the 3D geometrical primitive that was implemented as part of this research (this geometrical primitive supports internal topology, see Section 6.4).

All these 3D representations consist conceptually of polyhedrons.

Topology

In the prototype the topological structure of right-volumes is stored and a function has been written that generates the geometrical description of right-volumes using these geometrical primitives. The geometry of right-volumes can be made available with a view. The z-list is sufficient to generate the topological representation of right-volumes based on the realized geometry of parcel boundaries.

The spatial model of right-volumes should support topological structure within one parcel, which means that faces, edges, and nodes are shared within one parcel. In a more advanced implementation of right-volumes one could think of sharing nodes and faces between right-volumes that are established on neighboring parcels. The support of topology within a parcel makes it possible to query neighbors of right-volumes on top of each other using an explicit topological structure, which is more efficient than performing this topological query on geometry (as seen in Chapter 6). In addition since stratified right-volumes share the inbetween faces, data consistency is assured; e.g., two right-volumes on one parcel cannot overlap or during updates when a 3D boundary (3D face) is moved.

Current geo-DBMSs do not support topological structure in 3D. Therefore, the user-defined model as described in Section 6.2.4 (SSM) is used in the prototype implementations to implement topological structure for right-volumes.

A function in the prototype implements the topological structure of 3D right volumes within one parcel, based on the z-list and the realized geometry of parcel boundaries. This implementation could be improved by storing only the z-list and making the topology structure of right-volumes available with a view.

10.1.2 ADMINISTRATIVE DATA MODEL

Persons are entitled to right-volumes by means of limited rights. This right is established on a parcel and associated with the right-volume. Persons entitled to a right-volume can be found by the right to which the right-volume is associated (e.g., a right of superficies). The right-owner of this right is the person entitled to the right-volume. The full ownership of a parcel as well as the ownership of a parcel that remains after limited real rights have been established on the parcel is never related to right-volumes because "negative" right-volumes are not registered. In the case of right-volumes four cases can be distinguished:

1. The person using space above or below the surface is the full owner of the surface parcel or parcels. No right-volume can be registered, because no limited rights are established.
2. The person who holds a construction above or below the surface is the bare owner of the surface parcel or parcels: other persons are entitled to the space above and below the construction by means of limited rights, such as right of superficies, right of long lease, right of easement, etc. Right-volumes (related to third parties) are registered for space above and below the construction but not for the bare owner of the parcel who is the holder of the construction.
3. The person who holds a construction (e.g., tunnel) is entitled to space above or below the surface by limited rights on the parcel, such as right of superficies or a right according to public law. A right-volume will be registered for the space related to the construction. This is the case in Figure 11.7.
4. The person using space above or below the surface is not the owner of the surface parcel and has no rights on the parcel: the legal status of the space used is not explicitly registered.

Attributes

The person entitled to the right-volume with a right is important, as well as the type of right that entitles the person to that right-volume. This information can be accessed via a view on the administrative tables of cadastral registration. Both the list of persons and the list of types of rights are presented as Abstract Data Types (arrays) in the right-volume table. Because the number of persons entitled to parcels can vary, it would result in multicolumn or multirow representations to list the persons and rights that are related to one parcel in a full relational implementation. Using ADTs enables us to use a one-column structure for the list of persons as well as for the list of rights.

Whole 3D real estate objects are also maintained, with their ids. The right-volumes contain a reference to the id of the real estate object of which they are a part, by which it is possible to obtain a list of all right-volumes that refer to the same 3D real estate object. Spatial information on the whole 3D real estate object could be maintained, as well as non-spatial information (function, holder). Right-volumes are identified by unique numbers. The numbering is done in such a way that right-volumes are related to the affected parcel. The numbering could be done similarly to the numbering of apartments. For example, on a ground parcel 1234, two right-volumes are generated, one ranging from -20 to 0 to hold a tunnel and one from 0 to 13 to hold a building on top of the tunnel. The right-volumes could be numbered 1234 RV1 and 1234 RV2. The RV refers to a right-volume.

In general, the table (or view) containing the information on right-volumes must have at least the following columns:

- id: id of the right-volume
- type_of_right: The type of right that is associated with the right-volume; this information comes from the administrative part of the cadastral database via the association to "RightOrRestriction"
- subject: Person entitled to the right-volume; this information also comes from the administrative part of the cadastral database via two associations: first to RightOrRestriction and second to Person
- geometry: The 3D geometry of a right-volume of type polyhedron; the geometry is defined with a spatial function on the topological structure
- id_real_estate_object: The id of the whole real estate object where the right-volume is a part (a registration of these real estate objects should be started)
- tmin: The start time of the right-volume (see Section 10.4)
- tmax: The end time of the right-volume (see Section 10.4)

Relationships

Juridical relationships between right-volumes and parcels exist via the limited rights, apartment rights and restrictions. A limited right, an apartment right, or a restriction is established on a parcel and is associated with a right-volume. The possible relationships between right-volumes and parcels are m:1. Several right-volumes can be stacked on one parcel, while a right-volume cannot cross parcel boundaries; i.e., a right-volume belongs to exactly one parcel. The relationship between right-volumes and parcels is explicitly maintained via the different associations (association to RightOrRestriction and association to Parcel).

Because right-volumes are related to surface parcel, a subdivision of a parcel that contains a right-volume will cause a subdivision of the right-volume. This is a weak point in the concept of right-volumes, as the pattern of right-volumes is very much influenced by the surface configuration but also by what happens with the parcels on the surface (e.g., in case of a transfer of the surface parcel or in case of a subdivision).

10.1.3 Data Collection

Because geometry of right-volumes is relatively simple (compared to the geometry of 3D physical objects and the geometry of 3D parcels as in the full 3D cadastre case), data collection is not very complicated. For the 3D description of right-volumes the only information needed is the lower and upper level of the space to which the right applies. The issue of how to express the z-value (using absolute or relative values) was addressed in Chapter 8. The height and depth of right-volumes are precisely defined in deeds and survey plans (preferably using absolute values, possibly combined with relative values). The deed can also contain a more precise definition of the space to which the right refers, e.g., using a drawing or 3D survey plan. The 3D description in the deed is not necessarily the same as the factual boundary of the construction. For example, the right of superficies established for the tram/bus station as part of the building complex of The Hague Central Station (Section 3.1.3) can be defined just for the construction of the bus/tram station. However, it is also possible to establish a right of superficies that is higher than the bus/tram station, in view of future expansion on top of the bus/tram station by the municipality of The Hague.

10.1.4 Querying

The main improvement that the registration of right-volumes will provide is the insight into the vertical dimension of rights. The queries that will be possible in the registration of right-volumes are as follows:

- Is someone else entitled to space below (or above) my parcel?
- Who is the owner of the construction below this parcel?
- Who is the owner of this space?
- What is the distribution of properties in 3D, i.e., show the 3D cadastral map of the situation?
- What is the spatial extent (both in 2D and in 3D) of this right of superficies?
- Who is the owner of the right-volume above/next to this right-volume?

These queries can be performed when right-volumes are registered. In the prototype implementation the topological structure of right-volumes belonging to one parcel is implemented. Therefore, the query to find the right-volumes on top of another can be performed on the topological structure.

The query to find neighbors next to each other needs to be performed on the geometrical primitives. Another possibility to find next-to neighboring right-volumes is to use the 2D cadastral geographical data set since right-volumes coincide with parcel boundaries: Which right-volumes are established on the neighboring parcel and at what level?

10.2 3D PHYSICAL OBJECTS IN THE DBMS

This section describes considerations for the spatial model (Section 10.2.1), for the administrative model (Section 10.2.2) and for data collection in case of a registration of 3D physical objects (Section 10.2.3). CAD models are important sources for data collection in case of a 3D physical object registration, as seen in this section.

Therefore, this section also includes a discussion on how to link CAD with GIS (Section 10.2.4). Finally, Section 10.2.5 describes the queries that are supported in a registration of 3D physical objects.

The registration of physical objects primarily focuses on a registration of infrastructure objects (objects for public good). Therefore, the considerations described in this section do not especially address aspects in case of property units within building complexes.

10.2.1 SPATIAL DATA MODEL

A 3D description of a physical object consists of the outer boundary of the physical object. The 3D legal space of physical objects is not within the scope of this registration, but will be registered in the full 3D cadastre (see Section 10.3). Preferably, the description of a physical object is provided and maintained by the organization responsible for the object and accessible through the cadastral database within a Geo-Information Infrastructure (GII).

Geometry

The 3D geometry of physical objects can be stored in the DBMS within current techniques by using primitives that are supported (with minor support for topology or support only for internal topology).

From Chapter 6 it can be concluded that current mainstream DBMSs only support 3D objects by using flat faces. This is a limitation in storing 3D information concerning 3D physical objects, which often have complex geometries. For example, the tunnel in Figure 11.8 has been created in MicroStation by using the center-line (defined with x, y, and z coordinates) and using the cross section of the tunnel, which is a circle with a radius of 7.5 meters. The CAD software extrapolated the cross section along the length axis. To store this object in a DBMS, a conversion from the parametric description to a polyhedron representation is required in which the tunnel geometry is defined by many flat faces. This will decrease the quality of the 3D representation while storage space will increase. A better option would be to store the centerline and the cross section in the DBMS, whereupon the DBMS can generate the 3D representation. This is similar to the way circles are currently stored in 2D in the DBMS. Circles are not specified as a polyline consisting of many coordinates, but as a specific type of line or curve, i.e., a circle defined with three points on the circumference. Since current geo-DBMSs support only 3D geometries that are rather simple, the geometry of complex geometries is not taken into account in the prototypes.

Topology

A full topological structure in which relationships between 3D physical objects are maintained is not needed in a physical object registration, since the maintenance of 3D physical objects does not require a full partition of space. Therefore, a limited support of topology (only within objects and not between objects) as implemented in the geometrical primitives will be sufficient for 3D physical objects.

Topological relationships between two arbitrary objects (2D or 3D) can be derived by means of geometrical functions available in DBMS and can be used in constraints (e.g., to avoid overlaps). When topological relationships between objects are "derived on-the-fly" the accuracy of the data is very important (when are objects inside, touching, equal, overlapping?). This is complicated in 2D, but even more complicated in 3D.

10.2.2 ADMINISTRATIVE DATA MODEL

3D physical objects are registered, together with associations to holders of these objects. The holder of a 3D physical object is the person who has an economic ownership to the construction (right of exploitation). He benefits from the construction but also pays the costs for maintenance and replacements. In the case of physical objects below/above the surface, four similar cases as in the case of right-volumes can be distinguished:

- The holder of the object is the full owner of the surface parcel or parcels.
- The holder of the object is the bare owner of the surface parcel or parcels: other subjects have limited rights on intersecting parcels, such as right of superficies, right of long lease, right of easement, etc.
- The holder of the object is not the owner of the surface parcel. The holder has limited rights on the surface parcel, such as right of superficies or a right according to public law.
- The holder of the object is not the owner of the surface parcel and has no rights on the parcel: the legal status of the physical object is not explicitly registered.

These cases can vary for one physical object per intersecting parcel. The last case should be avoided or should be solved by other regulations (such as the Law on Telecommunications). However, when 3D physical objects are stored in the DBMS these "gaps" can be depicted. For the first case, the possibility to register 3D situations is not necessary from a legal point of view (the legal status of the space above/below the parcel is clear: the holder of the construction owns the whole parcel column). However, the information on the whole physical object might be needed for future or reference purposes (e.g., when the parcel intersecting with the construction is sold without selling the construction). Since the existence of a 3D physical object is the basis for registration, 3D physical objects will also be registered when the holder of such an object holds the intersecting parcels in full ownership. Both the parcel and the 3D physical object at the specific location will be registered and cadastral registration will be able to reflect the real situation at the location.

Attributes

The rights for physical objects are not established directly on a 3D physical object but on the intersecting parcels. To ease querying, these rights may refer to ids of the 3D physical object using primary and foreign keys. The set of rights that is associated with one physical object can also be found by finding the intersecting parcels of a construction and then find the rights established on these parcels of which the subject is the same as the holder of the construction.

In general, the table containing the information of 3D physical objects must have at least the following columns:

- id: id of the 3D physical object
- subject: Person who has a permit to exploit the physical object (accessible via association)
- geometry: The 3D geometry of a 3D physical object (which is explicitly stored)
- tmin: The start time of the 3D physical object
- tmax: The end time of the 3D physical object

The 3D physical object table could also contain a column with a list if all intersecting parcels. This list could be stored explicitly, but a better option would be to define this list in a view with a spatial function.

Relationships

When 3D physical objects are spatially described in the DBMS, it is not necessary to describe the relationships between parcels and physical objects explicitly, since these implicit relationships (n:m) can be obtained through spatial functions available in the DBMS or by visualizing all the spatial information in an integrated view. As a rule, when a physical object intersects with a parcel, a juridical relationship shall be established. This rule could be implemented as a constraint in the DBMS. When a 3D physical object is inserted in the DBMS, it is checked if there are parcels intersecting with the 3D physical object with no rights established for the 3D physical object.

Via rights the possible explicit relationships between 3D physical objects on the one hand and parcels on the other hand are m:n.

Juridical relationships between a 3D physical object and the surface parcels exist through the holders of the 3D physical object who should be subjects of rights or restrictions on intersecting parcels.

10.2.3 DATA COLLECTION

Spatial data models for 3D physical objects can be populated with data obtained by object reconstruction techniques. As seen in Section 7.1.2 the process of 3D object construction still needs to be done partly manually and is therefore time-consuming. In addition, underground constructions such as tunnels and pipelines cannot be captured using aerial laserscan and photogrammetric techniques. Therefore, it is useful to look at other possible sources. Since 3D data are available with designers, mostly as CAD models, this data could be used to populate the spatial data models of 3D physical objects in the DBMS.

The next step is to study how CAD designs can be used and what selections and generalizations are needed to obtain the relevant information from these designs, such as the outer boundary of objects. As part of this research a municipality (Rotterdam), two departments of the Ministry of Transport and Public Works (*Bouwdienst van Rijkswaterstaat* and *Projectorganisatie HSL-zuid*), and a designer (*Holland Railconsult*) were visited to look for usable CAD models. Based on this survey the conclusion

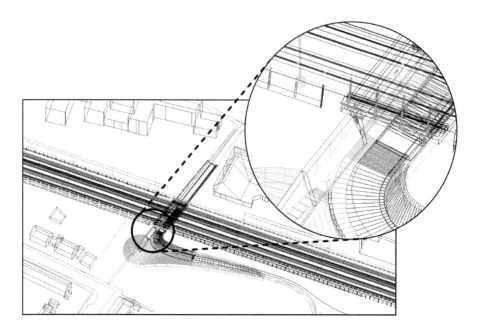

FIGURE 10.2 **(Color figure follows page 176.)** The CAD model designed for a cycle tunnel in Houten, the Netherlands. (Courtesy of Holland Railconsult.)

can be drawn that CAD models are not always created in the design process of 3D physical objects (and are therefore not available to be inserted into a 3D cadastre). Most tunnels are still designed on 2D drawings by using linear profiles and cross sections. Contractors and builders are used to the 2D drawings: understanding 3D drawings would require other skills and software. However, this information could be a very good basis for deriving a 3D model for the 3D cadastre.

There are plenty of examples in which 3D CAD models are generated in the design process, but mainly for visualization purposes (Figure 10.2).

A case study was carried out to see how 3D CAD models, mostly covering local environments, could be converted into a set of 3D geo-objects.[72] This study revealed that CAD models, mainly designed for visualization, are not directly suitable for 3D cadastre purposes for several reasons. The files can become unworkably large since they are most-often made not for interactive purposes, but for generating animations. Furthermore, they contain too much detail; objects can hardly be recognized in the file-based models and cannot be easily selected, and finally 3D spatial data in CAD models are defined by complex geometries, which are mostly parametrically described. At the moment these files cannot be automatically converted into a set of individual objects defined with (simple) geometrical primitives that are available in spatial DBMSs (point, lines, polygons, polyhedrons). Another problem is that CAD models are mostly defined in local coordinate systems, whereas 3D geo-information that needs to be combined with 2D geo-information should be defined within a national coordinate systems.

Although the use of 2D and 3D CAD models still seems to offer a lot of potential for the 3D cadastre (information on the third dimension is available in those models), generating relevant information from these models requires further study. Using CAD models to obtain spatial models for 3D physical objects touches the fundamental issue of bridging the gap between GIS and CAD. This requires further study on the fundamentals of GIS and CAD (see next subsection).

10.2.4 FUNDAMENTAL ISSUES WHEN LINKING GIS AND CAD

CAD systems were originally developed to design large-scale models (usually of relatively small size), without maintenance of attributes and not related to geographic coordinate systems. In contrast, GIS was able to manage geo-information obtained from some kind of measurement technique resulting in very large data sets, including attributes and supporting a variety of different geographic coordinate systems. Nowadays, large-scale geo-information is a topic of interest for both CAD and GIS users, although CAD and GIS are still two different worlds. For example, in ISO two different committees are responsible for standardization in GIS (TC 211 Geographic information/Geomatics geo-information) and in CAD (TC 184 Industrial automation systems and integration).

CAD designers are increasingly confronted with the request to provide and design geo-information, i.e., the geometry of identifiable objects, with fixed location with respect to the earth, to which information can be linked. These data may serve a variety of purposes, e.g., spatial analyses, spatial planning, decision support, updating existing geographical data sets with planned designed objects, etc.

The process of linking GIS and CAD raises some fundamental issues. The aim of CAD engineers is no longer to produce a geometric and visual representation of a local environment. These local environments are now part of the complete world, for which real coordinates are needed. Because the same information is reused and updated, a system is needed to maintain the integrity and consistency of the spatial, spatial-temporal, and thematic data, which used to be core business of GIS. These trends require a tighter connection between GIS and CAD to be able to harmonize the geometrical primitives common in CAD software with the geometrical primitives and topological structure as defined by the GIS community, e.g., as defined by OGS.

In 2D a great deal of progress has been observed in linking GIS and CAD during the last years; e.g., cadastral parcels can be designed in CAD systems (with some kind of geographic extension) and can be maintained in a DBMS. These are indications that the border between GIS and CAD is fading at least from the user's perspective. In 3D, CAD and GIS integration is even more challenging. CAD software provides all kind of primitives to create a geometrical model (and their visual attributes) close to reality; however, these primitives are not supported in the GIS world.

As seen in Section 7.1, CAD software and not GIS software contains a set of tools to design, edit, and update large-scale 3D geo-data. Therefore, a closer connection between GIS and CAD may be very beneficial for 3D GIS developments. Questions that need attention in this process are how to use CAD primitives (e.g., parametric primitives) in an OpenGIS-compliant environment (with possible extensions) and how OpenGIS primitives can be extended to use CAD functionalities (textures, shading, etc.) to represent a model close to reality. An extensive study of the challenges and opportunities in the integration of GIS and CAD can be found elsewhere.[231]

10.2.5   QUERYING

In the case that the description of 3D physical objects is available in the cadastral database, the queries that are supported are as follows:

- Is the owner of the parcel the same as the holder of the 3D physical object?
- Is this construction located above or below the surface?
- Which other 3D physical objects are located on top of or below a certain 3D physical object?
- Which surface parcels intersect with a projection of a 3D physical object, or vice versa, which 3D physical object intersects with a certain parcel?
- What rights are established on surface parcels intersecting with a 3D physical object? Which subjects (i.e., legal persons) hold these rights?
- What is the overlap area between a 3D physical object and a 2D parcel?
- What is the volume of a 3D physical object (may be relevant for tax purposes)?
- What is the area of the footprint of a 3D object?

When physical objects are maintained in the cadastral DBMS these queries are possible. Once one has detected which parcels intersect with a 3D physical object, the juridical relationships between parcels and the 3D physical object can be obtained by administrative queries on the tables that contain the juridical relationships between parcels, rights/restrictions, and subjects (persons).

10.3   VOLUME PARCELS IN THE DBMS

This section describes the issues of translating the selected conceptual model of a full 3D cadastre (containing infinite parcel columns and volume parcels) into a logical model concerning the following fundamentals: spatial data model (Section 10.3.1), administrative data model (Section 10.3.2), and how to populate the spatial data model (Section 10.3.3). Section 10.3.4 describes the queries that are possible if a full 3D cadastre is implemented.

10.3.1   SPATIAL DATA MODEL

The selected alternative of the full 3D cadastre maintains infinite and remainder parcel columns, volume parcels, restriction areas, and restriction volumes. The volume parcels are related to an amount of space that is bounded. In the database, the volume parcels are modeled in 3D, whereas the infinite and remainder parcels are defined by parcel boundaries described in 2D and by parcel surfaces in 2.5D. The 3D description of these infinite and remainder parcels will not be visualized or constructed in the cadastral registration itself, but can be conceptualized by the user by subtracting the intersecting volume parcels from the infinite parcel column. If the parcel column does not intersect with a volume parcel, the ownership to the surface parcel is defined as described in the Civil Code (including space above and below the surface and reaches as high and as low the user has interest; see Section 2.3.1).

Geometry

The volume parcels can be defined in a geometrical model in the same way as right-volumes: using the three data types that were tested and evaluated in Chapter 6:

1. Define a right-volume as a set of polygons defined in 3D (see Section 6.1.2).
2. Define a right-volume as one multipolygon defined in 3D (see Section 6.1.2).
3. Use the user-defined 3D geometrical primitive (Section 6.4).

The geometry of volume properties that are rather simple (defined with flat faces) can be modeled with these simple 3D primitives (using absolute z-values). However, complex volume properties need to be modeled using more complex primitives. In the prototype the internal topological structure of volume parcels is maintained whereupon a geometrical realization can be obtained. After the volume properties are described using the self-implemented 3D geometrical primitive they can be validated and queried in 3D. The geometry of infinite and remainder parcels are defined by parcel boundaries on the surface (based on the 2D topologcal structure), while the 2.5D surfaces of parcels are maintained in a TIN structure, preferably in an integrated view of parcel boundaries and point heights.

Topology

The infinite and remainder parcel columns, together with the volume parcels, form a full partition of space. In the proposed implementation, the full partition of space is not implemented as such, because the infinite and remainder parcels are not modeled with volumetric representations. However, in densely built-up areas a full 3D partition of space could be considered. A 2D (or preferably 2.5D) topological structure is maintained for the parcels that are defined by surface boundaries. To assure the full 3D topological model, the following constraints need to be implemented:

- Volume parcels should not intersect (touch is allowed).
- Volume parcels should not cross surface parcels that are not subdivided into the third dimension, i.e., parcels on which no volume parcels have been established (this constraint should be used when volume parcels are inserted in the cadastral database).

In the prototype, the 3D characteristics of volume parcels are inserted in the topological structure using the Simplified Spatial Model (SSM) (Section 6.2.4).

10.3.2 ADMINISTRATIVE DATA MODEL

In 3D property situations, only one case can be distinguished (instead of four as in both hybrid alternatives). The person who uses space above or below a surface belonging to another person is entitled with an ownership right to the volume parcel. In those cases the parcel column or columns are subdivided. When a person uses space above or below a surface from another person without establishing a volume parcel, the legal status of the situation is not established and cannot be registered in the cadastral registration.

Attributes

The principle of the full 3D cadastre is that the cadastral geographical data set (parcel columns and volume parcels) consists of a full partition of space. Every surface parcel or volume parcel is related to an amount of space. The attributes that are stored with these volume parcels do not differ much from those in the 2D case (except the attribute "area" is replaced by "volume").

Relationships

It is no longer necessary to project a 3D property situation on the surface, since it is possible to establish volume parcels that have no relationship with surface parcels. Only when the volume parcel is created (and subtracted from the infinite parcel column) should the owner of the surface parcel agree with the subdivision. The deed establishing the volume parcel needs in this case only to mention the surface parcel. The boundaries of the volume parcel are defined in a 3D survey plan. This procedure is quite similar to the horizontal subdivision of parcels in which a 2D survey plan is required. In the future the volume parcel can be sold, without relating it to the underlying surface parcel (with the exception that rights on the surface parcel might be necessary to assure the use of the volume parcel, e.g., when the surface parcel is needed to access the 3D property). Future transfer of a surface parcel that intersects with a volume parcel will transfer only the remainder of the parcel that is left after subtraction of the intersecting volume parcel. In contrast to the right-volumes, a subdivision of a surface parcel has no consequences for the volume parcels intersecting with the concerning parcel. The 3D cadastral geographical data set provides insight into the property situation for the parties involved; e.g., the transfer will not mean transferring an infinite parcel column. The legal status of the situation can be obtained by tracing the ownership back in history, as is current practice in the case of apartment rights: one deed is necessary to establish the apartment units (*splitsingsakte* in Dutch). After this transaction the apartment unit exists as a separate unit in the administrative part of the land registration. However, apartment units always keep a legal relationship with the other apartment units in an apartment complex in contrast to volume parcels, which are totally independent of other property units.

10.3.3 DATA COLLECTION

Every volume parcel should be established by means of a 3D survey plan, as in the case of volumetric parcels in the United States (Section 3.6) and Queensland, Australia (Section 3.4). Note that surveying in 3D might be difficult where the volume parcel does not relate to a built construction and also when the geometry of a volume parcel is complex. The 3D survey plan should define how the volume parcel is bounded by defining all the corner coordinates with x-, y-, and z-coordinates in the National Height Datum. The insertion of volume properties in the cadastral database makes it possible to check the volume property (is the property closed, are all faces planar?) and to check the constraints to assure the topological partition of space (the volume property should not intersect another volume property; the volume property should not intersect a parcel on which no volume properties have been established). After these checks, the geometry of the volume property can be inserted into the cadastral geographical data set.

A procedure should be developed and defined to convert the 3D survey into an internal topological structure and into geometrical primitives in the database. If this procedure is clear, the process from surveying to insertion in the cadastral database can be streamlined. In the prototype, the whole process from 3D survey plan to geometrical and topological representations in the cadastral database has been implemented.

10.3.4 QUERYING

The queries that are supported by the full 3D cadastre are as follows:

- Is this volume parcel valid (closed, planar faces)?
- Are these volume parcels overlapping? (This query can be used in constraints.)
- Is this volume parcel intersecting a parcel that is defined by an infinite parcel column (can be used in constraints)?
- To what space is this person entitled?
- Does this parcel refer to an infinite or a remainder parcel?
- What are the 3D neighbors of this parcel or volume parcel?
- What is the 3D cadastral map?

10.4 MAINTAINING HISTORY IN THE 3D CADASTRE

When the registration of 3D physical objects, right-volumes, or 3D volume parcel becomes practice, updates will occur by which version managing (old state before update and new state after update) is necessary, which is the 4D aspect of data modeling. An example of maintaining history at the database level is the cadastral database of the Netherlands Kadaster. This database maintains history as described in Reference 129 in a self-implemented extension, as history was not supported by mainstream DBMSs at the time of system development. History is currently maintained at the record level (only for the spatial part of the cadastral database). For every object (parcel and boundary) a start time and an end time are stored. When an object is created, the current time is set as the start time in the new record and a time in the faraway future is set as the end time. This is necessary to be able to reconstruct the correct situation at any given point in history, e.g., to show the cadastral map of October 10, 1988. The unique identifier for objects (key) is the pair (object-id, start-time). Only when a new object is created or an old object is drastically changed (e.g., subdivided) a new object-id will be used. For simple updates a creation of a new object version (with same object-id) is the way to capture full history.

This structure assures topologically consistent data. For topological references, only the object-id is used to refer to another object. In the situation that a referred object is updated and keeps its object-id, the reference does not change. This avoids, in a topologically structured data set, the propagation of changes for many objects when only one object is changed as all objects are somehow connected to each other. In case the object-id of a referred object is changed (becomes a different object), the referring object also has to be updated.

10.4.1 HISTORY FOR RIGHT-VOLUMES

History for right-volumes can be maintained in a similar way if we assume that the 3D internal topological structure is the basic structure that is maintained. When the face is the lowest-dimensional topological object, history can be maintained on faces similar to parcel boundaries. When a face is moved, the face is updated. If this is not seen as a major adjustment, no new object-id will be created for the face. The 3D geometrical description of right-volumes will change because the object consists of references to the faces that are updated. When nodes are the lowest-dimensional objects, the same apply for nodes (nodes are updated and the geometrical description of faces and right-volumes change through their references). When a new object-id is created for lower-dimensional objects, the change of such an object will propagate changes in the higher-dimensional objects that refer to these objects. Topological consistency of different time stamps should therefore always be checked while updating.

10.4.2 HISTORY FOR 3D PHYSICAL OBJECTS

History on the geometry of 3D physical objects can be maintained on the whole object. A start time and an end time are maintained as attributes for 3D physical objects. Updates work in the same way as updates of parcel boundaries. A new object version will be created and the old one will be ended when an update occurs. As no topology is maintained between 3D physical objects, updates of 3D physical objects do not affect other 3D physical objects. However, consistency checks should ensure that 3D physical objects, also in the new situation, do not overlap. In addition, changes will affect other objects that contain references to the physical objects (e.g., where the holder of a physical object refers explicitly to the physical object).

10.4.3 HISTORY IN A FULL 3D CADASTRE

In the case of a full 3D cadastre, the history can be maintained in the same way as in current registration if we assume that the 2.5D topological structure is maintained (for infinite and remainder parcels) as well as the internal 3D topological structure and a full partition of space in densely built-up areas. If the lower-dimensional objects are updated and no new object-ids are created, the geometrical description of the higher dimensional objects will change through the defined references. However, in the new situation consistency checks should ensure that 3D volume parcels do not overlap in 3D. Updates of an object (surface parcel or volume parcel) should result in a new object-id if the updates are major changes (e.g., in case of a subdivision) to avoid losing some of the history. In this process, topological consistency of different time stamps needs special attention.

10.5 CONCLUSIONS

This chapter described the issues that need to be considered when translating the conceptual models of right-volumes, 3D physical objects, and the full 3D cadastre into logical models for the selected object relational database structure.

Concerning spatial models, current techniques are appropriate for modeling the simple geometry of right-volumes and for modeling simple volume parcels, although the 3D topology structure needs more research (e.g., implementing consistency checks as part of the DBMS).

Geometry of both 3D physical objects and complex volume parcels is harder to model within current techniques. To define the geometry of 3D physical objects and complex volume parcels precisely in the DBMS, research is needed on storing complex 3D geometries (more complex than a polyhedron) in a DBMS.

Topology between objects in the proposed solutions for a 3D cadastre (right-volumes, 3D physical objects, and volume parcels) is not needed, only in the case of volume parcels in densely built-up areas. Spatial relationships between two 3D geo-objects can be obtained by spatial functions. Topology forms, therefore, no bottleneck for implementing the logical models of these objects, except when a full 3D partition is needed.

Populating the spatial data models with data was another issue that was considered in this chapter. The data collection in case of right-volumes and volume parcels requires 3D surveys instead of 2D surveys. Procedures should be set up that regulate the content of 3D survey plans (what data should be incorporated and how?). In case of 3D physical objects one could think of using the CAD designs of the physical objects. However, as described in this chapter, it is not straightforward to convert a CAD design into a GIS model (i.e., a collection of distinct objects geometrically defined in real-world coordinates, with both spatial and non-spatial attributes).

Apart from spatial and administrative modeling, 4D requirements for the logical models were also considered in this chapter: how to maintain history. History is not supported in current DBMSs. However, history in the three proposed logical models for the 3D cadastre could basically follow the same approach as history is currently implemented in the Dutch cadastral DBMS, where spatial objects have two additional attributes (tmin and tmax) to implement history.

Based on the considerations for the logical models as described in this chapter, the prototypes were implemented. The prototypes contain the basic aspects of the three selected alternatives. In Chapter 11 the prototypes are evaluated as well as the conceptual and logical models of the different alternatives, by applying the prototypes to the case studies introduced earlier in this research.

Part IV

Realization of a 3D Cadastre

11 Prototypes Applied to Case Studies

Three conceptual models of a 3D cadastre (the registration of right-volumes, 3D physical objects, and the registration of volume parcels) were completed in Chapter 9. The considerations for translating these conceptual models into logical models for an object relational database structure were described in Chapter 10. To evaluate the conceptual models, the two logical models of the hybrid cadastre (the registration of right-volumes and the registration of 3D physical objects) were populated with data concerning the Dutch 3D property situations introduced in Chapter 3. The logical model for the combined 2D/3D alternative of the full 3D cadastre (volume parcels and infinite parcel columns) was populated with data concerning the case study from Queensland, Australia introduced in Chapter 3. The reason for this was that the legal doctrine in Queensland provides already the possibility of establishing multilevel ownership, while in the Netherlands the property to real estate is still land (surface) oriented, as in the hybrid case. Therefore, the prototype implementations resulted in an evaluation of possibilities and constraints of all three proposed conceptual models.

In this way this chapter shows the 3D potentials for cadastral registrations that already have legal instruments to establish 3D property units, as well as for cadastral registrations that are still surface oriented.

Section 11.1 describes the prototypes of the hybrid cadastre applied to the Dutch 3D property situations. Section 11.2 describes the full 3D cadastre prototype applied to the case study in Queensland, Australia. The chapter ends with conclusions.

For the prototype implementations the technical framework that was explored in Part II of this book was used. The 3D data are maintained in Oracle Spatial 9i. The data are accessed with both MicroStation GeoGraphics and Web-based techniques (ArcGIS has not been used for accessing the data, as ArcGIS is not yet able to visualize vertical polygons via ArcSDE from an Oracle Spatial database). Because no height data were available for the Dutch case studies, height information on parcels was not used in the Dutch case studies. For the case study in Queensland, ArcView (ArcGIS) was used to generate an appropriate TIN structure containing a 2.5D representation of the cadastral base map.

11.1 PROTOTYPES OF THE HYBRID CADASTRE

This section describes the concepts of the two alternatives of the hybrid cadastre (right-volumes and a 3D physical object registration) applied to the Dutch case studies that were introduced in Chapter 3 (Sections 11.1.1 to 11.1.5). The case study of The Hague Central Station (Section 11.1.2) is used to show the process of creating 3D representations of right-volumes and their data structure in more detail. In Section 11.1.6 both alternatives of the hybrid cadastre are evaluated.

11.1.1 Case Study 1: Building Complex in The Hague

Right-Volumes

Figure 11.1 shows the implementation of the registration of right-volumes applied to the building complex in The Hague. It is not the building itself that is registered, but the right-volumes established for the building, together with their 3D representation. These right-volumes are the 3D descriptions of the space to which the building owner is entitled with the following limited rights: right of superficies on parcel 1719 (parcel in the middle) and right of long lease on parcel 1720 (parcel on the right) (the cadastral map of the building containing the parcel numbers was shown in Figure 3.4). No right-volume is maintained on parcel 1718 (parcel on the left) because no limited real right has been established on this parcel. The parcel owner is the same as the building owner. The ownership right on this parcel applies therefore to the whole parcel column.

Note that the 2D extent of the 3D representations (i.e., footprint) is the same as the parcel boundaries. The 3D descriptions give an indication of the space to which the owner of the building is entitled.

The right-volumes refer to non-spatial information such as the person who is entitled to the right-volume and the type of right. The legal status of the building can be obtained by querying the right-volumes via the right association (what is the right associated with the right-volume, who is the subject of the right?). The relationship between right-volumes and the whole real estate object (whole building in this case) may also be maintained (note that it is not maintained in this case).

The legal status of the space above and below the building complex is not explicitly registered. It is disputable who owns the space above the construction that is registered with a right of superficies, unless this is explicitly stated in the deed. However, in current deeds the exact location of the right of superficies is often not

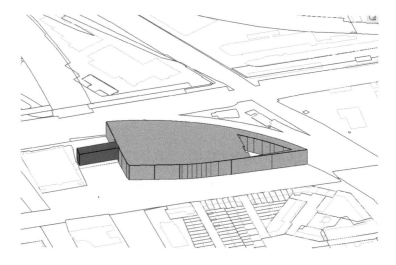

FIGURE 11.1 Registration of right-volumes: right of ownership on the left parcel, right of superficies on the middle parcel, and right of long lease on the right parcel.

clearly described. In the case of long lease, the long leaseholder has the right to use the whole parcel column within the conditions stated in the deed. The condition can restrict the long leaseholder in using the whole parcel column. In this specific case the deed did not contain conditions with respect to the spatial extent of the right established on parcel 1720.

In this case study the heights of the right-volumes (relative heights respective to surface parcel) are related to the construction as built and horizontal planes are used to define the right-volumes as the definition of right-volumes prescribes. If the space to which the rights apply is precisely defined in 3D in deeds and in 3D survey documents, this information can be used to construct the right-volumes. In that case it can happen that the visualization of the right-volumes differs from the actual built construction (e.g., when a right of superficies exceeds the actual construction in view of future plans).

3D Physical Objects

Figure 11.2 shows the implementation of the registration of 3D physical objects applied to the building complex in The Hague. The physical object in this figure represents the factual building. Also in this case, the z-values used to define the physical object are defined relative to the surface.

Although the building crosses parcel boundaries, the whole building is registered as one object in the cadastral database. Separated from the parcels and outlines of buildings, the 3D physical object is maintained in a table containing the id of the object and the 3D geometry of the object (in this case defined as one multipolygon defined in 3D). The 3D object can be visualized and queried with a front end in combination with 2D (preferably 2.5D) cadastral data. The legal status of the building is still registered by establishing rights on surface parcels. Consequently, the legal status of the building can be obtained only by examining the surface parcels. Because the 3D location of the building is integrated with the cadastral geographical data set, this information can be used in the querying process.

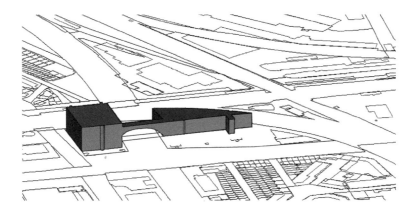

FIGURE 11.2   (Color figure follows page 176.) Registration of a 3D physical object.

11.1.2 CASE STUDY 2: THE HAGUE CENTRAL STATION

Generating Right-Volumes

The process of automatically generating right-volumes is described in this section using the case of The Hague Central Station.

For every parcel on which limited real rights or apartment rights are registered, a z-list is generated, which defines the upper and lower limits of (limited) rights (and apartment rights) established on the specific parcel. For example, in the case of The Hague Central Station, the vertical extents of the rights on the parcel that contains the tram and bus station and the railway platform are as follows (parcel 13295; see Figure 11.3b):

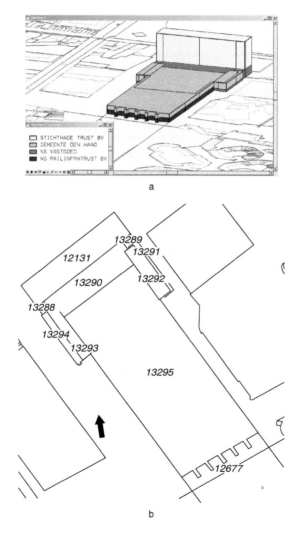

a

b

FIGURE 11.3 (Color figure follows page 176.) Cadastral representation (2D and 3D) for The Hague Central Station. (a) 3D representations of right-volumes; (b) 2D cadastral map.

- Railway platform (owned by NS Vastgoed): 0 to 6 meters
- Tram/bus station (right of superficies, holder municipality of The Hague): 6 to 12 meters

Because the notarial deed gave no information about the boundaries of the established right of superficies for the bus/tram station in 3D, the levels were obtained by measuring the building (construction).

The z-lists are inserted in the right-volume table, as described in Section 10.1.2. For The Hague Central Station the right-volume table is as follows (implying 15 right-volumes):

PARCEL	Z_LIST
12131	Z_ARRAY(0, 12, 40)
13290	Z_ARRAY(0, 12)
13288	Z_ARRAY(0, 12)
13289	Z_ARRAY(0, 12)
13294	Z_ARRAY(0, 3, 12)
13291	Z_ARRAY(0, 3, 12)
13293	Z_ARRAY(0, 3, 12)
13292	Z_ARRAY(0, 3, 12)
13295	Z_ARRAY(0, 6, 12)

Non-spatial information on the right-volumes (person who is entitled to the space; what right is established to entitle the person to the space) can also be obtained. The type of right and the person entitled to a right-volume (right-owner) are made available via views on the administrative base tables ("right" and "subject"). The list of owners and the list of types of right for every level are presented as arrays (Abstract Data Types) in the following view:

```
SELECT parcel, o_list, r_list FROM dh_input_3d_view;
```

PARCEL	O_LIST	R_LIST
12131	O_ARRAY(`NS VASTGOED', `STICHTHAGE')	R_ARRAY(`App', `App')
13290	O_ARRAY(`NS VASTGOED')	R_ARRAY('VE')
13288	O_ARRAY(`NS VASTGOED')	R_ARRAY('VE')
13289	O_ARRAY(`NS VASTGOED')	R_ARRAY('VE')
13294	O_ARRAY(`NS VASTGOED', `GEMEENTE DEN HAAG')	R_ARRAY(`EVOS', `OS')
13291	O_ARRAY(`NS VASTGOED', `GEMEENTE DEN HAAG')	R_ARRAY(`EVOS', `OS')
13293	O_ARRAY(`NS VASTGOED', `GEMEENTE DEN HAAG')	R_ARRAY(`EVOS', `OS')
13292	O_ARRAY(`NS VASTGOED', `GEMEENTE DEN HAAG')	R_ARRAY(`EVOS', `OS')
13295	O_ARRAY(`NS RAILINFRATRUST BV', `GEMEENTE DEN HAAG')	R_ARRAY(`EVOS', `OS')

```
9 rows selected.
```

A PL/SQL script has been written to generate the spatial description (topology and geometry) of the right-volumes, which is linked to the non-spatial information.

Topological structure   DBMSs do not yet support topology structure management (neither 2D nor 3D). Therefore, the topological structure has to be defined in a DBMS by means of user-defined references. For the prototypes we use the Simplified Spatial Model.[229] A 3D geometry object is therein defined as a polyhedron consisting of nodes and faces (see Section 6.2.4). A PL/SQL script was written to generate the

BODY, FACE, and NODE table based on the z-list containing height values of properties on one parcel and the geometry of the parcel. In this model the faces within one parcel are shared between bodies, as they should be in a full topological model, and nodes are shared between faces.

From topology to geometry To realize the geometry of the right-volumes, based on the topological tables, a function has been written. In this function the nodes of right-volumes are retrieved by the following query (see Section 6.2.4) whereupon the 3D geometry is reconstructed:

```
/* for the body bid=1*/
SELECT body.bid,face.fid, face.seqn, node.nid, node.x, node.y, node.z
FROM body, face, node WHERE body.bid=1;
```

Having the geometry it is possible to visualize, query, and edit the data in GIS and CAD software and to perform spatial queries in the DBMS.

Geometrical primitives The geometry of right-volumes is generated by means of the implemented realization function and is made available in a view. The realized geometry of right-volumes can be defined as a polyhedron data type as it is possible within current techniques: either as a set of polygons in 3D or as a multipolygon defined in 3D. For the prototypes, we used the multipolygon representation as this representation is recognized as one object by front ends. As part of the implementation of the 3D primitive in the DBMS (Section 6.4), a conversion tool has been written to convert 3D objects stored as multipolygons into the 3D polyhedron primitive. For the 3D right volume with bid=3 (which is the right-volume on parcel 13290), this appears as:

```
SELECT return_polyhedron(shape) FROM dh_multipol WHERE bid=3;

RETURN_POLYHEDRON(SHAPE)
------------------------
(SDO_GTYPE, SDO_SRID, SDO_POINT(X, Y, Z), SDO_ELEM_INFO,
SDO_GEOMETRY(3002, NULL, NULL,
-- 3002 refers to geometrytype: (fictive) 3D polyline
-- in sdo_elem_info_array the elements are listed, first triplet is a line,
-- followed by the (outer) faces (starting offset, e_type, interpretation code)
SDO_ELEM_INFO_ARRAY(1, 2, 1, 67, 0, 1006, 78, 0, 1006, 82, 0, 1006, 86, 0,
1006, 90, 0, 1006, 94, 0, 1006, 98, 0, 1006, 102, 0, 1006, 106, 0,
1006, 110, 0, 1006, 114, 0, 1006, 118, 0, 1006, 122, 0, 1006),
-- in the oridinate array, first the vertices are listed
SDO_ ORDINATE_ARRAY(82140054, 455389862, 0, 82124400, 455378306, 0,
82103103, 455361960, 0, 82036913, 455311156, 0, 82054915, 455287619, 0,
82063070, 455293838, 0, 82107247, 455327528, 0, 82151729, 455361401, 0,
82161846, 455369105, 0, 82159770, 455371792, 0, 82143723, 455392571, 0,
82140054, 455389862, 12000, 82124400, 455378306, 12000,
82103103, 455361960, 12000, 82036913, 455311156, 12000,
82054915, 455287619, 12000, 82063070, 455293838, 12000,
82107247, 455327528, 12000, 82151729, 455361401, 12000,
82161846, 455369105, 12000, 82159770, 455371792, 12000,
82143723, 455392571, 12000,
-- then the faces, defined by references to the nodes
1, 2, 3, 4, 5, 6, 7, 8, 9, 10, 11, 1, 12, 13, 2, 2, 13, 14, 3,
3, 14, 15, 4, 4, 15, 16, 5, 5, 16, 17, 6, 6, 17, 18, 7,
7, 18, 19, 8, 8, 19, 20, 9, 9, 20, 21, 10, 10, 21, 22, 11,
11, 22, 12, 1, 12, 13, 14, 15, 16, 17, 18, 19, 20, 21, 22))
```

Once a right-volume is defined with the 3D polyhedron primitive, the implemented 3D functions can be performed on it, such as a validation, 3D area, or 3D volume calculation (in this case in mm^3):

```
SQL> SELECT bid, volume(return_polyhedron(shape)) FROM dh_multipol;

BID        VOLUME(RETURN_POLYHEDRON(SHAPE))
---        ------------------------------
1          1.6024E+13
2          5.3413E+12
3          1.1888E+13
4          4.6118E+11
5          5.7372E+11
6          6.6815E+11
7          6.6815E+11
8          2.6769E+11
9          2.6769E+11
10         6.8230E+11
11         6.8230E+11
12         1.0486E+12
13         1.0486E+12
14         4.6100E+13
15         4.6100E+13
15 rows selected.
```

Topological model compared to geometrical primitives   The following are disadvantages of using a topological model:

- The data model needs three tables instead of just one (as in the multipolygon or polyhedron case).
- Because the DBMS does not recognize topology, inserting the data is one thing, but updates require a lot of effort and experience or other software. In addition, the consistency of the data must be checked by other software (until DBMSs offer topology structure management in 3D).
- Querying can be difficult at the SQL level (topology is not recognized by DBMSs); for geometrical queries it is always necessary to generate a realization of the object, instead of being able to use the spatial queries available in the DBMS directly.

Another problem mentioned in Chapter 6 is the required storage capacity of the topological structure compared with the storage capacity needed for the geometrical primitives. Every row in the tables defining the topological structure has its overhead, and the references require a lot of storage capacity. To illustrate this we queried the storage capacity needed for the tables of the topological structure in the case of The Hague Central Station, and found it to be twice the storage capacity needed for the polyhedron representation. An advantage of the topological structure is that topology structure management can be used in the storage and the retrieval of data. For example, by means of the shared (horizontal) faces one can easily find the upper and lower neighbors of a right-volume.

Evaluating Right-Volumes for The Hague Central Station

The visualization of the right-volumes for The Hague Central Station (Figure 11.3a) gives clear insight into the various rights in the building complex. It not only gives an indication of the spatial extent of the property rights on each of the parcels concerned, but it also shows the relation between the rights established on adjacent parcels. The 3D map of The Hague Central Station (classified on subject) clearly shows that the municipality of The Hague holds the right of superficies not only on parcel 13295 (the big parcel in the center, with the railway platforms on ground level), but also on parcels 13291, 13292, 13293, and 13294 (see the cadastral map in Figure 11.3b). At one glance one can see that the municipality is owner of the bus/tram station on the second floor, with the adjacent entrances at left- and right-hand side of the railway station. This is an advantage compared to the traditional 2D cadastral map. More advanced visualization techniques may be needed in complex clusters of right-volumes (e.g., make certain right-volumes are semitransparent).

In this case the space of the railway platforms on parcel 13295 is visualized; however, in the strict definition of right-volumes (right-volumes relate only to positive rights) this is not correct. The space occupied by the railway platform belongs to the space that is left for NS Railinfratrust BV after the space related to the right of superficies has been subtracted from the infinite parcel column. The right-volumes on parcel 13288, 13289, and 13290, which also have been related to the construction as built, are also visualized. This is also not correct according to the definition of right-volumes as these parcels are held in full ownership.

Again, the limits of the right-volumes are related to the construction as built. However, it would be better to define the limitations in the deed based on 3D survey plans. For example, who owns the space above the bus/tram station and the space below the railway platform? At the moment this seems not a relevant question. However, if a business company wants to build a business center on top of the bus/tram station, the ownership of the space above the tram/bus station will become an important issue.

3D Physical Objects

The 3D map in Figure 11.3 shows only the right-volumes and not the physical objects (although in this case the right-volumes are related to the physical objects). The physical objects in the case of a physical object registration are maintained to visualize constructions, not subdivided by parcels that are crossed. In some cases the physical object coincides with a conglomerate of right-volumes. This is also the case for The Hague Central Station where right-volumes are related to the construction as built due to lack of other information. For example, the physical object for the bus/tram station is a combination of the right-volumes established for the bus/tram station including the entrances on parcels 13291, 13292, 13293, 13294, and 13295. In the case of the railway platform, the whole platform (or railway) would be registered as one 3D physical object also crossing parcels that are in full ownership of NS Railinfratrust BV. Normally, the collection of right-volumes that refers to a whole real estate object embraces the space occupied by a 3D physical object, except on parcels for which no cadastral recording of the 3D situation has taken place since on those parcels no right-volumes are established. However the physical object will also be visual in those situations.

11.1.3 CASE STUDY 3: APARTMENT COMPLEX

Right-Volumes

The case of the apartment building is more complex, because on the ground floor there are three apartment units and on the first and second floor two units, all established on one parcel. Furthermore, the building does not cover the whole parcel. This is quite common for apartment complexes. Consequently, the footprints of the apartment units on every floor do not coincide with the parcel boundary.

To be able to apply the z-list with upper and lower limits of rights established on one parcel, the 2D boundaries of the individual apartment units are generated, which resulted in the 2D objects as shown in Figure 11.4, with object a (whole building minus b and c), objects b and c defined for the ground floor, and object d and e (both half of the building) defined for the first and second floor. The garden area (which belongs to the apartment unit on the ground floor) and the space above this area are not included in the right-volumes, although this could have been done.

The right-volume table for the whole apartment complex is as follows:

PARCEL	Z_LIST
6408_a	Z_ARRAY(0,3)
6408_b	Z_ARRAY(0,3)
6408_c	Z_ARRAY(0,3)
6408_d	Z_ARRAY(3, 10)
6408_e	Z_ARRAY(3, 10)

"Parcel" in this case does not refer to parcel numbers but to the 2D polygons of apartment units generated to define inner boundaries of the apartment units in order to be able to extract them in 3D. The drawings added to deeds of subdivision could be used to construct the 2D footprints, that is to say, when the spatial information on the drawings is defined in, or can be transferred into, world coordinates and in vector format. The visualization of the generated right-volumes is shown in Figure 11.5.

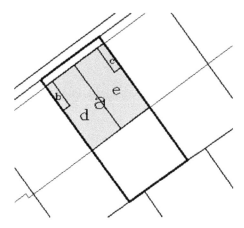

FIGURE 11.4 The generated 2D objects: footprints of individual apartment units on every floor. The gray part is the extent of the building.

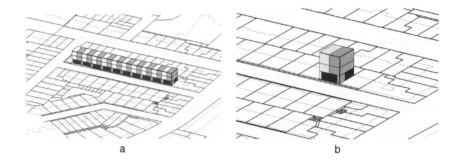

a   b

FIGURE 11.5 **(Color figure follows page 176.)** Apartments as right-volumes. The horizontal lines between the first and second floor are for visualization purposes. (a) All apartments in the street; (b) the apartment complex of the case study, which is the second complex from right.

This case shows some complications. Not only is a horizontal division of the parcel column needed to define the right-volumes, but a vertical division of the parcel area (dividing the parcel into smaller parts) is also required. If only one right-volume was to be established for the whole parcel area, it would not reflect the real situation and it would definitely not provide a clear picture of the situation (right-volumes would overlap), which is one of the main aims of 3D registration. However, the generation of the smaller parts (footprints of apartment units) is a change in the concept of right-volumes if they have the same legal status as "normal" (traditional) parcels.

3D Physical Objects

The physical object registration would in this case be the same as the right-volume registration in which the right-volumes are related to the apartment units. The apartment units are the physical objects that would be identified as registering objects in the physical object registration. It can be disputed if a 3D physical object registration is appropriate for apartment units, as the main objective of a 3D physical object registration is to provide more insight into locations containing infrastructure objects (crossing parcel boundaries and mostly meant for public good) rather than to improve insight into private property situations.

11.1.4 CASE STUDY 4: RAILWAY TUNNEL IN URBAN AREA

Right-Volumes

The right-volume table for the railway tunnel in Rijswijk is as follows (for the parcel numbers, see Figures 3.10 and 3.11):

```
7854:    Z_ARRAY(-20, 0, 4)  –tunnel with kiosk on top of it
7855:    Z_ARRAY(-20, 0, 4)  –tunnel with kiosk on top of it
7857:    Z_ARRAY(-20, 0, 12)  –tunnel with railway station on top of it
7944:    Z_ARRAY(-20, 0, 12)  –tunnel with public space on top of it
7949:    Z_ARRAY(-20, 0, 12)  –tunnel with public space on top of it
```

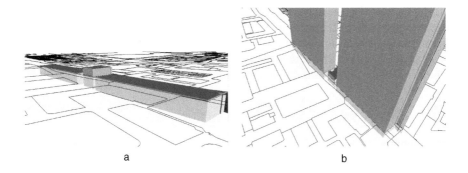

a   b

FIGURE 11.6 (Color figure follows page 176.) The first two alternatives for unrestricted right of superficies (third option is not displayed). Note that the lowest right-volumes (for the railway tunnel) are located below the surface (below the $z = 0$ plane).

For parcel 7945 and parcel 7946 no right-volumes are generated as NS Railinfratrust BV has a full ownership on these parcels. The right of superficies established for the municipality to hold the public area at street level is in this case supposed to be bounded on a level 12 meters above the surface level. This is not the real case. In the deed, the space of the right of superficies is defined as "above the surface level." How can one visualize this? There are basically three solutions:

1. Make a 3D description of just the street level (Figure 11.6a); this representation could be confused with a right-volume that is limited in height and is therefore not a good solution.
2. Make a very high ("unlimited") right-volume (Figure 11.6b); this also does not reflect the real situation correctly, because it looks as a very high building has been built.
3. Use an "open" polyhedron, without a top (and the side faces visualized until a reasonable height related to the height of the physical object).

All these alternatives are vague indications that something is happening above the surface. The question is if these alternatives are correct and clear representations of the real situation.

The 3D registration of this situation gives significantly more insight compared to the registration in the current cadastre. It is now possible to see not only which persons have a right on a parcel, but also where these rights are located in space. Although the gaps in the registration caused by the full ownership of NS Railinfratrust BV on some parcels as well as the undefined right-volumes when right-volumes are defined "above street level" make the situation unclear. The real situation might be better reflected when the tunnel itself is registered as 3D physical object.

3D Physical Objects

The registration of physical objects registers the tunnel as one whole object, together with the spatial extent of the tunnel and information on the tunnel. The station building could be registered as one physical object as well. The fragmented pattern of parcels could then be avoided. 3D information on the tunnel was not available for this research, but it would have been similar to the registration of the 3D physical object in the case of the railway tunnel in a rural area as described in the next case. The 3D location of the tunnel helps to understand the real situation.

11.1.5   CASE STUDY 5: RAILWAY TUNNEL IN RURAL AREA

Right-Volumes

Also in the case of the HSL, the right-volumes start with the surface parcel boundaries. To avoid the situation where part of parcels that do not cross the tunnel are encumbered with a right for the tunnel, the intersecting parcels need to be subdivided. As was seen in Section 3.1 most intersecting parcels were already subdivided but have not yet been surveyed. Therefore, we created fictive new parcel boundaries using the new parcel boundaries as shown in Figure 4.1. These new parcel boundaries were created by a spatial overlay in the database. First the tunnel axis, stored as a line, was buffered with 15 meters, based on the diameter of the tunnel (15 meters) and a safety zone of 7.5 meters at each side ("shape" in this query is the geometry column of the table in which the tunnel is represented with the centerline):

```
CREATE table hslbuffer AS
SELECT sdo_geom.sdo_buffer(shape,15000,1) shape
FROM hsltunnel;
```

Then a spatial overlay was carried out between the layer containing the tunnel buffer and the realized geometry of parcels:

```
CREATE TABLE hsl_parcel_new AS
SELECT parcel, municip, osection,
sdo_geom.sdo_intersection(hb.shape,hp.return_polygon(object_id),10) shape
FROM hsl_parcel hp,hslbuffer hb;
```

The newly created parcels (as well as the remainder parcels) received a unique parcel number. These new parcels were used to create the right-volumes. The spatial extent of the right-volumes is the spatial extent of the spaces where the rights established for the tunnel apply: in this case the same as the 3D spatial extent of the tunnel under the specific parcels extended with a safety zone of 7.5 meters (in all directions). The upper and lower limits of a right-volume for a specific parcel were derived from two sources: (1) the 3D centerline of the tunnel that intersects with the specific parcel and (2) information on the extent of rights established for the tunnel (diameter of 15 meters plus safety zone of 7.5 meters).

The obtained upper and lower limits of the right-space per parcel were inserted in the right-volume table and used for the generation of right-volumes for the tunnel (the z-values are in mm and in NAP, the Netherlands National Ordnance Datum):

MUNICIP	SECTION	PARCEL	Z_LIST
HZW00	E	740	Z_ARRAY(-28809, -13598)
HZW00	E	2396	Z_ARRAY(-28384, -12055)
HZW00	E	2397	Z_ARRAY(-26826, -3426)
HZW00	F	57	Z_ARRAY(-37501, -21069)
HZW00	F	58	Z_ARRAY(-35970, -20869)
HZW00	F	59	Z_ARRAY(-40100, -23997)
HZW00	F	60	Z_ARRAY(-38960, -23368)
HZW00	H	14	Z_ARRAY(-38129, -23116)
HZW00	H	15	Z_ARRAY(-38103, -22664)
HZW00	H	17	Z_ARRAY(-37857, -22819)
HZW00	H	21	Z_ARRAY(-37651, -22638)
HZW00	H	25	Z_ARRAY(-37625, -22586)

etc.......

From this right-volume table the topology structure (and geometry) of right-volumes was obtained, representing the space to which the ministry is entitled.

The right-volumes give insight into the vertical dimension of the rights established (Figure 11.7a). Now it is clear that the rights are established for an underground construction and not for a viaduct or a road. This solution also provides insight into the depth and height of the construction (if the height surfaces of parcels are also available), which is a considerable improvement of current registration.

The registration of a right for the tunnel will not take place, when the ministry owns the intersecting parcel. This leads to "gaps" in the 3D registration. This is clearly illustrated in Figure 11.7b and c. Figure 11.7b shows the situation when new parcels are created and some of these parcels are in full ownership with the Ministry of Transport and Public Works. For those parcels a right-volume will not be created (the ministry owns the whole parcel column). The situation is even less clear in Figure 11.7c. This will be the case when both new parcels and original parcels that are not divided are in full ownership of the ministry.

Special cases are the parcels that are held in bare ownership by the ministry, while other persons are entitled to use space above and below the tunnel via limited real rights. In that case, a right-volume (multivolume object) would need to be maintained for the space above and below the tunnel. The representation of such "open" right-volumes would meet the same complications as the right-volumes that refer to space "above street level" in the Rijswijk case.

3D Physical Objects

Figure 11.8 shows the implementation of a physical object registration applied to the HSL tunnel. The spatial description of the whole tunnel is maintained as one 3D object in the database. Although the tunnel is a round-shaped object, which can easily be modeled in CAD software, implementing it in a DBMS reduces the precision of

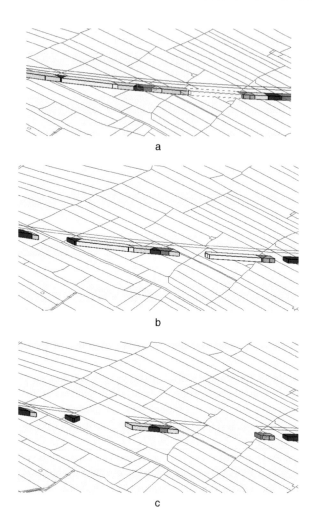

a

b

c

FIGURE 11.7 (Color figure follows page 176.) Three possible recordings of right-volumes in the case of a railway tunnel. (a) All the parcels are encumbered by right of superficies; new parcels are created for all intersecting parcels. (b) As a, but now three newly created parcels are in full ownership. (c) Three newly created parcels are in full ownership; two parcels that are not subdivided are in full ownership. All the other new parcels are encumbered by a right of superficies.

the data, since now the object needs to be approached by many flat faces to be able to use the spatial primitives available in DBMSs.

The rights for the tunnel are still registered on the intersecting parcels. However, since the exact location of the tunnel is also maintained, it is not necessary to create new parcel boundaries. The holder of the tunnel is stored in the DBMS and is in this case the same as the subject who has a right of superficies or the right of ownership on the intersecting parcels. Note that in this case the safety zone is not included since

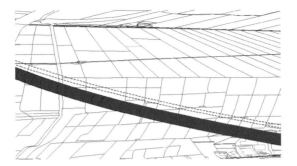

FIGURE 11.8 (Color figure follows page 176.) Registration of the 3D physical object in the case of the HSL tunnel. The dashed line is the projection of the tunnel on the surface. Note that the parcels are not divided into smaller parcels.

the 3D representation relates to the actual construction. The location of the tunnel helps to better understand the real situation.

11.1.6 EVALUATION OF HYBRID CADASTRE

When a legal framework cannot entitle persons to volumes independent from the surface, the 3D cadastre can be implemented within a hybrid environment introducing either right-volumes or a physical object registration.

Right-Volumes

From the prototype implementations it can be concluded that the introduction of right-volumes leads to a significant improvement of current cadastral registration in 3D situations. The inclusion of right-volumes in the cadastral geographical data set provides an overview of the distribution of 3D property units. The registration warns the user of the cadastral registration that something is located under or above the surface. It also gives information on what is located under or above the surface (relationship with whole real estate object can be maintained). For precise information, the deed in the land registration can be consulted. A 3D survey of the situation can be made and used to describe the situation in the deeds and to determine the upper and lower limits of right-volumes. From a technical point of view, the geometry of right-volumes is simple and can therefore be maintained in the DBMS within current techniques.

Basic disadvantages of right-volumes are as follows:

- Because parcels are still the basis for registration, gaps can occur when no rights have been established that require a cadastral recording, e.g., when the holder of the construction is the same as the owner of the intersecting parcel. In these cases the location of the construction is still not known in the cadastral registration.

- If rights are established on just a part of a parcel, new parcel boundaries need to be created. This leads to fragmentation of both parcels and right-volumes.
- When space to which the right applies is not precisely restricted in height or depth, registration of right-volumes does not give satisfying insight, as was seen in the Rijswijk case. This could be solved by a rule applying to 3D surveys that will allow open polyhedrons (not defined either in height or in depth).
- Horizontal boundaries restrict the spatial description of right-volumes. The concept of a right-volume could be improved when other than just horizontal boundaries could be defined.

Registration of 3D Physical Objects

From the experiments with the case studies it can be concluded that a registration of physical objects offers several improvements. The 3D description of the physical objects (extent of the object) can be used for reference purposes (i.e., to improve the reflection of the real situation) and to support cadastral tasks. When a 3D physical object is registered, parcels do not need to be divided into parcels matching the 2D projection of the physical object as the exact location of the physical object is known in the cadastral database. Only one object needs to be registered by which the registration for all intersecting parcels can be guaranteed. All parcels intersecting with the physical object can be found by a spatial query (by an overlay with the projection of the 3D object).

From a technical point of view, the geometry that has to be maintained for physical objects can become complex. It is therefore not easy and straightforward to insert and maintain the spatial information on 3D physical objects within current techniques.

Conclusion on Hybrid Cadastre

In both alternatives, rights to hold 3D property units are still registered on the intersecting parcels. Querying the legal status of 3D property units still needs to be done by querying the legal status of the intersecting parcels. However, maintenance of the 3D situation can assist considerably in understanding the real situation. Cadastral registrations that are not yet ready for a full 3D cadastre will benefit from a hybrid registration, for a number of reasons:

- The solutions give visual insight in 3D into the real situation. It is now clear from the cadastral registration that persons are entitled to space above or under the surface.
- Both solutions are implemented within the cadastral registration as part of the cadastral geographical data set and can therefore be queried with parcel surfaces in one integrated view.
- The proposed solutions show the persons responsible for setting up deeds (i.e., notaries and licensed surveyors) how the inclusion of spatial information in deeds can be used to visualize the 3D component of rights in the cadastral registration. The solution can make notaries and licensed

surveyors aware of the improvements of 3D registration and may motivate them to include well-defined 3D information in the deeds or to require a 3D survey plan.

- Registrations and databases outside the cadastral domain can benefit from the information on 3D situations that is available in the cadastral registration, and vice versa, via the Geo-Information Infrastructure (monument registration, building registration, taxes for immovable goods, management of soil pollution areas, management of cables and pipes, management of the subsurface).

The case studies were divided into building complexes and infrastructure objects. A physical object relating to a property unit within a building complex coincides with the legal space of a property unit. The main objective of cadastral registration in the case of building complexes is to give insight into property boundaries in all dimensions rather than to reflect the built constructions in the cadastral registration for reference purposes. On the other hand, the main objectives of a registration of 3D physical objects are, first, to be able to locate infrastructure objects to support cadastral tasks and, second, to register the person who holds an infrastructure object. Therefore, the registration of 3D physical objects will specifically be suitable for infrastructure objects. For registering property units in building complexes, right-volumes are more appropriate, because the spatial extent of property units can be easily and clearly defined with right-volumes, which refer to the legal space to which a person is entitled.

The two concepts of the hybrid cadastre (right-volumes and 3D physical object registration) have a different line of approach and therefore meet other needs of 3D cadastral registration. The right-volume is a considerable improvement of insight into 3D property units as part of the cadastral geographical data set, while the 3D physical object registration provides information on constructions that is available in the cadastral geographical data set to improve the reflection of the real situation. The concepts could be combined to take advantage of both solutions.

The main limitation of both hybrid solutions is that the property rights are still related to surface parcels.

11.2 PROTOTYPE OF THE FULL 3D CADASTRE

In the full 3D cadastre it is possible to entitle a person to a volume parcel that is no longer related to the surface parcel (only in the case when it is subdivided from the infinite parcel column defined by the surface parcel). Section 11.2.1 describes the results of the prototype applied to the case study in Queensland; Section 11.2.2 evaluates the prototype of the full 3D cadastre.

11.2.1 THE GABBA STADIUM IN QUEENSLAND

As seen in Section 3.4, the legal framework in Queensland, as in some other countries and states, provides a good basis for a full 3D cadastre. Within this framework it is possible to establish property rights to (1) standard, infinite parcels, (2) volumetric parcels (no longer related to the surface), and (3) remainder parcels that are left after a volumetric parcel has been subtracted from a standard parcel. In our model volumetric parcels are referred to as "volume parcels".

The cadastral framework in Queensland does not yet provide the possibility of maintaining the 3D geometry of the volumetric parcels in the cadastral registration. In Section 3.4 it was concluded that the current cadastral registration of volumetric properties, in which only the 2D geometry is registered, meets the following limitations:

- Because the 3D information is laid down on paper or scanned drawings (which is a 2D visualization of 3D information), the 3D information cannot be interactively viewed.
- The 3D properties are described only by coordinates and faces on drawings; i.e., no 3D primitive is used. Therefore, it is not possible to check if a valid 3D property has been established. Is the 3D property closed? Are the faces planar?
- The 3D information is not integrated with the cadastral map or with other 3D information; e.g., two or more neighboring parcels cannot be visualized in one view in 3D and it is also not possible to check how volumetric parcels spatially interact in 3D (overlap, touch, etc.).

To improve cadastral registration we applied the feasible concept of the full 3D cadastre (combination of volume parcels and infinite parcel columns) to the described case study in Queensland: the Gabba Stadium in Brisbane at the location of Vulture Street (in the north), i.e., parcel 100 (stratum parcel) and parcel 101 (volumetric parcel); see Figure 3.15 and Figure 11.9.

The required survey plans for the volumetric parcel and the stratum parcel contain 3D information that can be used to describe the 3D geometry and the 3D topological structure of these objects in the cadastral database. The following steps were followed to convert the spatial information on the scanned 3D survey plans into a 3D geometrical primitive in the DBMS:

- The field measurements, as indicated on the survey plan by distances and bearings between the successive points, were adjusted by traverse adjustment for each parcel in a local coordinate system.[200]
- The local rectangular coordinates were fitted to the (global) map coordinates by an overdetermined conformal (Helmert) transformation using three connection points in both coordinate systems.[200]
- The faces were constructed with references to nodes.
- This information was inserted in a 3D topological structure (SSM) in the DBMS.
- From the topological structure the geometry (as polyhedron primitive) can be realized, validated, and spatially queried using the self-implemented 3D primitive and 3D functions.

After these steps, the 3D geometries could be visualized and queried in one integrated view (Figure 11.10), which offers major improvements. It is now possible to see if and how the volumetric parcels interact and to view the 3D situation interactively.

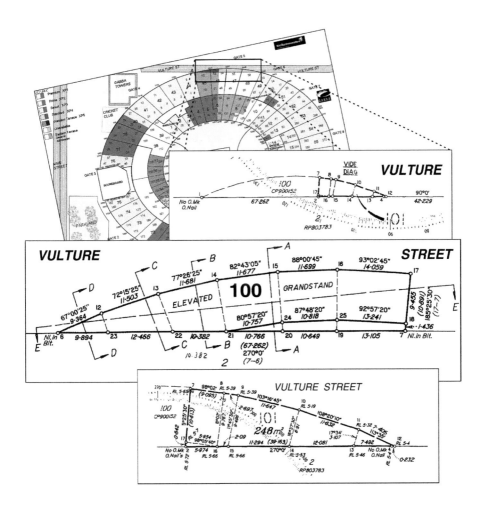

FIGURE 11.9 Volumetric parcel (101) and stratum parcel (100) used in the case study.

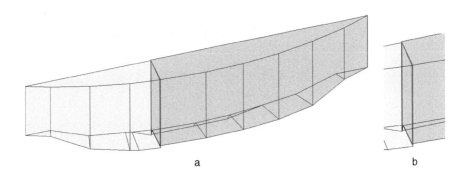

FIGURE 11.10 Visualization of 3D geometries of volumetric parcels, stored in DBMS. Zoom in on shared faces (b) shows that the shared faces do not coincide.

The neighboring polygons as defined do not match face to face; comparing the common boundary between parcels 100 and 101 shows a difference of about 30 centimeters (Figure 11.10b). This may indicate an error, but in this case it is correct. The two parts were determined at different times, and parcel 101 allows more space around the structure. The measurements define the space while there is no real object to mark the limits of the parcels. Therefore, the geometry of the volume parcels must by definition be correct. In the 2D map (Figure 3.16) there is no such error; the reason is that in the 2D map some topology processing (based on a certain tolerance) is performed during the update process. Small input errors are cleaned in 2D, but in 3D no topology processing is performed.

To validate the volumetric parcels and to perform 3D spatial functions on the volumetric parcels, the geometry of the volumetric parcels was represented using the self-implemented 3D geometrical primitive (Section 6.4). Therefore, we were able to query the 3D objects in an integrated DBMS environment:

```
/* validate of 3D geometries */
SELECT bid, validate_polyhedron(return_polyhedron(shape), 0.5) validate
FROM qld_3Dgeom;


BID    VALIDATE
----   ----------
100    True
101    True


/* calculate volumes of 3D geometries
SELECT bid, volume(return_polyhedron(shape)) volume
FROM qld_3Dgeom;

BID    VOLUME
----   ----------
100    12725.1989
101    5329.18583

/* check if two geometries intersect (1=TRUE and 0=FALSE) */
SELECT d1.bid, d2.bid FROM robject3dql d1, robject3dql d2
WHERE intersection(return_polyhedron(d1.shape), return_polyhedron(d2.shape),0.01) = 1
AND d1.bid < d2.bid;

BID       BID
------    ------
100       101
```

The 3D geometries can be incorporated in a cadastral geographical data set that contains surface parcels represented in 2.5D in order to get a 3D overview of the complete situation. For this purpose a conforming TIN was generated using ESRI software that incorporated the planar partition of the cadastral base map (see Chapter 8). The result is shown in Figure 11.11.

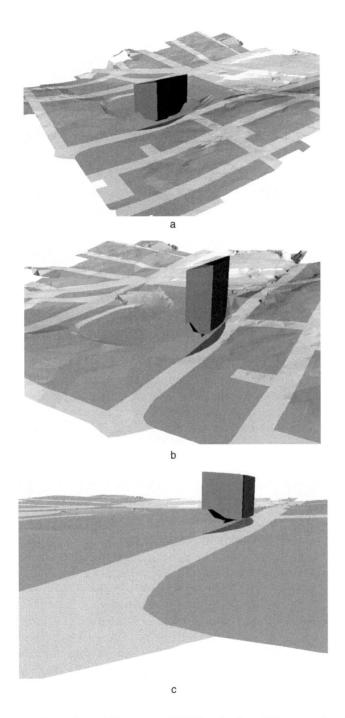

a

b

c

FIGURE 11.11 (Color figure follows page 176.) Visualization of 3D geometries of volumetric parcels together with the 2.5D cadastral base map, seen from different viewpoints.

11.2.2 Evaluation of Full 3D Cadastre

As can be concluded from this case study, the full 3D cadastre offers many improvements compared to traditional cadastral registrations:

- The real situation is no longer projected on the surface; i.e., volumetric parcels are not dominated by the parcel pattern on the surface.
- Persons can be entitled to space in a transparent way instead of establishing property rights on intersecting parcels to establish the legal status above and below the surface.
- The space is precisely described in a 3D survey document, which offers a uniform way of defining 3D property units.

The full 3D cadastre also offers improvements in countries and states that already establish 3D property units unrelated to the surface:

- The information from the 3D survey document can be used to insert the volume parcels in a topological structure and in geometrical primitives in the DBMS.
- The volume parcels can be viewed interactively.
- The geometry of volume parcels can be checked, e.g., are the faces planar, is the volume closed, are there no self-intersections?
- The 3D situation can be spatially queried in the DBMS (e.g., do volume parcels intersect?).
- The volume parcels can be visualized and queried in an integrated view with a 2.5D representation of the parcels that are defined by parcel boundaries on the surface, e.g., is the volume parcel located above or below the surface, or does it intersect the surface?

Having the 3D property unit in the same environment as the 2D parcels clearly offers great potential. However, even starting from one of the more advanced environments (Queensland, where both the legal aspects and the 3D survey documents are satisfactorily dealt with) quite a number of nontrivial issues still need to be addressed:

- In the survey plans both the 3D points and edges are specified (as required); however, there is no explicit listing of faces and the polyhedron itself. It is not trivial to reconstruct the faces and it is possibly ambiguous, especially in more complex cases (such as parcel 103 in Figure 3.15).
- The validation of the polyhedron is nontrivial (especially if it consists of other faces than horizontal, vertical, or triangular faces). Is the volume completely closed? Is the orientation correct? Are holes or cavities modeled correctly? Are all the faces planar within a certain tolerance? The points that define a face in 3D can be slightly out of the ideal flat plane because of the geodetic measuring methods and the finite representation of coordinates in a digital computer. Therefore, the faces of a polyhedron should be flat within a given tolerance.

- The footprints of the 3D objects do not fit perfectly in the cadastral map: a straightforward conversion from the local coordinates to global coordinates (rotate, translate) resulted in a mismatch of about 60 centimeters. Additional field measurements are required to solve these differences.
- The Queensland regulations also allow non-polyhedral 3D objects, such as rotated ellipsoids or cylindrical patches (Figure 11.12). Should these be converted to polyhedrons (approximation within given tolerance) to be modeled in the DBMS or should the DBMS be extended with complex 3D data types?
- Attention should be paid on how to make sure that two polyhedra do not overlap in 3D space (but at most touch in a common node, edge, or face) and on how to make sure that there is no 3D sliver between two polyhedra that are supposed to be touching neighbors.
- The cadastral registration should be organized in a uniform manner. In the case study (with only three 3D objects all related to the same construction) some differences are noticeable:

  Neighbor parcels 100 and 101 are both on the same side of the stadium, but parcel 100 is related to a stratum parcel, since it was established before 1997, and parcel 101 is related to a 3D volumetric parcel, which is only possible after 1997. Therefore, the available information for the parcels differs.

  Parcels 101 and 103 are both volumetric parcels, while parcel 101 is relatively rough; it seems that parcel 103 is defined quite tightly around the construction (making this object quite complex).
- Trivial registration errors should be avoided, such as the recording of the volume. It turned out that the recorded volume of parcel 101 in the cadastral registration was not correct (10,000 times too large), probably due to some typing error (because the survey plan was correct).

In addition to this, it is also a challenging task to integrate a terrain elevation model with the 2D surface parcels in order to obtain 2.5D surface parcels that can be combined with the 3D objects. This should preferably be implemented as an integrated view (in the DBMS sense) on the two data sets from the independent, distributed sources and not as a physical (permanent) integration with copies of the data sets (see Chapter 8).

In areas with high density of 3D volume parcels, a true space partitioning might be needed (defined in a full topological model).

11.3 CONCLUSIONS

In this chapter the concepts of the hybrid cadastre and the full 3D cadastre were applied to case studies in order to evaluate the concepts.

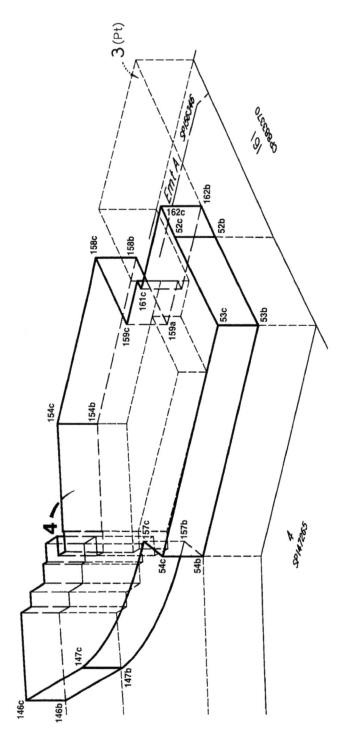

FIGURE 11.12 Volumetric parcel defined with more complex geometry than polyhedron.

11.3.1 HYBRID CADASTRE

Right-Volumes   The experiments with the case studies showed that right-volumes considerably improve insight into the property situation in 3D property situations. It is now clear how property units are distributed in 3D. Generating and maintaining spatial data are easy when registering right-volumes: the parcel boundaries that are already registered form the basis for the 3D representations while the geometry of right-volumes is simple. A disadvantage of right-volumes is when no limited rights have been established on a parcel in a 3D situation. These cases are not registered in the cadastral registration and therefore they lead to gaps in the 3D registration. Another disadvantage is when rights are not clearly restricted in the vertical dimension in the deed. In this case the right-volume does not necessarily yield more insight than the current cadastral registration; as seen in the Rijswijk case. In the case of apartment units, the 2D boundary of right-volumes did not coincide with the parcel boundary. However, if drawings added to deeds of divisions were available in vector format and in world coordinates the spatial information from the drawings can be used to automatically produce the 2D description of right-volumes for every floor. These polygons can then be used to generate the right-volumes. The right-volume concept could be improved if the boundaries between two right-volumes on top of each other were not restricted to horizontal boundaries. A nonhorizontal boundary can reflect more detail.

   3D physical object registration   The experiments with the case studies showed that the availability of physical objects in the cadastral geographical data set offers better means to reflect the real situation. In addition, parcels no longer need to be divided into parcels that match the 3D objects and "gaps" in the cadastral registration can easily be traced. Technical issues have to be solved to be able to maintain the complex geometry of physical objects in the cadastral DBMS. The geometries of physical objects will mostly have to be provided by third parties. As can be concluded from the experiments in this case study, the registration of 3D physical objects is specifically suitable for infrastructure objects, while the right-volumes are more appropriate for registering property units in building complexes. The two concepts of the hybrid cadastre (right-volumes and 3D physical object registration) have a different line of approach and therefore meet other needs of 3D cadastral registration. The concepts could be combined to take advantages of both solutions. The disadvantage of both hybrid solutions is that the rights to real estate are still related to land and not to volumes.

11.3.2 FULL 3D CADASTRE

In the full 3D cadastre, volume parcels can be established that no longer have a relationship with surface parcels. This concept was applied to the Gabba Stadium case study in Queensland, Australia. The legal framework in Queensland already provides the possibility of establishing volumetric parcels, as in the Gabba Stadium case; however, the cadastral framework does not provide the possibility of incorporating the precisely defined volumetric parcels as part of the cadastral geographical data set in 3D. The prototype applied to this case study showed that it is possible to use the 3D information from the 3D survey plans (needed to establish volumetric parcels) to insert the 3D geometrical and topological characteristics in the DBMS. This makes it possible to validate the volumetric parcels, to perform 3D functions on these parcels, and to query and visualize the 3D situation in one integrated view containing volumetric parcels and 2.5D surfaces of standard and remainder parcels. The prototype of the full 3D cadastre showed the very good potentials of a full 3D cadastre since insight into the 3D situation is considerably improved, while the concept is based on an integrated approach of the legal aspects (to allow volume parcels), cadastral aspects (to register volume parcels), and technical aspects (to define volume parcels in 3D survey documents and to incorporate this information in the cadastral database, followed by an integration of volume parcels and a 2.5D surface of the base map).

From the experiments with the case study in Queensland it can be concluded that, although the states and countries that already establish 3D property units have some remarkable differences (some require real constructions to be related to the 3D property registration, others not; some limit the 3D property to be within the column of one surface parcel, others not; some require quite detailed 3D survey plans to support the 3D property registration, others not), they all can be supported by a cadastral registration based on the proposed full 3D cadastre model, although there are some nontrivial aspects (in the conversion and use of a 3D cadastre), that require further attention.

An important condition of the full 3D cadastre is that the legal system is flexible enough to permit volume parcels. In other 3D cases the hybrid solution can be considered to improve traditional cadastral registration.

12 Summary, Conclusions, and Further Research

The main research question presented in this book was "how to record 3D situations in cadastral registration in order to improve insight into 3D property situations." To answer this question, this book was divided into four major parts. This chapter summarizes these parts and lists the main conclusions that can be drawn from the four parts:

- Analysis of the background (Section 12.1)
- Technical framework for modeling 2D and 3D situations (Section 12.2)
- Models for a 3D cadastre (Section 12.3)
- Realization of a 3D cadastre (Section 12.4)

Based on the conclusions, recommendations for future directions and future research can be outlined. Section 12.5 contains recommendations for future directions toward a 3D cadastral registration. Section 12.6 lists the recommendations for future research. This chapter ends by summarizing the most important results of this research.

12.1 ANALYSIS OF THE BACKGROUND

In the analysis of the background, the cadastral registration of 3D property units in several countries was studied to establish a clear overview of the requirements, the constraints, and the state of the art of 3D cadastral registration. The developments on 3D cadastral registration depend on the national legal system and on the type of cadastral registration (see Chapter 2). In Section 12.1.1 current 3D practice in cadastral registrations that are still surface oriented is summarized as well as the basic limitations. In Section 12.1.2 current practice of cadastral registrations that already provide the possibility to establish 3D property units is summarized. From the limitations and constraints of current practice, the basic needs and requirements for a 3D cadastre are summarized in Section 12.1.3.

12.1.1 SURFACE-ORIENTED CADASTRAL REGISTRATIONS

In countries that are still surface oriented, property rights to space are related to and dependent on the property of surface parcels. Persons can only be entitled to 3D property units by establishing rights and limited rights on intersecting surface parcels. The basic drawback of the land- or surface-oriented concept of property rights to real estate is that the 3D reality in which persons are entitled to volumes is projected on the surface.

The deeds concerning real estate archived in the land registration must always relate to land parcels. In the deeds it is possible to precisely define the space to which the concerning rights apply, for example, by adding an analog cross section. The basic drawback of these land registrations is that 3D property units are not known as individual property units in the land registration, except in the case of apartment rights or strata titles. In addition, how the legal status in 3D situations is established (what rights are used? are parcels subdivided? and also what information is added and included in the deeds?) depends very much on the choices in the deed and we may assume on the legal advice of the official charged with drawing up authentic deeds and legalizing documents. In general, there are no instructions for a 3D survey that could be added to a deed. Only in the case of apartment rights and strata titles is an analog or scanned drawing containing an overview of every floor (assuming that there are clearly identifiable floors) required, and only in the case of apartment units are there special requirements concerning the quality of the spatial information.

The surface parcel is always the entrance to a cadastral recording. Only in the case of apartment rights are individual 3D property units known as such in the administrative part of the cadastral registration. The 2D parcel as the basic and only real estate object in the cadastral registration meets several drawbacks. The legal status of the space above and below the surface can be obtained only by collecting information on the legal status of intersecting surface parcels. However, one first must determine which parcels intersect with the 3D construction. This is not always an easy query, as the construction itself is not available in the cadastral registration. In addition, the third dimension of rights and restrictions cannot be reflected in the cadastral registration, even if this information is available in deeds, drawings, or survey plans. Consequently, the current cadastral registration provides information on which persons have a right on a parcel but not on the spatial extent of these rights. Access to information in 3D property situations will soon be improved, because deeds and drawings archived in the land registration will be accessible in scanned format through the cadastral database in the near future, although this still will be limited to querying the static information (scanned document) per parcel, instead of visualizing the 3D situation of several parcels in one environment (similar to viewing the current cadastral map).

12.1.2   3D PROPERTY UNITS IN CADASTRAL REGISTRATIONS

Some cadastral registrations are based on a more advanced concept of the right of ownership and other property rights that is no longer always related to surface parcels but can be related to volumes. These solutions were found within legal frameworks that are able (or were able after some minor adjustments) to establish multilevel ownership, e.g., "volumetric parcels" in Queensland, "air-space parcels" in British Columbia, and "construction properties" in Norway and Sweden. These solutions to establish volume parcels differ per country; e.g., the footprints of 3D property units should be within the 2D surface parcels (British Columbia, apartments in the United States and Argentina) or not (Norway, Sweden, Queensland), the 3D property units have to relate to built constructions (Norway, Sweden, the United States, Argentina) or not (British Columbia, Queensland), the 3D property units have to be described in survey plans (British Columbia, Queensland, the United States, Argentina) or not (Norway, Sweden). From these new solutions it can be concluded that within some legal frameworks it is possible to explicitly entitle persons to volumes, which is an important precondition for a well-working 3D cadastral registration. The establishment of 3D property units that are no longer related to surface parcels provides better means to reflect the real property situation.

Although the 3D property units can be established within the legal framework in Queensland, British Columbia, the United States, Argentina, Norway, and Sweden and registered in the land registration and cadastral registration as individual property units, none of these solutions includes a complete 3D cadastral registration of 3D property units. This causes a few problems. First, a digital description of the 3D property unit in vector format is not maintained in the land registration (only scanned or paper drawings). Therefore, the 3D property unit cannot be viewed interactively and the geometry of the 3D property unit cannot be validated. Second, the 3D properties are still not incorporated in 3D in the geographical data set of the cadastral registration (only as footprints), by which it is not possible to query and view the 3D situation in the cadastral registration. These solutions therefore do not address technical issues, such as how to store, query, and visualize 3D property objects in 3D and how to make sure that 3D properties do not overlap (the condition that 2D parcels may not overlap assures complete and consistent registration in current cadastral registrations).

12.1.3 NEEDS AND REQUIREMENTS FOR A 3D CADASTRE

The basic needs for a 3D cadastre, focusing on improving insight into 3D situations, can be summarized as follows:

- To have a complete registration of 3D rights as such (rights which entitle persons to volumes). The current cadastre already registers rights that entitle persons to volumes, e.g., full ownership (applies to whole parcel column), right of superficies, etc.; however, a 3D cadastre should explicitly register the space to which rights apply.
- To have good accessibility on the legal status of 3D property units including 3D spatial information, as well as on public law restrictions.

It will be more effective (e.g., with respect data integrity and data consistency) if information on constructions and other objects of interest is maintained at their source (e.g., in databases of holders of constructions) and accessible within and from the 3D cadastre within a GII.

Based on these considerations, we can conclude that a 3D cadastre should incorporate the following functionalities:

- Register 3D information on rights (what is the space to which the person is entitled?) and make this information available in a straightforward way
- Establish and manage a link with external databases that contain objects that are of interest for the cadastre (infrastructure objects, soil pollution areas, forest protection zones)
- Use the information on these objects to support registration tasks, i.e., to detect and correct errors in cadastral registration or in the process of registering and viewing the legal status of 3D property (are all intersecting parcels encumbered with a right for the infrastructure object?)

12.2 FRAMEWORK FOR MODELING 2D AND 3D SITUATIONS

The 3D cadastre needs to be implemented using current and new techniques. The framework of modeling 2D and 3D situations was studied in Part II of this book. In Chapter 6 it was concluded the DBMS plays an important role in the new-generation GIS architecture. Consequently, to implement the 3D cadastre, in which a lot of information needs to be managed, a DBMS is needed for maintaining the cadastral spatial and non-spatial information concerning 3D situations.

Fitting this research in a technical framework required a study to determine what is technically possible with respect to maintaining, accessing, and analyzing 3D geo-information in DBMSs using standard products and additional developments. Current technologies were tested and concepts were designed and implemented into proto-types to improve current technology.

In Section 12.2.1, conclusions on possibilities of support of spatial data types in geo-DBMSs are drawn. Apart from geo-DBMSs, other developments of 3D GIS are important for the 3D cadastre research, since available 3D GIS functionalities in general impose constraints and provide possibilities on how to maintain, access, and analyze 3D geo-information. In Section 12.2.2, state of the art of 3D GIS is summarized. How to access spatial information stored in a DBMS with different front ends (as the new-generation GIS architecture is organized) is described in Section 12.2.3. Finally, in Section 12.2.4 the possibilities and problems of combining 2D and 3D geo-objects in one environment are described.

12.2.1 2D AND 3D GEO-OBJECTS IN GEO-DBMS

The 3D spatial component of constructions and rights, but also of parcels, has to be registered in the cadastral database. This raises the question of how to structure spatial objects in 2D and 3D in a DBMS. Concerning this, the following conclusions can be drawn.

2D and 3D Geometrical Primitives in DBMS

Geometrical primitives as defined by the OpenGeospatial Consortium (OGC) have been adapted by mainstream DBMSs and popular noncommercial DBMSs. The OGC Implementation Specification for SQL[140] is to date 2D. It also does not cover topologi-cal structure, although topological relationships can be obtained by spatial functions on the geometrical primitives. The ISO DIS 19107 standard[83] (adopted as Abstract Specification by OGC) does define 3D spatial objects and topological structure; how-ever, these Abstract Specifications still have to be transformed into Implementation Specifications by OGC and to be adopted by DBMSs.

Current DBMSs do not support 3D volumetric data types. To maintain 3D geometrically structured data within current techniques, 2D primitives defined in 3D embedding space can be used (polygon defined in 3D). 3D objects can be defined either as a body that consists of a set of faces or as a multipolygon defined in 3D. However, these 3D objects are not recognized as such by DBMSs or only in a limited way (e.g., to calculate the 3D length of a line). The z-coordinates are stored, and in nearly all spatial analyses and validation checks the 3D object is projected on the surface. To support true 3D in a DBMS, a 3D geometrical primitive (polyhedron) has been defined and implemented in the DBMS as part of this research. By using this primitive, 3D geometries can be defined consisting of flat faces including holes. This implementation shows the possibilities of maintaining 3D objects in a geometrical structure. As part of the implementation 3D spatial functions and a 3D validation function were implemented.

2D and 3D Topological Structures in the DBMS

Awaiting an Implementation Specification for 2D and 3D topological structure, there are already some user-defined (Section 6.2.2) and commercial implementations of 2D topological structures available (Laser-Scan and Oracle 10g; see Section 6.2.3). These implementations look promising when applying topological queries on the structures (good performance). However, geometrical queries are faster on the geometrical primitives since many tables need to be queried to obtain a geometrical realization of the topological structure before the geometrical query can be executed. At the moment, topological structure is therefore appropriate mainly for representing relationship operations and for checking the quality of the data. The topology structure offers better maintenance possibilities with respect to quality. Topological structure supports consistency of spatial data since because lower-dimensional objects are stored only once, in contrast to data defined with geometrical primitives. Topological structure management to maintain 3D geo-objects and 2D geo-objects for the 3D cadastre is preferred, but, as can be concluded from the research presented in this book, has to be implemented using self-defined extensions.

We experimented with a DBMS implementation of a 3D topological structure: SSM (Simplified Spatial Model), which is a topological structure described in Reference 229. This topological structure supports only flat faces (as the implemented polyhedron primitive). In an object relational DBMS, the relationships between the high-dimensional (3D body) and low-dimensional objects (face and node) can be stored. The implementation shows that storing a 3D object and generating a geometrical realization of the 3D object within the DBMS is not a problem. However, because the topological structure is not recognized by the DBMS, topological consistency has to be checked and guaranteed outside the DBMS, available spatial indexing cannot be used, and spatial functions have to be self-defined (intersection, distance).

Summarizing, the 2D geometrical primitive (including spatial operations) is well implemented in DBMSs; support for topological structure in 2D in DBMSs has just started but will most probably be available in DBMSs within a few years. However, none of the DBMSs has started with support for 3D volumetric objects (using either geometrical primitives or topological structure). In addition the OGC still has to decide on Implementation Specifications for a geometrical and a topological schema in 3D. Therefore, the 3D cadastre will have to be based on a combination of commercial products and user-defined extensions that showed potential in the experiments in this research.

12.2.2   3D GIS

In Chapter 8 an extended overview was given concerning other basic aspects (apart from DBMS aspects) of 3D GIS: organization of 3D data, 3D data collection, and object reconstruction, visualization, and navigation in 3D environments, and 3D analyzing and 3D editing. Based on this overview it can be concluded that 3D GIS still has to mature. 3D GIS developments are mainly in the area of visualization and animation. Bottlenecks for commercial implementation of 3D GIS are as follows:

- 3D editing in GIS is not yet possible and is traditionally a functionality that is well supported in CAD software but not in GIS.
- There is poor linkage between CAD, traditionally designers of 3D models, and GIS.
- Methods to automatically reconstruct 3D objects are lacking.
- Visualization of 3D information requires special techniques; characteristics such as physical properties of objects (texture, material, color), behaviour (e.g., on-click-open), and different levels of detail representations need to be maintained and organized in DBMSs;
- Virtual Reality and Augmented Reality techniques should be incorporated in GIS software to improve interaction with and visualization of 3D environments.

12.2.3   ACCESSING SPATIAL INFORMATION ORGANIZED IN A DBMS

Once 3D geo-objects are stored in a DBMS with current techniques, the next issue is how to access and query the geo-objects by front ends. Three front ends were analyzed to access 3D objects stored in 3D geometrical primitives in Oracle Spatial 9i: a CAD-oriented front end, a GIS front end, and a self-developed front end using Web-based techniques.

CAD-Oriented Front End

With the CAD-oriented software MicroStation GeoGraphics (MS GG) it is rather easy to visualize 3D objects stored as multipolygons in a DBMS; however, querying and editing 3D objects require more complex steps but is possible, while true 3D editing is supported in MS GG. The main disadvantage is that the database structure is altered. The Java applet Spatial Viewer that is delivered with MicroStation GeoGraphics requires less customization and is therefore easier to use.

GIS-Oriented Front End

To be able to access a spatial layer stored in Oracle Spatial with the GIS front end ArcGIS, one first needs to register the spatial layer with ArcSDE. After registering the spatial layer, querying of spatial objects is, apart from some small problems, straightforward and the tables structure is not altered. One major complexity of ESRI is that ArcSDE validates spatial objects before they are inserted into ArcGIS. This means that spatial layers containing invalid spatial objects cause problems. The main consequence of not being able to handle invalid objects is that "vertical" polygons

(polygons perpendicular to the surface) cannot be visualized in ArcScene, although ArcScene does support vertical polygons when they are stored in other formats. It should be emphasized that ESRI, as other GIS software, does not offer graphical functionality to edit in 3D and to perform spatial analyses in 3D. Both MicroStation GeoGraphics and ArcGIS are specifically based on Oracle Spatial 9i, which is not fully OGC compliant. MicroStation GeoGraphics and ArcSDE also support other DBMSs. However, all combinations (i.e., front end combined with back end) have their own architecture. If both the DBMS and the front end were fully OGC compliant, it should be possible to query any DBMS that supports OGC geometries with any front end that is based on OGC specifications.

Web-Based Front End

To look for a more open solution in the sense of interoperability but also in the sense of open source, a prototype was built using Web-based techniques. The Internet has become a major tool for disseminating information in today's society, in which information has gained a crucial place. We studied the use of Web technologies that were designed outside the GIS world. These techniques included Internet formats for displaying and querying 3D objects and techniques to query a DBMS via the Internet. Based on these techniques two prototypes were built. The experiences with the prototypes showed that it is possible to view and query 2D and 3D geo-objects that are stored in a DBMS using open source Web technology. Although Oracle is used as back end, the prototype that looks most promising uses an XSQL servlet, which also works on other DBMSs provided these DBMSs can be accessed via JDBC connections.

To make the prototype OGC compliant, we studied the possibilities to use the OGC Web services. OGC has defined several OGC Web services that can be used to disseminate 3D information via the Internet: Web Map Service,[145] Web Feature Service,[146] Web Terrain Service,[141] and Web Coverage Service.[150] Navigation, querying, and identifying 3D geo-objects require a 3D vector representation of 3D objects, which is offered only by the Web Feature Service that returns geo-information in GML (Geography Markup Language). GML 3.0[147] includes the ability to handle complex properties, to describe coordinates with x,y, and z (already possible in version 1 and 2), and to define 3D objects. With the Web Feature Service it is also possible to edit 3D objects via the Internet.

12.2.4 2D PARCELS AND 3D GEO-OBJECTS IN ONE 3D ENVIRONMENT

When integrating 3D geo-objects and 2D parcels in one environment, the height issue needs to be addressed: how to locate the 3D geo-objects with respect to 2D surface parcels in one 3D view. Basically, there are two solutions for this:

1. z-coordinates of 3D geo-objects are stored within a national reference system.
2. z-coordinates of 3D geo-objects are stored relative to the surface.

The most sustainable solution is to define 3D objects with absolute z-coordinates within a national reference system: First, because absolute z-coordinates are not influenced by surface changes; second, the definition of the surface level (the reference level used for values with respect to the surface) is sometimes not clear. Finally, when using z-coordinates with respect to the surface defining the actual geometry of 3D objects is complicated. Having 3D objects defined in absolute values, the next issue is how to combine the 3D objects with parcels defined in 2D. For this purpose the parcels need to be draped over a height surface. A case study was carried out using a DBMS approach in which laserscan data (point heights) on a density of one point per 16 square meters was integrated with parcel boundaries in order to be able to extract height surfaces of individual parcels. TINs (Triangular Irregular Networks) representing height models were created outside the DBMS because TINs are not yet supported within DBMSs. The planar partition of 2D parcels was included in the TINSs.

The main conclusions that can be drawn from this case study are as follows:

- Incorporating the planar partition of parcels into a height surface makes it possible to extract the 2.5D surfaces of parcels and to visualize the 2D cadastral geographical data set in a 3D environment.
- It is not easy and straightforward to create a good integrated elevation and object model. Several alternatives of a TIN structure were investigated: unconstrained Delaunay TIN, constrained TIN, conforming TIN, and finally refined constrained TIN. After some analysis, the refined constrained TIN was selected as most appropriate for the purpose of this research.

The large data volume as a result of a dense laseraltimetry data set led to poor performance, while not all point heights significantly contribute to the height surface. Therefore, a generalization method was described to come to an effective model of parcel surfaces. The first part of this generalization method was implemented and applied to a study area. From these experiences it can be concluded that an initial filtering of the point heights results in a much improved integrated model: about four times fewer points, but still within the epsilon tolerance of the same size as the quality of the original input data sets.

12.3 MODELS FOR A 3D CADASTRE

In Part III of this book, conceptual models and logical models for a 3D cadastral registration were developed to meet the cadastral and technical requirements for a 3D cadastre that were studied in Part I and Part II of this book.

12.3.1 CONCEPTUAL SOLUTIONS FOR A 3D CADASTRE

Based on the conclusions of Part I and Part II, three concepts with several alternatives were distinguished (the UML class diagrams were also drawn in Chapter 9):

- Full 3D cadastre:
  Alternative 1: Combination of infinite parcel columns and volume parcels
  Alternative 2: Only parcels are recognized that are bounded in three dimensions (volume parcels)

- Hybrid cadastre:
    Alternative 1: Registration of 2D parcels in all cases of real property
        registration and additional registration of 3D legal space in the case
        of 3D property units
    Alternative 2: Registration of 2D parcels in all cases of real property
        registration and additional registration of 3D physical objects
- 3D tags linked to parcels in current cadastral registration; additional infor-
    mation is available on analog or scanned documents and drawings.

12.3.2 THE OPTIMAL SOLUTION FOR A 3D CADASTRE

The full 3D cadastre showed the best potential, because the 2D parcel as the sole
basic concept of cadastral registration is abandoned. Consequently, the 3D cadastral
issue is solved at a fundamental level. In a full 3D cadastre it is possible to transfer a
volume parcel, without relating the property rights for this space to the surface parcels.
In this book we have presented two variants of a full 3D cadastre: one with both
volume parcels and infinite 3D columns (which are defined by the parcel boundaries
on the surface from which volume parcels may be subtracted) and one with only
bounded volume parcels. The advantage of the first alternative is that this variant
still has a strong link to the current 2D registration and conversion of the current
cadastral registration into this variant is more feasible. The first alternative also has
the advantage of being able to represent infinite (open) 3D parcel columns, which
still suffice in 2D property situations (where only one person is entitled to a parcel).
It was therefore decided to select and refine this model.

The first alternative of a full 3D cadastre can only become practice if the legal and
cadastral frameworks can be extended to establish a volume parcel that is no longer
related to the surface configuration. However, as seen in Part I, whether volume
parcels will be easily permitted is dependent on the type of cadastral registration and
the legal system of a specific country. Cadastral registration in many countries seems
to be still very land oriented (as in the Netherlands, but also in Argentina and in British
Columbia where air-space parcels have to be totally located within one parcel), and
the step toward a full 3D cadastre might be too ambitious for the medium-term future.
Therefore, concurrently with the full 3D cadastre, the possibilities and constraints of
the hybrid cadastre were studied.

In the hybrid cadastre, 3D situations (factual situations) are registered apart from
2D parcels (legal situations) in one integrated environment. This solution fits within
the traditional 2D legal framework and to some extent within traditional cadastral and
technical frameworks (changes needed in the cadastral and technical framework can
be achieved within a few years in standard products or self-developed software). In
the hybrid cadastre, property rights to real estate are still always registered on parcels
on the surface. This is the basic difference from the full 3D cadastre concept.

Two possible alternatives were introduced to effectuate the hybrid solution: registration of right-volumes and registration of 3D physical objects. In the registration
of right-volumes the limited rights registered on parcels form the starting point: what
rights are established on a parcel and what is the space where the rights are valid? A 3D
representation of this space is registered in the cadastral registration. In contrast, in a

3D physical object registration the 3D physical object is the starting point of registration, independent of the rights that have been established. Preferably 3D information on physical objects is maintained by organizations responsible for the objects and accessible in the cadastral database via the Geo-Information Infrastructure.

The solution of "3D tags in the current cadastral registration" is a solution that works, as current practice proves; however, it has some basic limitations. The solution cannot provide one 3D overview of the cadastral map integrated with 3D property situations: 3D situations can be examined only per parcel, i.e., isolated from each other. This solution, therefore, does not provide a base for efficient and sustainable registration in the future.

12.4 REALIZATION OF A 3D CADASTRE

The concepts of the first alternative of the full 3D cadastre and the two alternatives for the hybrid cadastre were translated into logical models and prototypes in Part IV. The prototypes were implemented within the legal, cadastral, and technical framework described in Part I and Part II. The aim of the prototypes was to evaluate the conceptual models. The concepts of the hybrid cadastre were applied to the Dutch case studies introduced in Chapter 3, because the Netherlands Kadaster still holds strongly to the 2D parcel concept as in the hybrid cadastre. The concept of the full 3D cadastre was evaluated by applying this concept to a case study in Brisbane, Queensland that was also introduced in Chapter 3, because the legal framework in Queensland provides the establishment of 3D property units independently from the surface configuration.

Conclusions based on the experiments with the full 3D cadastre prototype are listed in Section 12.4.1, while the conclusions for the hybrid cadastre are described in Section 12.4.2.

12.4.1 Full 3D Cadastre

In the full 3D cadastre prototype environment (based on a 3D polyhedron extended version of the Oracle spatial DBMS and ESRI and Bentley GIS/CAD software), the 3D property survey plans were converted into a spatial representation in the DBMS and the surface parcels were successfully merged with a terrain elevation model.

Conclusion on Full 3D Cadastre

The full 3D cadastre offers many improvements compared to traditional cadastral registrations:

- The real situation is no longer projected on the surface; i.e., volume parcels are not dominated by the parcel pattern on the surface.
- Persons can straightforwardly be entitled to space instead of establishing property rights on intersecting parcels.
- The space is precisely described in a 3D survey document, which offers a uniform way of defining 3D property units.

In addition, implementing the full 3D cadastre offers also improvements for cadastral registration in countries and states that already establish 3D property units as volume parcels in the legal framework:

- The information from the 3D survey document can be used to insert the geometrical and topological description of volume parcels in the DBMS.
- The volume parcels can be viewed interactively.
- The geometry of volume parcels can be checked, e.g., whether faces are planar or not.
- The 3D situation can be spatially queried in the DBMS (e.g., do volume parcels intersect?).
- The volume parcels can be visualized in an integrated view with a 2.5D representation of the parcels that are defined by parcel boundaries on the surface.
- The volume parcel and the 2.5D surface parcels can be queried in the DBMS (e.g., is a volume parcel located above or below the surface, or does it intersect the surface?).

The prototype environment of the full 3D cadastre offers the possibility to query, analyze, and visualize the true 3D situation of the properties. However, while the legal, organizational, and technical aspects of a 3D cadastre have been solved, some nontrivial aspects (in the conversion and use of a 3D cadastre) require further attention as was shown by the case study (e.g., how to model volume parcels with complex geometries).

12.4.2 HYBRID CADASTRE

When a legal framework cannot yet deal with the establishment of property rights to volumes independent of the surface, the 3D cadastre can be implemented within a hybrid environment introducing either right-volumes or a 3D physical object registration.

Right-Volumes

From the prototype implementations it can be concluded that the introduction of right-volumes means a significant improvement of current cadastral registration in 3D situations. The inclusion of right-volumes in the cadastral geographical data set provides an overview of the distribution of 3D property units. The registration warns the user of the cadastral registration that something is located under or above the surface. It also gives information on what is located under or above the surface (relationship with whole real estate object is maintained). For precise information the deed in the land registration can be consulted. A 3D survey of the situation can be made and used to describe the situation in the deeds and to determine the upper and lower limits of right-volumes.

From a technical point of view, the geometry of right-volumes is simple and can therefore be maintained in the DBMS within current techniques.

Basic disadvantages of right-volumes are as follows:

- Since parcels are still the basis for registration, gaps can occur when no rights have been established that require a cadastral recording, e.g., when the holder of the construction is the same as the owner of the intersecting parcel. In these cases the location of the construction is still not known in the cadastral registration.
- If rights are established on just a part of a parcel, new parcel boundaries need to be created. This leads to fragmentation of both parcels and right-volumes.
- When space where the right applies to is not precisely restricted in height or depth, registration of right-volumes does not give satisfying insight, as was seen in the Rijswijk case. This could be solved by a rule applying to 3D surveys that will allow open polyhedrons (either not defined in height or in depth).
- Horizontal boundaries restrict the spatial description of right-volumes. The concept of a right-volume could be improved when other than just horizontal boundaries could be defined.

Registration of 3D Physical Objects

The registration of 3D physical objects comprises the registration of physical objects as they occur in the real world. From the experiments with the case studies it can be concluded that such a registration offers several improvements. The 3D description of the physical objects (extent of the object) can be used for reference purposes (i.e., to improve the reflection of the real situation) and to support cadastral tasks. When a 3D physical object is registered, parcels do not need to be divided into parcels matching with the 2D projection of the physical object because the exact location of the physical object is known in the cadastral database. Only one object needs to be registered by which the registration for all intersecting parcels can be guaranteed. All parcels intersecting with the physical object can be found by a spatial query (by an overlay with the projection of the 3D object).

From a technical point of view the geometry that has to be maintained for physical objects can become complex. It is therefore not easy and straightforward to insert and maintain the spatial information on 3D physical objects in the DBMS within current techniques.

Conclusion on Hybrid Cadastre

Because both solutions of the hybrid cadastre are implemented within traditional legal and cadastral frameworks, rights to hold 3D property units are still registered on the intersecting parcels. Querying the legal status of 3D property units still needs to be done by querying the legal status of the intersecting parcels. However, the maintenance of the 3D situation can assist considerably in understanding the real situation.

Cadastral registration will benefit from a hybrid registration for a number of reasons:

- The solutions give insight into the real situation. It is now clear from the cadastral registration that persons are entitled to space above or below the surface.

- Both solutions are implemented within the cadastral registration as part of the cadastral geographical data set and can therefore be queried with parcel surfaces in one integrated view.
- The proposed solutions show authorities how the inclusion of spatial information in deeds can be used to visualize the 3D component of rights in the cadastral registration. The solution can make notaries aware of the improvements of 3D registration and may motivate them to include well-defined 3D information in the deeds.
- Registrations and databases outside the cadastral domain can benefit from the information on 3D situations that is available in the cadastral registration and via the Geo-Information Infrastructure (monument registration, building registration, taxes for immovable goods, management of soil pollution areas, management of cables and pipes, management of the subsurface).

A physical object relating to a property unit within a building complex coincides with the legal space of the property unit. The main objective of cadastral registration in the case of building complexes is to give insight into property boundaries in all dimensions rather than to reflect the built constructions in the cadastral registration for reference purposes. On the other hand, the main objectives of a registration of 3D physical objects are, first, to be able to locate infrastructure objects to support cadastral tasks and, second, to register the person who holds an infrastructure object. Therefore, the registration of 3D physical objects will specifically be suitable for infrastructure objects. For registering property units in building complexes, right-volumes are more appropriate, because the spatial extent of properly units can be easily and clearly defined with right-volumes, which refer to the legal space to which a person is entitled.

The two concepts of the hybrid cadastre (right-volumes and 3D physical object registration) have a different line of approach and therefore meet other needs of 3D cadastral registration. The right-volume is a considerable improvement of insight into stratified property as part of the cadastral geographical data set, while the 3D physical object registration provides information on constructions, which is available in the cadastral geographical data set to improve the reflection of the real situation. The concepts could be combined to take advantage of both solutions.

12.5 FUTURE DIRECTIONS FOR A 3D CADASTRE

In countries where the property rights to real estate are still very much land (i.e., surface) oriented within the existing legal doctrine and cadastral framework, the hybrid cadastre seems to be the best solution for the medium-term future. The proposed alternatives meet the requirements of 3D cadastral registration. Insight into the legal status of 3D property is improved, because the 3D extent of rights can be visualized and queried in the case of right-volumes. In the case of a 3D physical object registration the construction itself is available in the cadastral registration by which the real situation is much better reflected. However, one basic principle is not addressed. Since the legal status of constructions and 3D property units is still registered by means of land parcels, querying the legal status in 3D still means collecting information on the legal status of the intersecting parcels.

As seen in Chapter 1, the FIG Bathurst Declaration[50] concluded that "most land administration systems today are not adequate to cope with the increasingly complex range of rights, restrictions and responsibilities in relation to land." Many existing cadastres are still based on the paradigm of a land parcel that has its origin centuries ago. This paradigm needs to be reconsidered and adjusted to today's world. Although parcels are traditionally represented in 2D, someone with a right to a parcel always has been entitled to a space in 3D. This led to no disputes as long as only one person was entitled to a land parcel since the traditional cadastre was capable of reflecting such property situations. However, in recent times stratified property is common practice, and in many countries multifunctional use of space is official planning policy. The way humans relate to land has also changed drastically (value of private property has increased considerably). Today's cadastral registration should therefore reflect the true principle of property rights that entitle persons to volumes and not just to areas.

The ultimate ambition for 3D cadastral registration should be a full 3D cadastre in which it is possible to entitle persons both to unconstrained parcel columns that are defined by boundaries on the surface and to bounded amounts of space (volume parcels).

The optimal solution for such a full 3D cadastre starts with the regulations for 3D surveys in case of stratified property (volume parcels). The volume parcel is then inserted in the land registration and known as an individual property that can be transferred independently from other properties.

The information from the 3D survey plans (in which the height is defined in absolute z-values within the national reference system) can be used to register the volume parcels in the cadastral registration and can be used to insert the 3D geometrical and topological characteristics of the volume parcels in the cadastral database. To be able to query the volume parcels and the parcels that are defined by boundaries on the surface in one environment the surface parcels need to be represented by 2.5D surfaces. After this whole procedure is clearly defined, the process from 3D survey to insertion in the cadastral database can be streamlined.

The experiments of the case study in Queensland using the prototype of a full 3D cadastre showed that the legal, organizational, and technical aspects in a full 3D cadastre are solved and that the proposed alternative of a full 3D cadastre is realizable. In a technical sense, the basic conditions for a full 3D cadastre are met (although these technologies still need further development). However, the actual implementation of a full 3D cadastre in many countries may meet complications. It requires a change in the way of thinking about the right of ownership and other property rights because in the full 3D cadastre the basic paradigm of land-oriented real estate has to be abandoned. However, since a full 3D cadastre offers solutions for 3D cadastral registration at a fundamental level and since the 3D principle of property is appropriately reflected in the cadastral registration, steps toward a full 3D cadastre should be further studied and be taken in the future.

12.6 FURTHER RESEARCH

Future research concerning the different aspects of 3D cadastral registration should focus on a number of areas.

12.6.1 INSTITUTIONAL ASPECTS OF 3D CADASTRAL REGISTRATION

To create better methods to register 3D property units, legal frameworks should be further examined. 3D cadastral registration in case of a full 3D cadastre is possible only when the legal framework provides the possibility of establishing volume parcels.

Further research should therefore focus on the following questions. How flexible is the definition of ownership of land (surface) from both a legal and a cadastral point of view? Is it possible to establish volume parcels as in Queensland, British Columbia, the United States, Argentina, Norway, and Sweden without changing the Civil Code? Are the costs to establish volume parcels lower than the expected benefits? How easy is it to change cadastral registration to register real estate objects other than parcels and apartment units? In addition, further research is needed on what 3D information is needed in deeds and survey plans and how this information should be collected, structured, and offered to make a 3D cadastral registration possible.

12.6.2 GEO-INFORMATION INFRASTRUCTURE

In today's society, information is of growing importance, and information exchange via the Internet is especially vital. The Geo-Information Infrastructure (GII) facilitates the exchange of geo-information across the Internet. A distributed setup of registrations within a GII provides the possibility to link information maintained in different databases. In this way the geometry of objects such as infrastructure, soil pollution area, and monuments can remain and be maintained at their original source (in databases of organizations that are responsible for these objects), while this information can be used to improve cadastral registration in the case of 3D situations. In addition, other persons and organizations can benefit from a 3D cadastral registration within a GII, since information from the cadastre is much easier to access. Therefore, the establishment of a GII needs further research concerning both technical and organizational aspects. The research on GII is also pushed by a growing need to integrate data sets from different domains and different countries e.g. the INSPIRE program in the European Union. This requires specific research, including the development of formal semantics.

12.6.3 3D IN THE NEW-GENERATION GIS ARCHITECTURE

In the area of 3D in the new-generation GIS architecture, future research should include the following topics: 3D modeling in DBMS, accessing 3D objects organized in a DBMS, 3D data collection, and generating an effective integral model of point heights and parcels.

3D Modeling in DBMSs

To improve 3D modeling in DBMS, the following issues need special attention:

- 3D (volumetric) data types are future work for standard DBMSs. The polyhedron primitive as it is implemented and described in this reserach showed possibilities for maintaining 3D geo-objects in a DBMS and is a first step toward 3D support in DBMSs. Future research should focus on implementing a more complex geometrical primitive in the DBMS, e.g., using curved surfaces.

- At present, 3D implementations are focused on boundary representation. However Constructive Solid Geometry (CSG) may appear appropriate for designed large-scale human-made objects (traffic signs, buildings) and voxel representation (3D raster) for continuous phenomena. Therefore, future 3D GIS may ask for the support of CSG and voxels (or other 3D tessellations) in DBMS.

- Topological structure in 3D is not supported in DBMS within current techniques although topological structure can be stored in relational tables. As the DBMS does not support the 3D topological structure, future research should focus on full support for 3D topological structure, i.e., performing consistency checks and resolving topological errors inside the DBMS. An OGC Implementation Specification for 2D topological structure also still needs to be finalized. Recently, 2D topological structure is available in commercial products. Research is needed to assess these implementations.

- Current DBMSs support only spatial functions in 2D. Future research should focus on 3D spatial analyses (e.g., overlay, buffer, route planning, visibility) and 3D querying.

Accessing 3D Objects Organized in a DBMS

Concerning the Web-based solution to access 3D spatial objects that are organized in a DBMS, future work should focus on a number of research issues. Although first experiences with the prototypes look promising with respect to performance, serious tests on larger data sets need to be set up. Fast rendering of 3D objects is, of course, critical when displaying data via the Internet. Other issues that need attention when disseminating 3D geo-information via the Internet are how to access data stored in separate DBMSs and how to address 3D cartographic aspects (perspective, stereo, movement, transparency, sticks that indicate the distance between a subsurface construction and the surface level, etc.). This requires that not only spatial and non-spatial information for spatial objects be maintained in the DBMS, but also characteristics such as physical properties of objects (texture, material, color), behavior (e.g., on-click-open) and different levels of detail representations. To join the interoperability standards of the OGC Consortium the OGC Web Services (and especially the Web Feature Service) should be studied to see how these services can be used to disseminate 3D geo-information across the Internet.

3D Data Collection

Future research is needed to make the process of 3D object reconstruction automatic or semiautomatic. In general, the process of 3D object construction is nontrivial (even using advanced sensors and reconstruction software) and still needs to be done partly manually, which is relatively time-consuming. In addition, underground construction such as tunnels and pipelines cannot be modeled using aerial laserscan and photogrammetry techniques. Therefore, it is needed to look at the designed CAD models. To improve 3D object reconstruction for geo-purposes, future research should focus in detail on the interoperability problem between GIS and CAD in order to be able to use

CAD designs in 3D GIS and to use 3D edit functionality and advanced visualization techniques available in CAD in 3D GIS. A small number of such problems have already been investigated in the case studies that were carried out as part of the research presented in this book, i.e., lack of object definitions in the CAD models, different scale representations, transformation of the local (CAD) coordinates into a reference system for both the horizontal and vertical coordinates, parametric shapes that cannot easily be converted into simple geometries, different levels of detail that requires generalization. The use of detailed CAD models in GIS requires 3D generalization algorithms. Therefore, 3D generalization is a fundamental issue that needs special attention when bridging the gap between CAD and GIS.

In general, the basic problems of linking GIS and CAD need better understanding in order to close the gap between GIS and CAD.[135] For this at least two important developments are needed. The first one is a semantic analysis of the concepts of these 'different' worlds. A two-way translation is needed between these concepts. Second, both GIS and CAD should base their data management on the same technology, for example spatial DBMSs compliant with OGC (ISO) standards. Therefore, first the CAD standards and GIS standards need to be harmonized.

Generating Effective Integrated Model of Point Heights and Parcels

The integrated height and object model represented in a TIN could be improved. Future work with respect to the integrated model should include the following topics:

- The TIN computation should be performed inside the DBMS to avoid time-consuming conversions that may lead to a decrease in the quality of the data. The ideal case would be just storing the point heights and the parcel boundaries in one, or preferably in distributed, DBMS and to generate the TIN (available in a view) of the area of interest on user's request in the DBMS, without storing the TIN explicitly. This is more efficient because no data transfer (and conversion) is needed from DBMS to TIN software and back. Future research should therefore focus on supporting TIN data structure, TIN creation, and TIN data reduction methods within the DBMS.
- As indicated in Section 8.3, the current TIN computation takes place in the 2D plane. It may be better to compute the integrated height and object model in true 3D space, based on tetrahedra (and then finding the proper parcel surface within this tetrahedron network; see also Reference 210).
- In the implemented data reduction method, the reduction is based on reducing only the number of point heights. In the future, the generalization of the integrated model should also take into account the 2D objects, especially the boundary line generalization and the object aggregation. This will lead to an integrated data reduction procedure of 2D objects and point heights, taking the constraints defined by the 2D objects into account. In the data reduction process, the planar partition of the 2D objects should always be part of the TIN structure, by which it is possible to extract height surfaces for individual 2D objects.
- It is important to maintain the result of the generalization in a multiscale data structure, as the costs of the computations are significant. This requires further research on multiple representations at different scales in DBMSs.

12.7 MAIN RESULTS OF THE RESEARCH PRESENTED IN THIS BOOK

The main objective of this book focused on how to record 3D situations in a cadastral registration in order to improve insight into 3D property situations. Based on the findings of the background study, the study on technological possibilities, and the experiments with the case studies, it can be concluded that a full 3D cadastre that both registers surface parcels and volume parcels offers the best potential and is realizable. Cases in Queensland, British Columbia, the United States, Argentina, Norway, and Sweden have already shown that it is possible to establish volume parcels within legal frameworks. The research presented in this book showed that it is possible to register volume parcels together with 2.5D surfaces of parcels within a cadastral and technical framework. It can therefore be expected that in the near feature more countries and states will implement further steps toward the full 3D cadastre model as described in this book.

In a technical sense this research contributed to 3D developments within the new-generation GIS architecture in general, as the prototypes showed that it is possible to maintain 3D objects using 3D geometrical primitives in a DBMS, to perform spatial functions in 3D within the DBMS on the 3D objects, to access the geo-DBMS containing 2D and 3D geo-objects using GIS/CAD front ends and Web-based technology, and, finally, to combine 2D geo-objects and 3D geo-objects in one 3D environment by generating an effective structure of a 2.5D surface that incorporates the planar partition of 2D geo-objects.

The research presented in this book showed and implemented major preconditions to establish a full 3D cadastre within a legal, cadastral, and technical framework. However, a number of technical limitations still need to be tackled to have commercially available tools to support a full 3D cadastre that can operate as part of a GII. Apart from technical aspects, many cadastral, legal, and institutional issues also need to be addressed to accomplish a full 3D cadastre. For instance, the cadastral registration should be organized in a uniform manner, and the legal framework should allow the establishment of independent 3D property units. Other questions must be answered: What will be the value of the 3D information provided (mere informative, or also binding to third parties)? Who is responsible for the information to the cadastre, e.g., do we need survey plans made by certified surveyors? What type of 3D real estate objects can be dealt with (building complexes, infrastructure objects)? Countries that are still based only on the 2D parcel paradigm especially need to reconsider the traditional ownership concept in order to better reflect the object of property that is a volume and not just a surface.

References

1. A. Aguilera. Orthogonal Polyhedra: Study and Application. Ph.D. thesis, Universitat Politècnica de Catalunya, Barcelona, Spain, 1998.
2. T.M. Aldridge and A. van Velten. Apartment ownership in the European Union. *Notarius International*, pp. 17–30, 1997.
3. C. Arens. Maintaining reality: modelling 3D spatial objects in a Geo-DBMS using a 3D primitive. Technical Report M.Sc. thesis, Delft University of Technology, Delft, the Netherlands, 2003.
4. C. Arens, J.E. Stoter, and P.J.M. van Oosterom. Modelling 3D spatial objects in a GeoDBMS using a 3D primitive. *Computer & Geosciences*, March: 165–177, 2005.
5. C. Arens, J.E. Stoter, and P.J.M. van Oosterom. Modelling 3D spatial objects in a GeoDBMS using a 3D primitive. In *Proceedings AGILE 2003*, Lyon, France, April 2003.
6. Argentinean Federal Government. Ministrio de Justicia y Derechos Humanos. Ley 13.512. Regimen de la Propiedad Horizontal, Buenos Aires, Boletin oficial (in Spanish), 1948. Available online at www.saij.jus.gov.ar/download/ley13512/13512. html/.
7. S. Aronoff. *Geographic Information Systems: A Management Perspective*. WDL Publications, Ottawa, Canada, 1989.
8. B.G. Baumgart. Winged-edge polyhedron representation for computer vision. In *Proceedings of National Computer Conference*, Stanford, California, USA, May 1975, 589–596.
9. L.A. Belfore. An architecture supporting live updates and dynamic content in VRML based virtual worlds. In *Proceedings of Symposium on Military, Government and Aerospace Simulation 2002 (MGA 2002)*, San Diego, California, April 2002, 138–143.
10. M. Benhamu and Y. Doytsher. A multilayer 3D cadastre: problems and solutions. In *Proceedings FIG, ACSM/ASPRS*, Washington, D.C., USA, April 2002.
11. Bentley. Bentley MicroStation GeoGraphics. Spatial edition, 2004. Available online at www.bentley.com/products.
12. E. Bignone, O. Henricsson, P. Fua, and M. Stricker. Automatic extraction of generic house roofs from high resolution aerial imagery. In *Proceedings of European Conference on Computer Vision—ECCV'96*, Vol. 1, Cambridge, U.K., April 1996, 85–96.
13. T. Blaschke and D. Tiede. Bridging GIS landscape analysis, modelling and 3D simulation. Is this already 4D? In *Proceedings of CORP 2003 Geo Multimedia*, Vienna University of Technology, Austria, February 2003.
14. L. Bodum. 3D mapping for urban and regional planning. In *Proceedings of URISA Annual Conference 2002*, Chigaco, Illinois, USA, 2002, 472–479.
15. P. Bottelier, R. Haagmans, and N. Kinneging. Fast reduction of high density multibeam echosounder data for near real-time applications. *Hydrographic Journal*, 98:23–28, 2000.
16. P.W. Bresters. 3D visualisations with the height model of the Netherlands (AHN). In *Proceedings of EuroSDR Commission V Workshop on Visualisation and Rendering*, Enschede, the Netherlands, January 2003.
17. British Columbian Government. Land Title Act. British Columbia, Canada, 1996.
18. British Columbian Government. Land Title Regulations. British Columbia, Canada, 1996.

19. R. Brügelmann. Automatic breakline detection from airborne laser range data. *International Archives of Photogrammetry and Remote Sensing*, 33(B3/1):103–110, 2000.

20. B. Cambray. Three-dimensional modelling in a geographical database. In *Proceedings Auto-Carto'11: 11th International Conference on Computer Assisted Cartography*, Minneapolis, Minnesota, USA, 1993, 338–347.

21. E. Carlson. Three-dimensional conceptual modelling of subsurface structures. In *Proceedings of ASPRS/ACSM Annual Convention*, Vol. 4, Baltimore, Maryland, USA, 1987, 188–200.

22. CGAL Consortium. CGAL Basic Library, 2004. Available online at www.cgal.org.

23. P.P. Chen. The entity relationship model: toward a unified view of data. *ACM Transactions on Database Systems*, 1(1):9–36, 1976.

24. E. Clementini, P. Di Fellice, and P. van Oosterom. A small set of formal topological relationships suitable for end-user interaction. In *Proceedings of the Third International Symposium on Advances in Spatial Databases, SSD'93*, Vol. 692 of *Lecture Notes in Computer Science*, Singapore, June 23–25. Springer-Verlag, Berlin, 1993, 277–295.

25. V. Clerc, R. van Lammeren, A. Ligtenberg, H. Kramer, and A. Ligtenberg. Virtual reality in the landscape design process. In *Proceedings International Conference on Landscape Planning*, Portoroz, Slovenia, November 2002.

26. S. Cockroft. Towards the automatic enforcement of integrity rules in spatial database systems. In *Proceedings of 8th Colloquium of the Spatial Information Research Centre*, University of Otago, Dunedin, New Zealand, July 1996, 33–42.

27. E.F. Codd. A relational model of data for large shared banks. *Communications ACM*, 13(6):377–387, 1970.

28. D. Comer. The ubiquitous B-tree. *ACM Computing Surveys*, 11(2):121–137, 1979.

29. V. Coors. Resource-adaptive 3D maps for LBS. In *Proceedings UDMS 2002*, Prague, Czech Republic, October 2002.

30. V. Coors. 3D GIS in networking environments. *Computers, Environments and Urban Systems (CEUS)*, 27(4):345–357, 2003.

31. V. Coors and V. Jung. Using VRML as an interface to the 3D Data Warehouse. In *Proceedings of the Third Symposium on VRML*, New York, USA, 1998, 121–129.

32. D.J. Cowen. GIS versus CAD versus DBMS: What are the differences? *Photogrammetric Engineering and Remote Sensing*, 54(11):1551–1555, 1988.

33. D.H. Douglas and T.K. Peucker. Algorithms for the reduction of points required to represent a digitized line or its caricature. *Canadian Cartographer*, 10(2):112–122, 1973.

34. S.E. Dowson and V.L.O. Sheppard. Land registration, page 47. Colonial Research Publications No. 13, 2nd ed. Her Majesty's Stationary Office, London, U.K., 1952.

35. S. Doyle, M. Dodge, and A. Smith. The potential of web-based mapping and virtual reality technologies for modelling urban environments. *Computers, Environments and Urban Systems (CEUS)*, 22(2):137–155, 1998.

36. Dutch Government. Belemmeringenwet Privaatrecht. Wet van 13 mei 1927, tot opheffing van privaatrechtelijke belemmeringen, Staatsblad 2001, 548 (in Dutch), 1927.

37. Dutch Government. Monumentenwet. Wet van 23 December 1988, to vervanging van de Monumentenwet, Staatsblad 1997, 291 (in Dutch), 1988.

38. Dutch Government. Dutch Civil Code (Burgerlijk Wetboek), Boek 5: Zakelijke rechten (in Dutch). The Hague, 1992.

39. Dutch Government. Wet Bodembescherming. Staatsblad 1994, 331 (in Dutch), 1994.

40. Dutch Government. Uitspraak van de Hoge Raad m.b.t. de status van kabels voor telecommunicatie. Hoge Raad 6 juni 2003, nr. 36.076, Jurisprudentie Onderneming & Recht 2003/222 (in Dutch), June 2003.

41. M. Egenhofer, M.J. Sharma, and D. Mark. A critical comparison of the 4-intersection and 9-intersection models for spatial relations: formal analysis. In *Proceedings of Autocarto 11*, Minneapolis, Minnesota, USA, October 1993.

42. M.J. Egenhofer. Spatial SQL: a query and presentation language. *IEEE Transactions on Knowledge and Data Engineering*, 6(1):86–95, 1994.

43. M.J. Egenhofer, E. Clementini, and P. Di Felice. Evaluating inconsistencies among multiple representations. In *Proceedings of 6th International Symposium on Spatial Data Handling*, Edinburgh, Scotland, 1994, 901–920.

44. M.J. Egenhofer and J. Herring. Categorizing binary topological relationships between regions, lines and points in geographic databases. Technical report, Department of Surveying Engineering, University of Maine, Orono, USA, 1991.

45. ERDAS, 2004. Available online at www.erdas.com.

46. ESRI. ESRI, ArcGIS, 2004. Available online at www.esri.com.

47. C. Faloutsos and S. Roseman. Fractals for secondary key retrieval. In *Proceedings of the Eighth ACM Symposium on Principles of Database Systems*, Philadelphia, Pennsylvania, USA, March 1989, 247–252.

48. FIG. The FIG Statement on the Cadastre. Technical Report Publication 11, Federation International des Géomètres, Commission 7, 1995.

49. FIG. Cadastre 2014, a vision for a future cadastral system. Technical report, Federation International des Géomètres, Commission 7, J. Kaufmann and D. Steudler, 1998.

50. FIG. The Bathurst Declaration on Land Administration for Sustainable Development. Technical Report Publication 21, Federation International des Géomètres, October 1999.

51. S. Flick. An object-oriented framework for the realisation of 3D Geographic Information Systems. In *Proceedings of the Second Joint European Conference and Exhibiton on Geographical Information*, Barcelona, Spain, 1996, 187–196.

52. J. Forrai and G. Kirschner. Transition from two-dimensional legal and cadastral reality to a three-dimensional case. In *Proceedings of International Workshop on 3D Cadastres, FIG*. Delft, the Netherlands, 28–30 November 2001, 9–24.

53. J. Forrai and G. Kirschner. An interdisciplinary 3D cadastre development project in practice. In *Proceedings FIG Working Week 2003*, Paris, France, April 2003.

54. W. Förstner. A framework for low level feature extraction. In *Proceedings of Computer Vision—ECCV94*, Vol. 2, Stockholm, Sweden, 1994, 383–394.

55. C. Fowler and E. Treml. Building a marine cadastral information system for the United States—a case study. *Computers, Environments and Urban Systems (CEUS)*, 25(4–5): 493–507, 2001.

56. J. Gerremo and J. Hannson. Ownership and real property in British Columbia: a legal study. Technical Report M.Sc. thesis 48, Royal Institute of Technology, Department of Real Estate and Construction Management, Division of Real Estate Planning and Land Law, Stockholm, Sweden, 1998.

57. B. Gorte. Segmentation of TIN-structured surface models. In *Proceedings of Joint Conference on Geo-spatial Theory, Processing and Applications*, Ottawa, Canada, July 2002.

58. K. Gray and S.F. Gray. *Elements of Land Law*. Oxford University Press, Oxford, U.K., 2004.

59. A. Grinstein. Different aspects of a 3D cadastre in the new town Modi'in, Israel. In *Proceedings of International Workshop on 3D Cadastres, FIG*. Delft, the Netherlands, 28–30 November 2001, 25–34.

60. R. Grinstein. A real-world experiment in 3D cadastre: mapping of underground parking for registration rights. *GIM International*, September 2003, 65–67.

61. R. Groot and J. McLaughin, Eds. *Geospatial Data Infrastructure—Concepts, Cases, and Good Practice*. Oxford University Press, Oxford, U.K., 2000.

62. M. Gruber, M. Pasko, and F. Leberl. Geometric versus texture detail in 3D models of real world buildings. In *Proceedings of Ascona Workshop 95 on Automatic Extraction of Man-Made Objects from Aerial and Space Images*, Basel, Switzerland, 1995. Birkhäuser Verlag, 189–198.

63. A. Grün and X. Wang. CC-modeller: a topology generator for 3D city models. *ISPRS Journal*, 53(5):286–295, 1998.

64. A. Guttman. R-trees: a dynamic index structure for spatial searching. In *Proceedings ACM International Conference on Management of Data*, Boston, Massachusetts, USA, June 1984, 188–196.

65. N. Haala. Combining multiple data sources for urban data acquisition. In *Proceedings of Photogrammetric Week 1999*, Stuttgart, Germany, September 1999, 329–339.

66. P. de Haan. Eigendom, beheer en registratie van ondergrondse infrastructuur (in Dutch). *Nederlands Juristenblad*, 79:564–570, 2004.

67. P. de Haan. Eigendomsverhoudingen bij privatisering van energiebedrijven (in Dutch). *Bouwrecht*, 41(4):283–292, 2004.

68. R.M. van Heerd et al. Productspecificatie AHN 2000. Technical Report MDTGM 2000.13, Rijkswaterstaat, Meetkundige Dienst, 2000.

69. J. Henssen. Basic principles of the main cadastral systems in the world. In *Proceedings of One Day Seminar Held During the Annual Meeting of Commission 7, Cadastre and Rural Land Management, FIG*, Delft, the Netherlands, May 1995.

70. I. Heywood, S. Cornelius, and S. Carver. *An Introduction to Geographical Information Systems*. Prentice Hall, Englewood Cliffs, New Jersey, USA, 1998.

71. F. Hobbs and C. Chan. AutoCAD as a cartographic training tool: a case study. *Computer Aided Design*, 22(3):151–159, 1990.

72. M. Hoefsloot. 3D Geo-Informatie uit bestaande CAD modellen (in Dutch). Technical Report M.Sc. case study report, TU Delft, Section GIS Technology, 2003.

73. E. Hoel, S. Menon, and S. Morehouse. Building a robust relational implementation of topology. In *Proceedings of 8th International Symposium on Spatial and Temporal Databases*, Santorini, Greece, July 2003, 508–524.

74. A.D. Hofmann, H.-G. Maas, and A. Streilein. Knowledge-based building detection based on laser scanner data and topographic map information. In *Proceedings of Symposium on Photogrammetric Computer Vision, ISPRS Commission III*, Graz, Austria, September 2002, 169–174.

75. R. Hoinkes and E. Lange. 3D for free—toolkit expands visual dimensions in GIS. *GIS World*, 8(7):54–56, 1995.

76. T. Höllerer, S. Feiner, T. Terauchi, G. Rashid, and D. Hallaway. Exploring MARS: developing indoor and outdoor user interfaces to a mobile augmented reality system. *Computers and Graphics*, 23(6):779–785, 1999.

77. M. Huml. Legal view, conditions and experiences in the Czech Republic. In *Proceedings International Workshop on 3D Cadastres*, Delft, the Netherlands, November 2001.

78. IBM. IBM DB2 Spatial Extender User's Guide and Reference. Special Web release edition. Technical report, IBM, 2000.

79. Informix. Informix Spatial DataBlade Module User's Guide. Technical Report Part no. 000-8441, Informix, 2000.

80. Ingres. CA-OpenIngres, INGRES/Object Management Extension User's Guide, Release 6.5. Technical report, 1994.

81. Intergraph. Geomedia, 2004. Available online at www.intergraph.com.

82. ISO. ISO/TC 211, Geographic information/Geomatics, Revised report of the secretariat to the 17th plenary meeting of ISO/TC 211 in Berlin, Germany, 2003-10-30/31. Technical report, 2003.

83. ISO. ISO/TC 211, ISO International standard 19107:2003, Geographic Information—Spatial Schema. Technical report, 2003.

84. K. Jacobsen and P. Lohmann. Segmented filtering of laser scanner DSMs. In *Proceedings of ISPRS Working Group III/3 Workshop 3D Reconstruction from Airborne Laserscanning and InSAR Data*, Dresden, Germany, October 2003, 87–93.

85. J. de Jong. *Erfpacht en opstal* (in Dutch). Kluwer, Deventer, the Netherlands, 1995.

86. J. de Jong. Juridische aspecten van ondergronds bouwen (in Dutch). *Bouwrecht*, 35(6):453–459, 1998.

87. B. Julstad and A. Ericsson. Property formation and three-dimensional property units in Sweden. In *Proceedings of International Workshop on 3D Cadastres, FIG*, Delft, the Netherlands, 28–30 November 2001, 173–190.

88. A.P. Kap and J.A. Zevenbergen. Valkuilen en kansen bij de opzet van landelijke registraties: een (inter)nationale vergelijking (in Dutch). Technical report, Department of Geodesy, Delft University of Technology, 2000.

89. H. de Kluijver and J.E. Stoter. Noise mapping and GIS: optimising quality and efficiency of noise effect studies. *Computers, Environment and Urban Systems (CEUS)*, 27(1):85–102, 2003.

90. M. Kofler. R-trees for Visualizing and Organizing Large 3D GIS Databases. Ph.D. thesis, Institute for Computer Graphics and Vision (ICGV), Graz University of Technology, Austria, 1998.

91. T.H. Kolbe, G. Gröger, and L. Plümer. CityGML—interoperable access to 3D city models. In *Proceedings of the First International Symposium on Geo-Information for Disaster Management GI4DM*, Delft, the Netherlands, March 21–23 2005. LNCS, Springer Verlag, Berlin.

92. M.J. Kraak and F.J. Ormeling. *Cartography, Visualization of Spatial Data*. Addison-Wesley, London, 1996.

93. S. Landes. Funktionalität des internetbasierten 3D Campus Informations Systems der Universität Karlsruhe. Ph.D. thesis, Institute für Photogrammetrie und Fernerkundung, Karlsruhe, Germany, 1999.

94. Laser-Scan Radius. Laser-Scan Radius Topology, 2004. Available online at www.radius.laser-scan.com.

95. Laser-Scan Radius Topology. Radius Topology Database Administrator's Guide, Issue 1.0 for Radius Topology Version 2.0. Technical report, Laser-Scan Limited, Cambridge, U.K., April 2003.

96. R. Laurini. *Information Systems for Urban Planning, a Hypermedia Cooperative Approach*. Taylor & Francis, New York, USA, 2001.

97. S.H. Lee and K. Lee. Partial Entity Structure: a compact non-manifold boundary representation based on partial topological entities. In *Proceedings Sixth ACM Symposium on Solid Modeling and Applications*, 159–170. Ann Arbor, Michigan, USA, June 2001.

98. C.H.J. Lemmen, P. van der Molen, P.J.M. van Oosterom, H. Ploeger, C.W. Quak, J.E. Stoter, and J.A. Zevenbergen. A modular standard for the Cadastral Domain. In *Proceedings of Digital Earth*, Brno, Czech Republic, September 2003.

99. U. Lenk. Strategies for integrating height information and 2D GIS data. In *Proceedings of Joint OEEPE/ISPRS Workshop from 2D to 3D, Establishment and Maintenance of National Core Spatial Databases*, Hannover, Germany, October 2001.

100. C. Lindenbeck and H. Ulmer. Geology meets virtual reality: VRML visualisation server applications. In *Proceedings of WSCG'98 (Sixth International Conference in Central Europe on Computer Graphics and Visualization)*, Vol. III, Plzen, Czech Republic, February 1998, 402–408.

101. P.D. Lindstrom, W. Koller, W. Ribarsky, L. Hodges, N. Faust, and G.A. Turner. Real-time, continuous level of detail rendering of height fields. In *Proceedings of SIGGRAPH'96 (23rd Annual Conference on Computer Graphics and International Techniques)*, New Orleans, Louisiana, USA, August 1996, 109–118.

102. T.L. Logan and N.A. Bryant. Spatial data software integration: merging CAD/CAM/ Mapping with GIS and image processing. *Photogrammetric Engineering and Remote Sensing*, 53(10):1391–1395, 1987.

103. M.A. de Lòpez. Country Report 2003 Argentina—Based on the PCGIAP-Cadastral Template, 2003. Available online at www.cadastraltemplate.org/.

104. L. Louwman. De inschrijving van netwerktekeningen (in Dutch). *De Stichting tot Bevordering der Notariële Wetenschap*, 6547:720–722, 2003.

105. J. Louwsma, T.P.M. Tijssen, and P.J.M. van Oosterom. A comparison between topologically structured and "plain" spatial data. *Geoconnexion*, June 2003.

106. D.W. Lowe. Fitting parameterized three-dimensional models to images. *IEEE Transactions on Pattern Analysis and Machine Intelligence*, 13(5):441–450, 1991.

107. H. Luttermann and M. Grauer. Using interactive, temporal visualizations for WWW-based presentation and exploration of spatio-temporal data. In *Proceedings of Spatio-Temporal Database Management 1999, International Workshop STDBM'99*, Edinburgh, Scotland, September 1999, 100–118.

108. D.J. Maguire. Improving CAD-GIS Interoperability, 2003. Available online at www.esri.com/news/arcnews/winter0203articles/improving-cad.html.

109. D.J. Maguire, M.F. Goodchild, and D.W. Rhind. *Geographic Information Systems: Principles and Applications*. Longman Scientific and Technical, Harlow, Essex, U.K., 1991.

110. MapInfo. Mapinfo, 2004. Available online at www.mapinfo.com.

111. H. Mattsson. Towards three dimensional properties in Sweden. In *Proceedings of European Faculty of Land Use and Development, 32nd International Symposium*, Strassbourg, France, October 2003.

112. E. Mitrofanova. The needs and possibilities for three-dimensional determination of real estate in Ukraine. In *Proceedings International Workshop on 3D Cadastres*, Delft, the Netherlands, November 2001.

113. M. Molenaar. Single valued vector maps: a concept in geographic information systems. *Geo-Informations-systeme*, 2(1):18–26, 1989.

114. M. Molenaar. A topology for 3D vector maps. *ITC Journal*, 1992(1):25–33, 1992.

115. M. Mortenson. *Geometric Modelling*, 2nd ed. John Wiley & Sons, New York, USA, 1997.

116. S.M. Movafagh. GIS/CAD Convergence Enhances Mapping Applications. *GIS World*, 8(5):44–47, May 1995.

117. MySQL, 2004. Available online at dev.mysql.com.

118. S. Nebiker. Support for visualisation and animation in a scalable 3D GIS environment: motivation, concepts and implementation. In *Proceedings of ISPRS Commission V Working Group 6, Workshop on Visualization and Animation of Reality-Based 3D Models*, Engadin, Switzerland, February 2003.

119. H. Netzel and F. Kaalberg. Settlement risk management with GIS for the North/South Metroline in Amsterdam. In *Proceedings of World Tunnel Congress*, Oslo, Norway, 1999.

120. R.G. Newell and T.L. Sancha. The difference between CAD and GIS. *Computer-Aided Design*, 22(3):131–135, 1990.

121. S. Ng'ang'a, M. Sutherland, S. Cockburn, and S. Nichols. Towards a 3D marine cadastre in support of good ocean governance. In *Proceedings of International Workshop on 3D Cadastres, FIG*, Delft, the Netherlands, 28–30 November 2001, 99–114.

122. E. van Nieuwburg. Visualisatie van 3D geo-informatie met VRML/X3D (in Dutch). Technical Report MSc case study report, TU Delft, Section GIS Technology, 2003.

123. P.R. van Nieuwenhuizen and F.W. Jansen. Computer graphics lecture notes. Technical report, Delft University of Technology, Delft, the Netherlands, 2002.

124. NIST, 2004. Available online at cic.nist.gov/vrml/vbdetect.html.

125. D. Nebert (Technical Working Group Chair of Global Spatial Data Infrastructure). Developing Spatial Data Infrastructures: The SDI Cookbook, version 1.1, May 2001. Available online at www.gsdi.org/pubs/cookbook/cookbook0515.pdf.

126. H. Onsrud. Making laws for 3D cadastre in Norway. In *Proceedings FIG, ACSM/ASPRS*, Washington, D.C., USA, April 2002.

127. H. J. Onsrud. The land tenure system of the United States. *Forum (Zeitschrift des Bundes der Öffentlich Bestellten Vermessungsingenieure)*, (January), 1989.

128. P.J.M. van Oosterom. The GAP-tree, an approach to "On-the-Fly" map generalization of an area partitioning. In J.C. Müller, J.P. Lagrange, and R. Weibel, Eds., *GIS and Generalization, Methodology and Practice*, Taylor & Francis, New York, USA, 1995, 120–132.

129. P.J.M. van Oosterom and C.H.J. Lemmen. Spatial data management on a very large cadastral database. *Computers, Environments and Urban Systems (CEUS)*, 25(4–5): 509–528, 2001.

130. P.J.M. van Oosterom, B. Maessen, and C.W. Quak. Spatial, thematic and temporal views. In *Proceedings of 9th International Symposium on Spatial Data Handling*, Beijing, China, 10–12 August 2000.

131. P.J.M. van Oosterom, H.D. Ploeger, and J.E. Stoter. Analysis of 3D property situations in the USA. In *Proceedings of FIG Working Week*, Cairo, Egypt, April 2005.

132. P.J.M. van Oosterom, C.W. Quak, and T.P.M. Tijssen. Testing current DBMS products with real spatial data. In *Proceedings UDMS 2002*, Prague, Czech Republic, October 2002.

133. P.J.M. van Oosterom, C.W. Quak, and T.P.M. Tijssen. Polygons: the unstable foundation of spatial modeling. In *Proceedings ISPRS Joint Workshop on Spatial, Temporal and Multi-dimensional Data Modelling and Analysis*, Quebec, Canada, October 2003.

134. P.J.M. van Oosterom and V. Schenkelaars. The development of an interactive multi-scale GIS. *International Journal of Geographical Information Science*, 9(5):489–507, 1995.

135. P.J.M. van Oosterom, J.E. Stoter, and F. Jansen. *Bridging the Worlds of CAD and GIS*, chap. 1 Taylor & Francis, New York, USA, 2005.

136. P.J.M. van Oosterom, J.E. Stoter, W.C. Quak, and S. Zlatanova. The balance between geometry and topology. In D. Richardson and P.J.M. van Oosterom, Eds., *Proceedings of 10th International Symposium on Spatial Data Handling*, Ottawa, Canada, July 2002.

137. P.J.M. van Oosterom, J.E. Stoter, E. Verbree, and S. Zlatanova. 3D GIS komt er wel, maar 't zal wel even duren (in Dutch). *VI Matrix*, 10(3):20–23, 2002.

138. P.J.M. van Oosterom, J.E. Stoter, E. Verbree, and S. Zlatanova. Onderzoek brengt 3D GIS in gangbare geo-informatie naderbij (in Dutch). *VI Matrix*, 10(5):26–29, 2002.

139. P.J.M. van Oosterom, W. Vertegaal, M. van Hekken, and T. Vijlbrief. Integrated 3D modelling within a GIS. In *Proceedings of Advanced Geographic Data Modelling:*

Spatial Data Modelling and Query Languages for 2D and 3D Applications, Delft, the Netherlands, September 1994, 80–95.

140. OpenGIS Consortium. OpenGIS Simple Features Specification for SQL. Technical Report Revision 1.1, OpenGIS Project Document 99-049, OpenGIS Consortium, Wayland, Massachusetts, USA, 1999.

141. OpenGIS Consortium. OGC Web Terrain Server (WTS), version 0.3.2. Technical Report OGC 01-061, Wayland, Massachusetts, USA, 2001.

142. OpenGIS Consortium. OpenGIS Abstract and Implementation Specifications, 2001. Available online at www.opengis.org/techno/specs.htm.

143. OpenGIS Consortium. Request Number 12, Geometry Working Group, A Request for Proposals: OpenGIS Feature Geometry. Technical report, Wayland, Massachusetts, USA, 2001.

144. OpenGIS Consortium. The OpenGIS Abstract Specification, Topic 1: Feature Geometry (ISO 19107 Spatial Schema), Version 5. Technical Report OpenGIS Project Document 01-101, Wayland, Massachusetts, USA, 2001.

145. OpenGIS Consortium. Web Map Service Implementation Specification, version 1.1.1. Technical Report OGC 01-068r2, Wayland, Massachusetts, USA, 2001.

146. OpenGIS Consortium. Web Feature Service Implementation Specification, version 1.0.0. Technical Report OGC 02-058, Wayland, Massachusetts, USA, 2002.

147. OpenGIS Consortium. OpenGIS Geography Markup Language (GML) Implementation Specification. Technical Report 02-023r4, Wayland, Massachusetts, USA, 2003.

148. OpenGIS Consortium. OpenGIS Reference Model. Technical report, Wayland, Massachusetts, USA, 2003.

149. OpenGIS Consortium. OpenGIS Web Map Server Cookbook. Technical Report 03-050r1, Wayland, Massachusetts, USA, 2003.

150. OpenGIS Consortium. Web Coverage Service (WCS), version 1.0.0. Technical Report OGC 03-065r6, Wayland, Massachusetts, USA, 2003.

151. OpenGIS Consortium. OGC, 2004. Available online at www.opengis.org.

152. Oracle. Oracle Spatial User's Guide and Reference Release 9.2 part number a96630-01. Technical report, ORACLE, March 2002.

153. Oracle Spatial 10g. Oracle Spatial User's Guide and Reference Release 10.1, 2004. Available online at www.oracle.com/pls/db10g/db10g.show_toc.

154. A. Osskó. Problems in registration in the third vertical dimension in the unified Land Registry in Hungary, and possible solution. In *Proceedings International Workshop on 3D Cadastres*, Delft, the Netherlands, November 2001, 305–314.

155. W. Pasman and F.W. Jansen. Scheduling level of detail with guaranteed quality and cost. In *Proceedings of 7th International Conference on 3D Web Technology*, Tempe, Arizona, USA, February 2002, 43–51.

156. W. Pasman, A. van der Schaaf, R.L. Lagendijk, and F.W. Jansen. Low latency rendering and positioning for mobile augmented reality. In *Proceedings of Vision Modeling and Visualization '99*, Erlangen, Germany, November 1999, 309–315.

157. R. Passini and D. Betzner. Filtering of digital elevation models. In *Proceedings FIG, ACSM/ASPRS*, Washington, D.C., USA, April 2002.

158. PCIGEOMATICS, 2004. Available online at www.pcigeomatics.com.

159. W. Peng. Automated Generalization in GIS. Ph.D. thesis, Wageningen University, ITC, the Netherlands, 1997.

160. F. Penninga. Detectie van kenmerkende hoogtepunten in TIN's voor iteratieve datareductie (in Dutch). In *Proceedings of Geo-Informatiedag Nederland 2002*, Ede, the Netherlands, February 2002.

161. F. Penninga. Oracle 10g Topology: Testing Oracle 10g Topology Using Cadastral Data. Technical Report GISt report 26, Delft, the Netherlands, 2004.

162. S. Pigot. A Topological Model for a 3-Dimensional Spatial Information System. Ph.D. thesis, University of Tasmania, Australia, 1995.

163. M. Pilouk. Integrated Modeling for 3D GIS. Ph.D. thesis, Wageningen University, ITC, the Netherlands, 1996.

164. E. Pogorelčnik and M. Korošec. Land cadastre and building cadastre in Slovenia: current situation and potential of 3D data. In *Proceedings International Workshop on 3D Cadastres*, Delft, the Netherlands, November 2001, 79–90.

165. PostGIS. PostGIS Manual, 2002. Available online at postgis.refractions. net/docs.

166. PostgreSQL, 2004. Available online at www.postgresql.org.

167. C.W. Quak, T.P.M. Tijssen, and J.E. Stoter. Topology in spatial DBMSs. In *Proceedings of Digital Earth*, Brno, Czech Republic, September 2003.

168. Queensland Government. Land Title Act 1994, reprinted as in force on 16 May 2003. Queensland, Australia, 2003.

169. Queensland Government. Registrar of Titles Directions for the Preparation of Plans. Queensland, Australia, 2003.

170. R. Ramakrishnan and J. Gerhke. *Database Management Systems*. McGraw-Hill Higher Education, New York, 2003.

171. S. Rana and J. Sharma, Eds. *The Role of DBMS in New Generation GIS Architecture*. Springer, New York, USA, 2005.

172. D.E. Richardson. Automated Spatial and Thematic Generalization Using a Context Transformation Model. Ph.D. thesis, Wageningen University, the Netherlands, 1993.

173. P. Rigaux, M. Scholl, and A. Voisard. *Spatial Databases with Applications to GIS*. Morgan Kaufmann, San Diego, California, USA, 2001.

174. R. Rikkers, M. Molenaar, and J. Stuiver. A query oriented implementation of a 3D topological data structure. In *Proceedings of EGIS'93*, Genoa, Italy, 1993, 1411–1420.

175. G. Roberts, A. Evans, A. Dobson, B. Denby, S. Cooper, and R. Hollands. Look beneath the surface with augmented reality. *GPS World*, February 2002.

176. J. Ruppert. A Delaunay refinement algorithm for quality 2-dimensional mesh generation. *Journal of Algorithms*, 18(3):548–585, 1995.

177. M. Saadi Mesgari. Topological Cell-Tuple Structures for Three-Dimensional Spatial Data. Ph.D. thesis, University of Twente, ITC, the Netherlands, 2000.

178. A. Schutzberg. Bringing GIS to CAD: a developer's challenge. *GIS World*, 8(5):48–54, 1995.

179. S. Shekhar and S. Chawla. *Spatial Databases, A Tour*. Prentice Hall, Englewood Cliffs, New Jersey, USA, 2003.

180. I.D.H Shepherd. Mapping with desktop CAD: a critical review. *Computer Aided Design*, 22(3):136–150, 1990.

181. J.R. Shewchuk. Triangle: engineering a 2D quality mesh generator and Delaunay triangulator. In *Proceedings of First Workshop on Applied Computational Geometry*, Philadelphia, Pennsylvania, USA, May 1996, 124–133.

182. J.R. Shewchuk. The Quake Project, 2004. Available online at www-2.cs.cmu.edu/ quake.

183. W.Z. Shi, B.S. Yang, and Q.Q. Li. An object-oriented data model for complex objects in three-dimensional geographic information systems. *International Journal of Geographical Information Science*, 17(5):411–430, 2003.

184. U. Shoshani, M. Benhamu, E. Goshen, S. Denekamp, and R. Bar. A multi layers 3D cadastre in Israel: a research and development project recommendations. In *FIG Working Week*, Cairo, Egypt, 21–26 April 2005.

185. S.R. Simpson. *Land Law and Registration, Book 1*. Surveyor Publications, London, 1984.

186. J.W.N. van Smaalen. Automated Aggregation of Geographic Objects, A New Approach to the Conceptual Generalisation of Geographic Databases. Ph.D. thesis, TU Delft, ITC, the Netherlands, 2003.

187. A.P. Smith, M. Dodge, and S. Doyle. Visual Communication in Urban Planning and Urban Design, report to the Advisory Group on Computer Graphics. Technical Report CASA paper 2, 1998.

188. H.J. Snijders and E.B. Rank-Berenschot. *Goederenrecht* (in Dutch). Kluwer, Deventer, the Netherlands, 2001.

189. M. Stonebraker. *Object-Relational DBMSs: The Next Great Wave*. Morgan Kaufmann, San Francisco, California, USA, 1996.

190. J.E. Stoter. 3D Cadastre. Ph.d. thesis, Delft University of Technology, Netherlands Geodetic Commission, Delft, the Netherlands, 2004.

191. J.E. Stoter and B.Gorte. Height in the cadastre, integrating point heights and parcel boundaries. In *Proceedings FIG Working Week*, Paris, France, April 2003.

192. J.E. Stoter and P.J.M. van Oosterom. Incorporating 3D geo-objects into a 2D geo-DBMS. In *Proceedings FIG, ACSM/ASPRS*, Washington, D.C., USA, April 2002.

193. J.E. Stoter, F. Penninga, and P.J.M. van Oosterom. Generalization of integrated terrain elevation and 2D object models. In *Proceedings of 11th International Symposium on Spatial Data Handling*, Leicester, U.K., August 2004.

194. J.E. Stoter and M. Salzmann. Where do cadastral needs and technical possibilities meet? *Computers, Environments and Urban Systems (CEUS)*, 27(4):395–410, 2003.

195. J.E. Stoter and S. Zlatanova. 3D GIS where are we standing? In *Proceedings ISPRS Workshop on Spatial, Temporal and Multi-Dimensional Data Modelling and Analysis*, Quebec, Canada, October 2003.

196. J.E. Stoter and S. Zlatanova. Visualising and editing of 3D objects organised in a DBMS. In *Proceedings of EuroSDR Com V Workshop on Visualisation and Rendering*, Enschede, the Netherlands, January 2003.

197. M. Sun, J. Chen, and A. Ma. Construction of complex city landscape with the support of CAD. In *Proceedings International Workshop on Visualization and Animation of Landscape*, Kunming, China, February 2001.

198. I. Suveg and M.G. Vosselman. Automatic 3D reconstruction of buildings. In *Proceedings of SPIE Photonics West, Electronic Imaging, Three-Dimensional Image Capture and Applications V, conference 4661*, San Jose, California, USA, 2002, 59–69.

199. Swedish Government. Tredimensionell fastighetsindelning (in Swedish), 2004. Available online at justitie.regeringen.se/sb/d/1917/a/12244.

200. P.J.G. Teunissen. *Adjustment Theory, An Introduction*. Delft University Press, Deventer, the Netherlands, 2003.

201. W.J.M. Teunissen and P.J.M. van Oosterom. The Creation and Display of Arbitrary Polyhedra in HIRASP. Technical Report 88-20, University of Leiden, Leiden, the Netherlands, 1988.

202. M.D. Thomas. *Oracle XSQL: Combining SQL, Oracle Text, XSLT, and Java to Publish Dynamic Web Content*. Wiley Europe, West Sussex, U.K., 2003.

203. L. Ting and I.P. Williamson. Cadastral trends: a synthesis. *The Australian Surveyor*, 4(1):46–54, 1999.

204. D.C. Tsichritzis and F.H. Lochovsky. *Data Models*. Prentice-Hall International, Englewood Cliffs, New Jersey, USA, 1982.

205. J. Ullman and J. Widom. *A First Course in Database Systems*, 2nd ed. Prentice-Hall, Englewood Cliffs, New Jersey, USA, 2001.

206. UML. OMG Unified Modeling Language Specification, Version 1.5. Technical Report formal/03-03-01, an adopted formal specification of the Object Management Group, 2003.

207. Valetta Convention. European Convention on the Protection of the Archaeological Heritage, Council of Europe, 1992. Available online at conventions.coe.int/Treaty/en/Treaties/Html/143.htm.

208. T. Valstad. The Oslo Method: a practical approach to register 3D properties. In *Proceedings FIG Working Week 2003*, Paris, France, April 2003.

209. E. Verbree, G. van Maren, F. Jansen, and M. Kraak. Interaction in virtual world views, linking 3D GIS with VR. *International Journal of Geographical Information Science*, 13(4):385–396, 1999.

210. E. Verbree and P.J.M. van Oosterom. The STIN method: 3D surface reconstruction by observation lines and Delaunay TENs. In *Proceedings of ISPRS Workshop on 3D-Reconstruction from Airborne Laserscanner and InSAR Data*, Dresden, Germany, October 2003.

211. T. Vijlbrief and P.J.M. van Oosterom. GEO++: an extensible GIS. In *Proceedings of 5th International Symposium on Spatial Data Handling*, Charleston, South Carolina, USA, August 1992.

212. G. Vosselman. Building reconstruction using planar faces in very high density height data. In *ISPRS Conference on Automatic Extraction of GIS Objects from Digital Imagery*, Munich, Germany, September, 1999, 87–92.

213. G. Vossen. *Data Models, Database Languages and Database Management Systems*. Addison-Wesley, Wokingham, U.K., 1991.

214. J. de Vries. 3D GIS en grootschalige toepassingen, de opslag en analyse in een geïntegreerde drie-dimensionale GIS (in Dutch). Technical Report M.Sc. thesis, Delft University of Technology, Delft, the Netherlands, 2001.

215. M.E. de Vries and J.E. Stoter. Accessing a 3D geo-DBMS using Web technology. In *Proceedings ISPRS Workshop on Spatial, Temporal and Multi-Dimensional Data Modelling and Analysis*, Quebec, Canada, October 2003.

216. M.E. de Vries and S. Zlatanova. Interoperability on the Web: the case of 3D geo-data. In *Proceedings IADIS International Conference e-Society*, Avila, Spain, July 2004.

217. X. Wang and A. Grün. A hybrid GIS for 3-D city models. *International Archives of Photogrammetry and Remote Sensing*, (4/3):1165–1172, 2000.

218. J. Warmer and A. Kleppe. *The Object Constraint Language: Precise Modeling with UML*. Addison-Wesley, Boston, Massachusetts, USA, 1998.

219. Web3D Consortium. X3D, Open Standards for Real-Time 3D Communication, 2004. Available online at www.web3d.org/fs_specifications.htm.

220. J.D. Wees, R.W. Versseput, H.J. Simmelink, R.R.L. Allard, and H.J.M. Pagnier. Shared Earth system models for the Dutch subsurface. In *Proceedings GIN 2002, Geoinformatiedag Nederland 2002*, Ede, the Netherlands, February 2002.

221. M. Werner. Integrating GIS with inundation models for flood extent mapping. In *Proceedings UDMS 2002*, Prague, Czech Republic, October 2002.

222. M.S. Widodo. The needs for marine cadastre and supports of spatial data: infrastructures in marine environment-a case study. In *Proceedings FIG Working Week 2003*, Paris, France, April 2003.

223. M.F. Worboys. Object-oriented models of spatio-temporal information. In *Proceedings of GIS/LIS'92*, San José, California, USA, 1992, 825–834.

224. M.F. Worboys. *Geographical Information Systems: A Computing Perspective*. Taylor & Francis, London, 1995.

225. L. Yaolin. Categorical Database Generalization in GIS. Ph.D. thesis, Wageningen University, ITC, the Netherlands, 2002.

226. J.A. Zevenbergen. Are cadastres really serving the landowner. In *Proceedings UDMS 1999*, Venice, Italy, April 1999.

227. J.A. Zevenbergen. Systems of Land Registration, Aspects and Effects. Ph.D. thesis, TU Delft, the Netherlands, 2001.

228. A. Zipf and R. Leiner. Mobile GIS based flood warning and information system. In *Proceedings of Symposium on LBS and TeleCartography*, Vienna, Austria, January 2004.

229. S. Zlatanova. 3D GIS for Urban Development. Ph.D. thesis, Institute for Computer Graphics and Vision (ICGV), Graz University of Technology, Austria, ITC, the Netherlands, 2000.

230. S. Zlatanova. Augmented Reality Technology, Report for SURFnet. Technical Report GISt 18, Department of Geodesy, Delft University of Technology, Delft, the Netherlands, 2002.

231. S. Zlatanova and D. Prosperi, Eds. *Large-Scale 3D Data Integration: Challenges and Opportunities*. Taylor & Francis, London, 2005.

232. S. Zlatanova, A. Rahman, and M. Pilouk. 3D GIS: current status and perspectives. In *Proceedings of ISPRS*, Ottawa, Canada, July 2002.

233. S. Zlatanova, A.A. Rahman, and W. Shi. Topology for 3D spatial objects. In *Proceedings International Symposium and Exhibition on Geoinformation 2002*, Kuala Lumpur, Malaysia, October 2002.

234. S. Zlatanova and E. Verbree. A 3D topological model for augmented reality. In *Proceedings Mobile MultiMedia Systems and Applications, MMSA*, Delft, the Netherlands, 2000.

Appendix A
Visualizing Attributes in VRML

Appendix A contains an example of VRML code representing a box. The box can be queried on its attributes. The attributes of the selected object are shown using a "Billboard."

The interaction with 3D objects in a VRML file has to be explicitly described in the VRML. This can be organized by two additional VRML nodes. First, a particular sensor (e.g., TouchSensor) has to be attached to the object (a Shape), which will monitor whether the cursor interacts with the object. Second, a billboard node has to be introduced to visualize the attributes in text format. In this example, a new "proto" node has been designed. The node is a TouchSensor extended with a Javascript code (included in the VRML file), which controls the text that is visualized (in this case, attribute information). The code provides a link between the attributes and the geometry. This link needs to be defined for every object using the specific code. The VRML code for the example of Figure 7.5 is shown below.

```
#VRML V2.0 utf8

Background {
        skyColor [
           0.0 0.2 0.7,
           0.0 0.5 1.0,
           1.0 1.0 1.0
        ]
        skyAngle [ 1.009, 1.571 ]
     }

PROTO TOUCH [
   field        SFInt32        object 0
   eventOut     MFString       string_changed
        ]
{
   DEF SENS TouchSensor {}
   DEF NODE Script {
     url "javascript:
        function set_boolean (bool)
           {
           if ((bool == true)&&(object == 30))
                   string_changed [0] = 'BUILDING /34';
           if ((bool == false)||(object == 0))
                   string_changed [0] = '';

           }"
   eventIn SFBool set_boolean
   eventOut MFString string_changed IS string_changed
   field SFInt32 object IS object
   }
```

```
ROUTE SENS.isOver TO NODE.set_boolean
}#TOUCH

Transform {
    translation 0 0 -30
    children [
        DEF Box30 TOUCH {
          object 30
        }
        Shape {
            appearance Appearance {
                material Material {
                        diffuseColor 0.60 0 0
                }
            }
            geometry Box {
                size 4 4 4
            }
        }
        Transform {
            translation 4 4 4
            children[
                Billboard {
                    children [
                        Shape {
                            appearance Appearance {
                                material Material {
                                        diffuseColor 0 0 0

                                }
                            }
                            geometry DEF TEXT30 Text {
                                length [5,120]
                            }
                        }
                    ]
                    axisOfRotation 0.0 1.0 0.0
                }
        ]}
]}

ROUTE Box30.string_changed TO TEXT30.string_changed
```

Appendix B
XSLT Stylesheet to Transform XML to X3D

Appendix B contains an example of an XSLT stylesheet to transform an Oracle table with geometries to X3D format.

When a Web server receives a request for an XSQL document, the page is passed to the XSQL servlet. The page is processed by the servlet: a connection to the database is made and the select statement is sent to the DBMS. The result set that comes back from the database is in XML format. The second step is then to transform the "raw" XML stream into a X3D or VRML output stream. Because of the XML syntax of X3D, the transformation from Oracle to X3D can easily be handled by XSLT stylesheets. The XSLT stylesheet below shows how to convert the XML output of Oracle in case of multipolygons to an X3D format.

```
<?xml version="1.0" encoding="iso-8859-1" ?>
<xsl:stylesheet version="1.0" xmlns:xsl="http://www.w3.org/1999/XSL/Transform" >
<xsl:output method="xml" indent="no" media-type="model/x3d+xml"
encoding="iso-8859-1" />

<!-- arguments for construction of getfieldinfo url -->
<xsl:param name="table" />
<xsl:param name="idcol" />
<xsl:param name="con" />

<!-- other variables -->
<xsl:variable name="apos">'</xsl:variable>
<xsl:variable name="zdummy">0</xsl:variable>


<xsl:template match="mymap">

  <!-- print outer x3d elements -->
  <X3D version='3.0' profile="Interactive">
    <Background skyColor="1 1 1" />
    <Scene>

      <!-- transform data in inputstream from XML to x3d -->
      <xsl:apply-templates />

    </Scene>
  </X3D>

</xsl:template>
```

```xsl
<xsl:template match="ROW/*[SDO_GTYPE='3003' or SDO_GTYPE='3004'
or SDO_GTYPE='3007']">

    <!-- get start positions in sdo_ordinates array of exterior and interior
    rings -->
    <xsl:variable name="startAt" >
        <xsl:for-each select="SDO_ELEM_INFO/SDO_ELEM_INFO_ITEM[position()
        mod 3 = 1]" >
           <xsl:value-of select="concat(';',..,';')" />
        </xsl:for-each>
    </xsl:variable>

    <!-- construct Anchor, Shape and IndexedFaceSet -->
    <Anchor parameter='target=new'>
      <xsl:attribute name="url">
          <xsl:if test="../ID and ../ID!='dummy'">
             <xsl:variable name="oid" select="../ID" />
             <xsl:text>fieldinfo.xsql?table=</xsl:text>
             <xsl:value-of select=
                 'concat($table, "&amp;idcol=", $idcol, "&amp;id=", ../ID,
                 "&amp;con=" ,$con)'
             />
          </xsl:if>
      </xsl:attribute>

      <Shape>
        <IndexedFaceSet>
          <xsl:attribute name="convex">false</xsl:attribute>
          <xsl:attribute name="solid">false</xsl:attribute>

          <!-- construct coordIndex= -->
          <xsl:attribute name="coordIndex">
              <xsl:call-template name="creCoordIndex" >
                <xsl:with-param name="startAt" select="$startAt" />
                <xsl:with-param name="triplet" select="0" />
                <xsl:with-param name="total"
                select="count(SDO_ORDINATES/SDO_ORDINATES_ITEM)" />
              </xsl:call-template>
          </xsl:attribute>

          <!-- construct Coordinate point= -->
            <Coordinate>
               <xsl:attribute name="point">
               <xsl:call-template name="MLSegment3d" >
                   <xsl:with-param name="startAt" select="$startAt" />
               </xsl:call-template>
               </xsl:attribute>
            </Coordinate>
        </IndexedFaceSet>

        <!-- print rest of x3d tags -->
        <Appearance>

          <xsl:variable name="difcolor">
```

```
            <xsl:choose>
              <xsl:when test="SDO_GTYPE='3003'">1 0 0</xsl:when>
              <xsl:when test="SDO_GTYPE='3004'">0 1 0</xsl:when>
              <xsl:when test="SDO_GTYPE='3007'">0 0 1</xsl:when>
              <xsl:otherwise>0 0 0</xsl:otherwise>
            </xsl:choose>
          </xsl:variable>

          <Material>
            <xsl:attribute name="diffuseColor"><xsl:value-of
             select="$difcolor"/></xsl:attribute>
            <xsl:attribute name="ambientIntensity">1</xsl:attribute>
            <xsl:attribute name="specularColor">0.8 0.8 0.8</xsl:attribute>
            <xsl:attribute name="transparency">0.0</xsl:attribute>
          </Material>

        </Appearance>
      </Shape>
    </Anchor>

</xsl:template>


<xsl:template name="MLSegment3d">
  <xsl:param name="startAt" />

    <xsl:for-each select="SDO_ORDINATES/SDO_ORDINATES_ITEM">
        <!-- ... routine to print xyz-coordinates ... -->
    </xsl:for-each>

</xsl:template>

<xsl:template name="creCoordIndex">
  <xsl:param name="startAt" />
  <xsl:param name="triplet" />
  <xsl:param name="total" />

    <xsl:variable name="ordinate" select="($triplet*3)+1" />
    <xsl:variable name="lookFor" select="concat(';',$ordinate,';')"/>

    <!-- ... routine to create coordinate index ... -->

</xsl:template>

<xsl:template match="ROW/*[count(SDO_GTYPE)=0]" >
  <!-- skip non-spatial properties -->
</xsl:template>

<xsl:template match="text()" >
  <!-- skip text output -->
</xsl:template>

</xsl:stylesheet>
```

Index

A

B

C

T - #0216 - 111024 - C0 - 234/156/17 - PB - 9780367577896 - Gloss Lamination